Frank Hoffmann
Solid-State Chemistry

Also of interest

Solid State Physics.
Hunklinger, Enss, 2022
ISBN 978-3-11-066645-8, e-ISBN 978-3-11-066650-2

X-Ray Structure Analysis.
Siegrist, 2021
ISBN 978-3-11-061070-3, e-ISBN 978-3-11-061083-3

Reticular Chemistry and Applications.
Metal-Organic Frameworks
Belmabkhout, Cordova (Eds.), 2023
ISBN 978-1-5015-2470-7, e-ISBN 978-1-5015-2472-1

Crystallography in Materials Science.
From Structure-Property Relationships to Engineering
Schorr, Weidenthaler (Eds.), 2021
ISBN 978-3-11-067485-9, e-ISBN 978-3-11-067491-0

Complementary Bonding Analysis.
Grabowsky, 2021
e-ISBN 978-3-11-066007-4

Frank Hoffmann

Solid-State Chemistry

DE GRUYTER

Author
Dr. Frank Hoffmann
Wissenschaftlicher Oberrat
Institut für Anorganische und Angewandte Chemie
Universität Hamburg
Martin-Luther-King-Platz 6
20146 Hamburg
Germany

ISBN 978-3-11-065724-1
e-ISBN (PDF) 978-3-11-065729-6
e-ISBN (EPUB) 978-3-11-065750-0

Library of Congress Control Number: 2023935968

Bibliographic information published by the Deutsche Nationalbibliothek
The Deutsche Nationalbibliothek lists this publication in the Deutsche Nationalbibliografie;
detailed bibliographic data are available on the Internet at http://dnb.dnb.de.

© 2023 Walter de Gruyter GmbH, Berlin/Boston
Cover image: Mordolff/iStock/Getty Images Plus
Typesetting: Integra Software Services Pvt. Ltd.
Printing and binding: CPI books GmbH, Leck

www.degruyter.com

Preface

Dear reader,

You might ask: another book about solid-state chemistry! Why? The first answer could be: Because the author just could not resist the publisher's invitation to write a textbook on that subject. But that is not the whole story. It is not that there is a shortage of books on this topic, but the cross-cutting nature and the vast amount of research make it a worthy subject.

It might be also justified to ask whether textbooks are still needed at all. Aren't Wikipedia and the latest generation of AI tools like ChatGPT, Dall-e, and Midjourney sufficient? If you prompt ChatGPT for a short definition of solid-state chemistry, you will get the following answer:

> Solid-state chemistry is the study of the chemical and physical properties of solids, including their synthesis, structure, bonding, and reactivity, with a focus on understanding the relationship between these properties and their applications.

That's an excellent answer, no question! But if you ask Midjourney to generate a picture of a "fancy crystal structure composed of various types of atoms, mainly colored in blue and orange, render look", this is what you get:

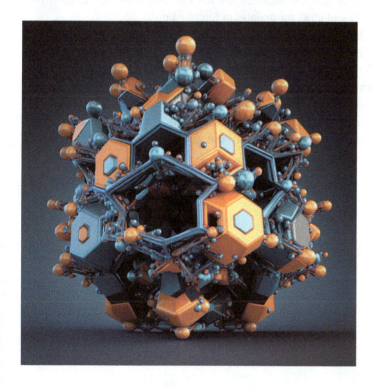

An interesting and beautiful image, indeed, but not a picture of a crystal structure. Maybe in the near future, the fusion of AI language tools with AI art creation tools is the answer to all questions! But I think that the decisive question is: Who will train the students to ask the right questions and to be able to evaluate the quality and correctness of the answers given by ChatGPT? And in this sense, this book should serve as a contribution to this goal.

You are invited to delve into fascinating structures and astonishing phenomena that occur in the solid state. You will gain a basic understanding of the systematics and principles of structures and their electronic, magnetic, and optical properties. In any case, special emphasis was placed on clear, high-quality illustrations and figures.

This small compendium offers a special selection of subjects, covering relevant topics, but without repeating all the basics from *general chemistry* that are already very well covered in many textbooks. The choice of the presented topics and the extent of their presentation have been made according to merely subjective assessments and may please some experts in the field as well as disappoint others – this cannot be avoided. And, last but not least, two topics in the area of solid-state chemistry that may be regarded as essential by others are not covered at all. Silicates except for zeolites are left out, because on the one hand it is a very comprehensive topic that deserves its own presentation, but is, on the other hand, from a conceptual point of view not very productive. Furthermore, lithium-ion batteries are not covered. However, I am always happy to receive wishes and ideas for updates in future editions. 😊

I feel deeply indebted to a number of people who have been very supportive in the preparation of the manuscript: First of all, of course, Ria Sengbusch from De Gruyter, whom I thank for her great supervision of the project and her patience, then Michael Fischer, because he is simply the best content checker, proofreader, and discussion partner in the world, then Ninja Reineke for her support and for making sure that the author's thoughts were put into a better linguistic form, and finally my boss, Michael Fröba, who allowed me to partially use my working hours with textbook writing.

Frank Hoffmann	Hamburg, Germany, March 2023

Contents

Preface — V

1 Introduction — 1

2 Categorizing and description of crystal structures — 3
2.1 Symmetry — 3
2.2 Sphere (and rod) packings — 4
2.3 Coordination polygons/polyhedra and the five Pauling rules — 5
2.3.1 The five Pauling rules — 6
2.3.2 Effective coordination numbers, linking coordination polyhedra, and Niggli notation — 11
2.4 Prototypical compounds – structure types — 12
2.5 Networks (nets) — 13

3 Symmetry of crystals — 15
3.1 The concept of the unit cell — 16
3.1.1 The metric and the stoichiometry — 16
3.1.2 Lattice points and motifs/bases — 17
3.1.3 The seven crystal systems — 19
3.2 The 14 Bravais lattices – centring of unit cells — 20
3.3 Crystallographic point groups (crystal classes) – the morphology of crystals — 24
3.3.1 Point symmetry elements — 25
3.3.2 The 32 crystal classes — 30
3.4 Further (micro-) translations – glide planes and screw axes — 40
3.4.1 Glide planes — 41
3.4.2 Screw axes — 42
3.5 The 230 space groups — 46
3.5.1 Nomenclature of space groups — 46
3.5.2 Deriving point groups from space groups — 47
3.5.3 International tables for crystallography — 47

4 Densest sphere packings — 50
4.1 The hexagonal closest packing — 52
4.2 The cubic closest packing — 54
4.3 Voids in the hcp and ccp packing — 55
4.4 Locations of interstitial sites along the packing direction — 57
4.5 Stacking variants/polytypes — 58

5	**Some important structure types —— 60**
5.1	Metal structures – sphere packings in action —— 60
5.1.1	Cu (ccp or fcc) (*Strukturbericht* type A1) —— 60
5.1.2	Mg (hcp) (*Strukturbericht* type A3) —— 62
5.1.3	Lanthanoids and actinoids – stacking variants —— 63
5.1.4	W (bcc) (*Strukturbericht* type A2) —— 65
5.1.5	α-Polonium —— 66
5.1.6	Special metal structures —— 66
5.2	(Ionic) compounds based on densest sphere packings —— 69
5.3	Compounds based on a cubic closest packing —— 72
5.3.1	NaCl (*Strukturbericht* type B1) —— 72
5.3.2	$CdCl_2$ (*Strukturbericht* type C19) —— 77
5.3.3	$CrCl_3$, $AlCl_3$, and $RhBr_3$ —— 78
5.3.4	CaF_2 (*Strukturbericht* type C1) —— 79
5.3.5	$MgAl_2O_4$ (spinel) (*Strukturbericht* type $H1_1$) —— 88
5.3.6	$CaTiO_3$ (perovskite, *Strukturbericht* type $E2_1$) —— 91
5.4	Compounds based on a hexagonal closest packing —— 102
5.4.1	NiAs (*Strukturbericht* type B8) —— 102
5.4.2	CdI_2 (*Strukturbericht* type C6) —— 105
5.4.3	β-V_2N (*Strukturbericht* type $L'3_2$) —— 107
5.4.4	$CaCl_2$ (*Strukturbericht* type C35), TiO_2 (rutile) (*Strukturbericht* type C4), and FeS_2 (marcasite) (*Strukturbericht* type C18) —— 108
5.4.5	BiI_3 (*Strukturbericht* type $D0_5$) —— 112
5.4.6	β-$TiCl_3$ and $MoBr_3/ZrI_3/RuBr_3$ —— 113
5.4.7	α-Al_2O_3 (corundum, *Strukturbericht* type $D5_1$) and $FeTiO_3$ (ilmenite, type $E2_2$) —— 114
5.4.8	ZnS (wurtzite) (*Strukturbericht* type B4) —— 115
5.4.9	Olivine $(Mg,Mn,Fe)_2SiO_4$ (*Strukturbericht* type $S1_2$) —— 117
5.5	Other important structure types not based on densest packings —— 118
5.5.1	CsCl (*Strukturbericht* type B2) —— 118
5.5.2	MoS_2 (*Strukturbericht* type C7) —— 119
	Further reading —— 120
6	**Defects in solids —— 121**
6.1	Point defects —— 122
6.1.1	Point defects in crystals of elements —— 122
6.1.2	Point defects in ionic compounds —— 123
6.1.3	Kröger-Vink notation for point defects —— 125
6.1.4	Colour centres – a special kind of point defects —— 126
6.1.5	Swapping places – order-disorder phenomena and the relation between superstructures and sublattices —— 128
6.1.6	Non-stoichiometric compounds and defect clusters —— 130

6.1.7	Substitutional and interstitial solid solutions —— 132	
6.2	Line defects – dislocations —— 138	
6.2.1	Edge dislocations —— 138	
6.2.2	Screw dislocations —— 140	
6.2.3	Movement of dislocations – plastic deformation —— 141	
6.2.4	Crystallographic shear planes —— 144	
6.3	Planar defects —— 146	
6.3.1	Stacking faults, turbostratification, and interstratification —— 146	
6.3.2	Internal boundaries in single crystals – mosaicity —— 148	
6.3.3	Grain boundaries —— 149	
6.3.4	Twin boundaries —— 150	
6.4	Volume defects —— 150	
	Conclusion and further reading —— 153	
7	**Phase diagrams —— 154**	
7.1	One-component systems —— 154	
7.2	Two-component systems —— 155	
7.2.1	Complete miscibility – solid solutions —— 156	
7.2.2	Singly eutectic systems —— 157	
7.2.3	Systems with compound formation that melt congruently —— 159	
7.2.4	Systems with compound formation but incongruent melting —— 160	
7.2.5	Systems with compound formation with an upper or lower limit of stability —— 161	
	Further reading —— 162	
8	**Electronic structure of solid-state compounds —— 163**	
8.1	Bloch functions, Bloch's theorem, the quantum number k, and crystal orbitals —— 166	
8.2	Bandwidth, density of states, and the Fermi level —— 172	
8.3	The Peierls distortion —— 174	
8.4	Band structures in two- and three-dimensional systems —— 178	
8.5	Examples of real band structures —— 178	
8.5.1	ReO_3 – a d^1 compound with metallic properties —— 180	
8.5.2	MoS_2 – a d^2 semiconductor with an indirect band gap —— 181	
8.6	Direct and indirect band gaps —— 183	
	Further reading —— 184	
9	**Magnetic properties of solid-state compounds —— 185**	
9.1	Diamagnetism and paramagnetism —— 185	
9.1.1	Quantifying the magnetic moments of paramagnetic substances —— 188	
9.1.2	Pauli paramagnetism —— 191	
9.2	Cooperative magnetic phenomena —— 193	

9.2.1	Ferromagnetism —— 199	
9.2.2	Antiferromagnetism and superexchange interactions —— 203	
9.2.3	Ferrimagnetism and double exchange —— 205	
9.3	Some magnetic materials —— 208	
9.3.1	Cubic spinel ferrites —— 208	
9.3.2	Garnets —— 208	
9.3.3	Hexagonal ferrites – magnetoplumbites —— 209	
	Further reading —— 211	
10	**Phosphors, lamps, lasers, and LEDs —— 212**	
10.1	Phosphors —— 212	
10.1.1	Fluorescent lamps —— 215	
10.1.2	Phosphors for CRTs of TVs and computer screens —— 217	
10.2	(Solid-state) lasers —— 218	
10.2.1	Operation conditions —— 219	
10.2.2	The ruby and the He-Ne laser —— 220	
10.2.3	Classification of lasers —— 222	
10.3	LEDs —— 224	
	Further reading —— 226	
11	**Superconductivity —— 227**	
11.1	From the metallic to the superconducting state —— 227	
11.1.1	The Meissner-Ochsenfeld effect and type I and type II superconductors —— 228	
11.1.2	The BCS theory —— 230	
11.2	Superconducting materials —— 232	
11.2.1	The elements —— 232	
11.2.2	Binary compounds, alloys, and intermetallics —— 232	
11.2.3	Oxo cuprates – the high-temperature superconducting revolution —— 234	
11.2.4	Further superconducting compounds —— 236	
	Further reading —— 241	
12	**Ceramics —— 242**	
12.1	Definition and classification of ceramics and the ceramic method —— 242	
12.2	(Alumo)silicate ceramics (traditional ceramics) —— 245	
12.3	Binary oxide ceramics —— 247	
12.3.1	α-Al_2O_3 —— 247	
12.3.2	ZrO_2 —— 247	
12.3.3	TiO_2, MgO, and BeO —— 249	
12.4	Mixed oxide ceramics —— 249	
12.4.1	Aluminium titanate —— 249	
12.4.2	Barium titanate and lead zirconate titanate —— 250	

12.5	Boride ceramics —— 253
12.6	Carbide ceramics —— 253
12.6.1	Boron carbide —— 254
12.6.2	Silicon carbide —— 254
12.7	Silicide ceramics —— 256
12.8	Nitride ceramics —— 257
12.8.1	Boron nitride —— 258
12.8.2	Aluminium nitride —— 259
12.8.3	Silicon nitride and SiALONs —— 260
12.8.4	Aluminium oxynitride —— 262
12.9	Glass-ceramics —— 263
	Further reading —— 264

13	**Intermetallic phases —— 265**
13.1	Classification scheme —— 266
13.2	Ordered solid solutions – superstructures —— 268
13.2.1	Superstructures of the bcc packing —— 269
13.2.2	Superstructures derived from densest packings (ccp or hcp) —— 270
13.3	Hume-Rothery phases —— 273
13.3.1	The γ-brass structure —— 275
13.4	Zintl phases —— 277
13.5	Packing-dominated phases (Frank-Kasper and Laves phases) —— 279
13.5.1	The geometrical principles in FK phases and the FK coordination polyhedra —— 280
13.5.2	Laves phases —— 281
13.5.3	Variants of Laves phases —— 285
13.5.4	σ-, μ-, M, P, and R phases —— 288
13.5.5	The Cr_3Si structure (*Strukturbericht* type A15) —— 290

14	**Porous crystals, reticular chemistry, and the net approach —— 295**
14.1	Zeolites and zeotypes —— 295
14.1.1	Structural chemistry and building units of zeolites —— 297
14.1.2	Technically important zeolites —— 303
14.2	Metal-organic frameworks —— 310
14.2.1	Carbon capture and sequestration —— 315
14.2.2	Hydrogen storage —— 316
14.2.3	Methane storage —— 318
14.2.4	Water harvesting from air —— 320
14.2.5	Other (potential) applications —— 322
14.3	Networks and topology —— 322
14.3.1	Graphs and nets —— 323
14.3.2	Turning crystal structures into their underlying nets —— 324

14.3.3	Topology and net descriptors —— **325**	
14.3.4	Characterizing and identifying nets —— **328**	
	Further reading —— **331**	

Appendix A
***Strukturbericht* designations and Pearson symbols —— 333**

References —— 345

Subject index —— 351

Formula index —— 363

1 Introduction

The question, what solid-state chemistry actually is, cannot readily be answered, but the term already implies some obvious characteristics: solid state refers to the aggregate state of the matter that is involved and chemistry is defined as the study of the structure, composition, properties, and behaviour of substances, in particular, their changes during chemical reactions.

It is not easy to discriminate solid-state chemistry from related disciplines or from certain subdisciplines of other fields. There is considerable overlap with solid-state physics, inorganic structural chemistry, metallurgy, materials science, mineralogy, and crystallography. A sharp distinction between solid-state chemistry and these disciplines is not possible.

There are two types of matter in the solid state: *crystalline* and *amorphous* substances. Crystals are homogeneous, anisotropic solid-state bodies that show a *strictly periodic order* of their constituents. Different approaches of categorizing and describing crystal structures are presented in Chapter 2, while the most important aspects of the symmetry of crystals are described in Chapter 3. Amorphous substances also show a certain order on the short range but lack the strict periodic order (long-range order). Examples of amorphous materials are glasses, wood, some types of ceramics, or plastics, although some types of plastics are at least partially crystalline (for instance, polyethylene terephthalate, PET). Wood also has a structure that is not completely chaotic but shows a certain repetition of some of its underlying constituents (in particular, the biopolymer lignin) but they do not possess a strictly periodic order.

In this introduction to solid-state chemistry, we will deal mainly with crystals, because we are interested in basic structure-property relations, for which a regular and, to some extent, also idealized structure is a prerequisite. However, some important properties of real crystalline materials depend on the kind and number of so-called *defects*, i.e., deviations from the ideal structure. Therefore, the most important types of defects will be presented in Chapter 6.

In our considerations of crystals, we will focus even more: we assume that they are *single crystals*, although many solid-state bodies, in particular metals and alloys, are *polycrystalline*, i.e., consist of an aggregate of smaller and differently oriented *crystallites*, which are also called *grains*. The properties of many technically important materials are in particular dependent on this *microstructure* and are rather the subject of material science and metallurgy.

Why is there a dedicated chemical discipline that refers to the solid state of matter, while no explicit subcategory like "liquid chemistry" or "gas-phase chemistry" exists? The author would argue that these subcategories exist, but that they are not called that way: chemical reactions are happening usually in a fluid-like medium; in solution or in the gas phase! Only in fluid-like media it is easy to ensure an intimate mixing between the reaction partners. This is obviously different in a solid-state body: we should be aware of the fact that in an ideal, defect-free solid body, reactions can take place only at the *interface* and that inside the body there are usually no *transport pathways*. This often requires special synthesis methods, the oldest of which have been known for thousands of years and are used in the production of porcelain and ceramics; ceramics are discussed in Chapter 12. In this context, the renaissance of *mechanochemistry* is worth mentioning, which has meanwhile fallen into oblivion, and which is also leading to new synthesis products and materials in the field of solid-state chemistry. A good introduction to this topic can be found in Ref. [1].

In order to understand the processes taking place during a chemical reaction and to explore new pathways to hitherto unknown compounds or materials, knowledge about the structures of the compounds involved is a prerequisite. Therefore, a large body of the book is devoted to the presentation of prototypical structure types (Chapter 5) and forms the basis for the understanding of more advanced structures. The preparation and examination of new solid-state compounds has led to technical advances as well as enormous progress in understanding of the properties of materials. The basics of electronic and magnetic properties of solid-state compounds are presented in Chapters 8 and 9, respectively. Optical properties and devices to generate light are the topic of Chapter 10. It is fascinating to observe that there are still new developments in the field of superconductivity and how different the classes of compounds that show this phenomenon can be. Superconductors are discussed in Chapter 11.

The book closes with a more detailed description of two further material classes. Intermetallics are covered in Chapter 13, a topic that is often only very briefly presented, although they represent a very large class of compounds and comprise those with equally complex, astonishing, and appealing structures. Finally, Chapter 14 deals with porous crystals, which, in addition to classical zeolites, comprise the class of metal-organic frameworks that could not be ignored in view of the rapid developments taking place in this field.

2 Categorizing and description of crystal structures

Crystals, i.e., chemical entities that show a strictly periodic order of their constituents in two or three dimensions, can be described and categorized in many different ways, all of which emphasize a certain aspect of the crystal structure and complement each other. The most important ones are the following: (a) to describe the symmetry of the crystal, (b) to look at the structure if it can be derived from a rather geometrical aspect, if we assume that atoms or ions behave like hard spheres, i.e., to emphasize the way the spheres (or rods) are packed together, (c) to explore the neighbouring species around a certain atom in the form of their *coordination polyhedra*, (d) to categorize crystals according to *prototypical compounds* of a certain type of structure (*structure types*), and (e) to specify the way in which the atoms are connected (bonded) to other atoms, i.e., to describe the *network* that a structure forms.

2.1 Symmetry

In crystallography, the most important aspect of a crystal structure is its symmetry. The presence (or absence) of certain symmetry elements leads to one of the 17 possible *plane groups* (for two-periodic systems) or to one of the 230 possible *space groups* (for three-periodic structures). A plane/space group is, in principle, nothing more than stating which *symmetry elements* are present in a structure. While the symmetry aspect is completely covered with this specification, it ignores all other aspects, like the nature of the constituents of a structure and how they are held together. Therefore, also other objects, even everyday objects, can be classified according to their symmetry: from a symmetry point of view, a hexagonal tiling in your bathroom is identical to a single graphene sheet of graphite. Disparate substances like the pure metal copper, held together by *metallic bonds*, the prototypical ionic compound NaCl (i.e., simple rock salt), held together by *electrostatic interactions*, noble gases (except helium) at very low temperatures, and the solid Buckminsterfullerene (at least at room temperature [2]), consisting of discrete C_{60} molecules (the famous football molecule), which are interacting only via *dispersion forces*, all have the same cubic symmetry: they all belong to the space group $Fm\bar{3}m$ (space group no. 225; for an explanation of the nomenclature of space groups, see below). What is different is, of course, (i) the *base* (also called *motif*) and (ii) the dimension of the unit cell; but the symmetry elements that are acting on this base are the same. The crystallographic description then will be completed by specifying the (i) *lattice parameters* and (ii) the nature and position of the atomic species of the base, i.e., their atomic (fractional) coordinates and sometimes their *Wyckoff letter*, this means their *multiplicity* and *site symmetry*. We will recap the most important aspects of the symmetry framework of crystals in Chapter 3.

2.2 Sphere (and rod) packings

Nature generally tends to avoid structures with pronounced voids. The reason for this is that the creation of surfaces or interfaces requires energy. Therefore, there is a trend to realize structures that are as compact as possible by minimizing the surface area and thus reducing energy. In particular, for systems composed of spheres (or rods, i.e., infinite cylinders; for an early overview of crystal structures based on rod packings, see, for instance, Ref. [3]) structures are realized by the guiding principle of how efficient and to which degree space can be filled, so that a minimum of unfilled space is left. In the case of only one type of spheres there are two equivalent densest sphere packings: the cubic closest packing (**ccp**) and the hexagonal closest packing (**hcp**).

A large number of metals crystallize in one of these two structures (**ccp**, for instance, by Rh, Ir, Ni, Pd, Pt, Cu, Ag, Au, and **hcp**, for instance, by Mg, Ti, Zr, Re, Ru, Os, Co, Cd). In both cases the volume is filled by approx. 74%. And in both arrangements, there are tetrahedral and octahedral voids, which can potentially be occupied by some other, smaller atoms or ions; depending on the composition of the (binary/ternary...) compound and the size and charge of the counterions, a fraction or all of these voids are filled. The structure of many ionic compounds can be described by sphere packings, in which usually the smaller cations are located in the voids of a **ccp** or **hcp** arrangement of the larger anions. However, interestingly, there are also a significant number of (simple) structures – although based on spheres – where *not densest* packings are formed and the principle of realizing structures that are as compact as possible seems to be violated: in elements with a body-centred cubic structure (**bcc**), for instance, V, Nb, Ta, Cr, Mo, W, α-Fe, the packing density is only 68%, and in primitive cubic (**pcu**) arrangements, like in α-Po, only 52% of the space is filled. And this means, of course, that atoms are not hard spheres that behave like billiard balls.

The packing density and specifications of the coordination of different kinds of sphere packings are gathered in Table 2.1. A more detailed view on densest sphere packings and their crystallographic descriptions are given in Chapter 4.

Table 2.1: Packing density and coordination in different sphere packings.

Sphere packing	Packing density	Coordination number of shell				Relative distance of the coordination shells		
		1	2	3	4	d_2/d_1	d_3/d_1	d_4/d_1
Primitive cubic (**pcu**)	0.52	6	12	8	6	1.41	1.73	2
Body-centred cubic (**bcc**)	0.68	8	6	12	8	1.15	1.63	1.91
Cubic closest packing (**ccp**)	0.74	12	6	24	12	1.41	1.73	2
Hexagonal closest packing (**hcp**)	0.74	12	6	8	24	1.41	1.63	1.73

2.3 Coordination polygons/polyhedra and the five Pauling rules

A slightly extended and complementary approach to categorizing crystal structures, mostly for compact and ionic crystalline compounds, is based on the specification of coordination polyhedra. Here, the coordination environments of the cations and/or anions are specified together with information on how these regular or also slightly distorted polyhedra are connected to each other (i.e., if they are corner- or/and edge- or/and face-connected). The advantage is that structures, which are not based on a densest packing of one of the components, can also be categorized and that polyhedra of any kind can be considered (not only tetrahedral or octahedral ones as in the case of sphere packings). Furthermore, this concept also allows, at least partly, predicting and rationalizing the crystal structure of ionic compounds. As early as 1929, Linus Carl Pauling (1901–1994, American chemist, awarded the Nobel Prize in chemistry in 1954 and the Nobel Peace Prize in 1962) published his seminal work on the principles on which the structure of complex ionic crystals is based, known as the five Pauling rules (see below).

The most important coordination environments, their names, and short symbols are shown in Figure 2.1.

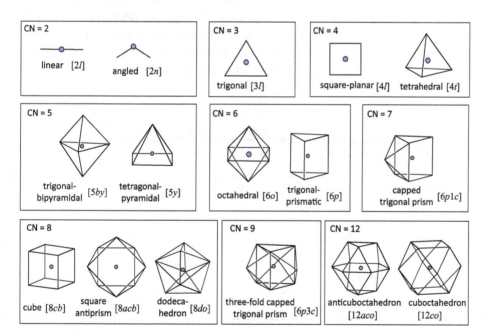

Figure 2.1: Important coordination polyhedra; l = collinear/coplanar, n = non-collinear/non-coplanar (redrawn and adapted from Ref. [4]).

2.3.1 The five Pauling rules

In 1929, Pauling published five rules for predicting and rationalizing the crystal structure of ionic compounds [5].

First rule: the critical radius ratio

Pauling's first rule is about the fact that the coordination number and geometry of a coordination polyhedron that is formed around a cation or anion by the respective other species can be deduced from the cation-anion radius ratio, i.e., r_c/r_a, assuming that the cation is smaller than the anion, which is usually the case. (Of course, if the cation is larger than the anion the quotient has to be inverted). Stable arrangements result when cations or anions are surrounded by a number of the other species, where the cations and anions are *in contact* with each other. Pauling mathematically derived the *minimum* or critical radius ratio for which the cation is in contact with a given number of anions (considering the ions as rigid spheres). If the cation and therefore the radius ratio is smaller, it will not be in contact with the anions, resulting in instability and leading to a lower coordination number. The critical radius ratios that Pauling derived for different coordination numbers (CN) and corresponding regular coordination polyhedra are shown in Table 2.2. Note that more irregular coordination environments are not possible if only electrostatic forces and rigid spheres are assumed. Note further that the limiting radii have to be considered in such a way that the coordination number must decrease if they are *below* the given value, but not in such a way that the coordination number must necessarily increase if the value is larger than the next limiting value.

Table 2.2: Overview of the critical radius ratios for selected coordination numbers and polyhedra in ionic compounds, as geometrically derived by Pauling.

Coordination number	Coordination polygon/polyhedra	Critical radius ratio
3	Triangle	0.155
4	Tetrahedron	0.225
6	Octahedron	0.414
7	Capped octahedron	0.592
8	Square antiprism	0.645
8	Cube	0.732
9	Tricapped triangular prism	0.732
12	Cuboctahedron	1.000

In Figure 2.2, the first Pauling rule is illustrated for the coordination number 6 starting with a cation-anion arrangement in which the radius ratio is actually larger than 0.414. This arrangement is more stable than at the critical value of 0.414 since the cation is still in contact with six anions, but the anions are further away from each other, so that their mutual electrostatic repulsion is reduced. If, however, the ratio rises above 0.732, a cubic geometry with a coordination number of 8 is favourable. If the radius ratio is below the critical value of 0.414, the stability limit is reached and a tetrahedral coordination environment will be realized.

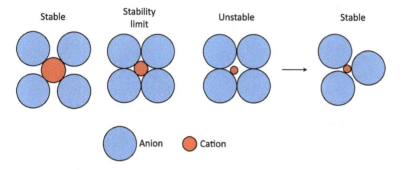

Figure 2.2: Diagram illustrating the stability range for an octahedron with a coordination number of six that will turn into an arrangement of lower coordination number, i.e., 4, in the form of a tetrahedron once the critical value of $r_c/r_a = 0.414$ is reached. Only the four anions that are lying in the plane are shown for the octahedron, one further anion is located above and below that plane, and only three anions of the tetrahedron are shown, one further is above (or below) that plane.

Despite its simplicity, the first Pauling rule shows some success in predicting the coordination number and geometry of many ionic compounds. However, there are also a significant number of exceptions, and the limitations are based (among other reasons) on the intrinsic assumption of the model, i.e., that the ionicity is 100% and that ions are hard spheres. Another problem is that the ionic radius itself is dependent on the nature of its counterion. If we take the alkali metal halides as an example, we find the following exceptions: For LiBr and LiI a tetrahedral coordination environment is predicted, but they are octahedrally coordinated as in NaCl, and KF, KCl, RbF, and CsF should prefer a cubic arrangement but also crystallize in the NaCl structure type.

Second rule: the electrostatic valence rule

The second Pauling rule is concerned with the electroneutrality of ionic compounds, specifically, in which sphere around an ion species this electroneutrality is given. Pauling says that ionic structures are stable, if electroneutrality is already achieved in the first coordination sphere of a given cation, or in other words, when *local* electroneutrality is achieved. Or to put it yet another way: there should be no local accumulation of charge of same sign. To express this more clearly in mathematical terms, Pauling defined for a given cation the electrostatic bond strength s:

$$s = \frac{z_c}{n_c}$$

with z_c being the charge of the cation and n the coordination number. A stable ionic crystal is obtained, if the sum of the bond valence strength of all cations surrounding an anion is equal or approximately equal to the (negative) charge of that anion:

$$z_a = -\sum_i s_i$$

Let us look at some examples, starting with the simplest one. In NaCl, the sodium ions are surrounded by six chloride atoms. Thus, the electrostatic bond strength is 1/6. The chloride atoms are surrounded by six Na^+ ions, giving $-(6 \times 1/6) = -1$, the charge of the Cl anion. Second example: suppose in a compound MX_2, the M^{2+} cations have the coordination number 6. What will be the coordination number of the anion? Well, $s = 2/6 = 1/3$, and therefore, as the charge of X is −1, it should be 3. This is the case for $CdCl_2$ and CdI_2. Third example: the electrostatic bond strength of Sc in Sc_2O_3 is $s = 3/6 = 1/2$. Therefore, the oxide anions should have a coordination number of 4, as $-(4 \times ½) = -2$, the charge of the oxide ion – this is indeed the case in Sc_2O_3. Last example: in $SrTiO_3$ (perovskite structure type, see Section 5.3.6), the titanium ions are surrounded by 6 oxide ions, hence $s_{Ti} = 4/6 = 2/3$. The Sr^{2+} ions are surrounded by 12 oxygen atoms, thus $s_{Sr} = 2/12 = 1/6$. The oxide ions are coordinated by 2 Ti ($2 \times 2/3 = 4/3$) and 4 Sr ions ($4 \times 1/6 = 2/3$) – the sum over both cations is exactly 2, the negative charge of the oxide ions. The electrostatic valence rule is surprisingly well obeyed for many compounds, even when significant covalent bonding is present, as long as one takes the formal oxidation state of the species involved as their charge.

Third rule: stability of linked polyhedra

The coordination polyhedra, which form around a cation (or anion), can be linked in different ways. They can be isolated, or they can be linked by common corners, edges (= 2 corners), or faces (= 3 or more corners), also combinations of these linkage types are possible, of course. Pauling's third rule now makes statements about how the way polyhedra are connected affects the stability of ion crystals: common edges and, to an even greater extent, common faces of polyhedra reduce the stability of an ion crystal (see Figure 2.3). This is even more valid the higher the charge and the smaller the coordination number of the cation, i.e., for example, face-linked tetrahedra of highly charged cations reduce the stability much more than face-linked octahedra of relatively low-charged cations. The reason for this is clear: the like-charged cations that are located at the centre of the polyhedra come closer together, if the polyhedra share common edges or faces and in this way the electrostatic repulsion between the cations is increasing. For a more quantitative estimate, of course, the anion-cation-anion bond angle must be considered for a corner linkage and the anion-cation-cation-anion torsion angle must be considered for an edge linkage, so no fixed values for relative distances can be given here. The standard example used to illustrate this principle is that given by Pauling himself, namely the three mineral modifications of TiO$_2$, in each of which TiO$_6$ octahedra are present. In the most stable modification of rutile, the TiO$_6$ octahedra have two common edges; in the less stable modifications brookite and anatase, which both irreversibly transform into rutile at higher temperature, three and four common edges are present, respectively.

In general, Pauling's third rule is considered to be satisfied well for purely ionic or highly polar crystals, less well for compounds with more covalent contributions, and there are exceptions, namely for those compounds involving transition metals that have d electrons and can form metal-metal bonds. There, preferentially common faces are occasionally encountered. This is, for instance, the case for ZrI$_3$, which forms strands of face-linked octahedra in which the metal atoms form metal-metal bonds in pairs between two neighbouring octahedra (see Section 5.4.6).

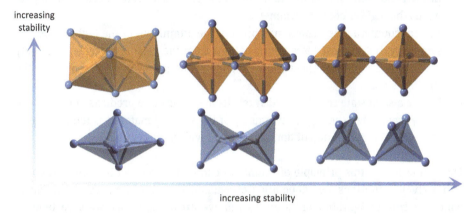

Figure 2.3: Illustration of Pauling's third rule concerning the stability of linked polyhedra. Cation polyhedra that share common edges or faces lead to decreased stability of ionic crystals. This is more pronounced, the smaller the coordination number and the higher the charge of the cation.

Fourth rule: linked polyhedra in crystals with different cations

Pauling's fourth rule is in principle derived from the third rule and concerns structures in which cations of different valency and coordination number are simultaneously present. It states that cations of high valency and low coordination number tend to form polyhedra that do not share common corners, edges, or faces, i.e., they tend to be *isolated*. The background – as with the third rule – is that this causes these cations to be further apart, minimizing their electrostatic repulsion. One example that Pauling gives is that of olivine, (Mg,Fe)$_2$SiO$_4$ (see Section 5.4.9). Olivine contains isolated SiO$_4$ tetrahedra and belongs to the nesosilicates; the cations with lower charge and higher coordination number, i.e., Mg/Fe, form Mg/FeO$_6$ octahedra that share common edges.

Fifth rule: the principle of parsimony

While Pauling's first four rules are essentially based on electrostatic considerations, the fifth rule formulates rather a principle of symmetry. It states that the number of different kinds of constituents in a crystal tends to be small. Constituents here do not mean chemical species, but coordination polyhedra. An interesting example here is the amphibole mineral group, where individual members can contain up to 13 different cations, but these are distributed over just four different cation positions, one tetrahedral, one octahedral, one cubic antiprismatic, and one cuboctahedral. This sparsity principle can also be expressed in such a way that a relatively small number of different symmetry positions are occupied in a crystalline solid-state structure. This symmetry principle has been recognized and refined by several researchers throughout the history of crystal chemistry. Hartmut Bärnighausen, in his paper from the year 1980, phrased it as follows [6]:

- In the solid state the arrangement of atoms reveals a pronounced tendency towards the highest possible symmetry.
- Several counteracting factors may prevent the attainment of the highest possible symmetry, but in most cases the deviations from the ideal arrangement are only small, and frequently observed symmetry reduction even corresponds to the smallest step.
- During a solid-state reaction, which results in one or more products of lower symmetry, very often the higher symmetry of the starting material is indirectly preserved by the orientation of domains formed within the crystalline matrix.

The first aspect of this principle of symmetry can be rationalized in such a way that under given conditions, i.e., depending on chemical composition, type of chemical bond, electron configuration of the atoms, relative size of the atoms, pressure, temperature, etc., there will be exactly *one* energetically most favourable environment for each of the atomic species. In a crystalline assembly, however, these sites are identical and thus have the same environment only if they lie on symmetry-equivalent sites.

Overall, Pauling's five rules are very plausible – they concern facts that are on the border of being trivial – and therefore it is not surprising that very many ionic compounds have structures that are in accordance with these rules, especially since the rules are formulated as tendencies and not as rigid laws. As the rules are rules, there are exceptions. In this context, a recent study in which approx. 5,000 oxides were examined to see to what extent all five rules apply to them simultaneously is interesting. The result, according to which only 13% of these 5,000 compounds fulfil rule no. 2–5 simultaneously, led the authors to the conclusion that Pauling's rules are only of limited predictive power [7].

2.3.2 Effective coordination numbers, linking coordination polyhedra, and Niggli notation

Effective coordination number

The coordination polyhedra given in Figure 2.1 are regular in the sense that all atoms that are located at the corners of these polyhedra have an *equal distance* to the atom which they enclose, and they are the nearest neighbouring atoms around that central atom. These geometrically ideal coordination environments are indeed very often encountered in real structures. However, there are also many compounds, where the situation is a bit more complicated; either in the sense that a particular coordination environment only approximately resembles the ideal polyhedra, meaning that the coordination polyhedron is distorted, or in the sense that it is not completely clear how to determine the distance limit from the central atom to count other atoms as neighbouring atoms. This situation is already encountered in the prototypical structure of the **bcc** packing, the structure of the element tungsten. Here, each W atom has eight nearest neighbours in a distance of 2.7412 Å forming a cube. But there are six next-nearest neighbouring atoms that are only 15% further away (3.1652 Å), meaning that the coordination number can also be described as 8 + 6, with the first number referring to the number of the atoms at the shortest distance.

The disadvantage of such integer notation is that the quantitative information about the distance to the next-nearest neighbours is lost. Therefore, several approaches have been proposed in order to apply schemes that take more quantitative information about roughened coordination shells into account: this leads to a so-called *effective* coordination number, often abbreviated as ECoN. In this ECoN, the atoms surrounding a central atom are counted either as full atoms or as fractions of atoms with a number between 1 and 0. However, there is still the problem of how to clearly demarcate coordination spheres from each other. Atoms that "clearly" belong to the second or third coordination sphere should be excluded. Investigations by Brunner and Schwarzenbach [8] on the basis of a large number of inorganic crystalline structures have now shown the following: if the number of neighbours of an atom are arranged as a function of their distances in a histogram, in approx. 90% of the structures there is a clear "gap" in the histogram, i.e., a range of distances without corresponding atoms. If the shortest distance to neighbouring atoms is set equal to 1, frequently this "gap" appears at distances larger than 1.3. After their approach, an atom at the distance of 1 obtains the weight 1, the first atom beyond the gap obtains zero weight, and all intermediate atoms are included with weights that are calculated from their distances by linear interpolation:

$$\text{ECoN} = \sum_i \frac{d_{\text{gap}} - d_i}{d_{\text{gap}} - d_1}$$

where d_{gap} is the distance to the first atom beyond the gap, d_1 is the distance to the closest atom, and d_i is the distance to the ith atom in the region between d_1 and d_{gap}.

If we apply the formula to tungsten with $d_{gap} = 4.4763$ Å, then we get an ECoN of 12.53.

Linked coordination polyhedra and Niggli notation

Extended inorganic structures can often appropriately be described by the way in which coordination polyhedra are connected. Two polyhedra can be joined by a common corner (vertex), a common edge, or a common face. The common atoms of two connected coordination polyhedra are called bridging or ligand atoms. A useful notation that expresses the coordination numbers of the species involved in the compound in the form of fractional numbers was introduced by Paul Niggli (1888–1953, a Swiss crystallographer, mineralogist, and petrologist). For instance, the formula $CdCl_{6/3}$ means that the Cd ions are surrounded by six Cl ions (here in the form of an octahedron), each of which is coordinated to three Cd ions, or in other words, the Cl ions belong to three octahedra simultaneously. This notation scheme is, of course, not limited to binary compounds. To give another example: the Niggli formula of niobium oxide chloride is $NbO_{2/2}Cl_{2/2}Cl_{2/1}$; here the coordination environment of niobium consists of 6 atoms, i.e., the sum of the numerators, the coordination number is 2 for the oxygen atoms and 2 and 1 for the two different kinds of Cl atoms. Provided that all atoms are taken into account, adding the individual fractions gives the chemical formula of the compound, here, $NbOCl_3$.

The given examples are the short notation form of the Niggli formulas. There are further extended symbols and notation schemes to express the way the structural fragments are organized. For instance, when coordination polyhedra are connected to chains, layers/sheets or a three-dimensional network, this can be expressed by the preceding symbols $^1_\infty$, $^2_\infty$, and $^3_\infty$, respectively; more details for recommendations concerning the nomenclature of inorganic structure types can be found in Ref. [9].

2.4 Prototypical compounds – structure types

For inorganic crystals there is another and even further extended approach to categorize crystal structures, which is based on a similarity principle. In this approach, certain compounds are selected to be a prototypical compound of a particular structure. This can be a well-known and abundant compound. Often also compounds that have been described for the first time serve as prototypes. Based on the structure of these prototypes, compounds which have the "same", or to be precise, an analogous structure can then be classified as belonging to the same structure type. One of the most prominent prototypes is NaCl, rock salt. Compounds that have essentially the same structures are then characterized by having a NaCl-like structure or crystallize in the structure type of NaCl.

To belong to the same structure type, the compounds must have the same symmetry (i.e., space group) and the relative positions of their atoms within the unit cell, expressed by the sequence of Wyckoff positions, must be identical. This implies that both the stoichiometry and the coordination polyhedra of the species involved must be the same. This also requires that the axial ratio of the unit cell is similar. Crystalline substances belonging to the same structural type are called isotypic.

The advantage of the classification according to structure types is that the type of atoms, the character of the interactions between the atoms (ionic, covalent, H-bonds, etc.) and the atomic distances play no role. Furthermore, structure types are helpful to derive certain structural relationships, when going from an idealized structure with highest possible symmetry to structures with less and less symmetry, so-called group-subgroup relations [10]. For some types, whole genealogical trees exist (e.g., perovskites [11]), also known as Bärnighausen trees [6]. The type with the highest possible symmetry is called the aristotype, while any crystal structure with a lower symmetry that is related to an aristotype is called a hettotype [12]. After Buerger, aristotypes are also known as *basic structures* and hettotypes as *derivative structures* [13].

We will refer to structure types throughout the book. In Chapter 5, the most important structure types are described. A note on the historical development of structure types and a list of structure types can be found in Appendix A.

2.5 Networks (nets)

The classification schemes based on sphere packings and coordination polyhedra essentially presuppose that the forces between the interacting chemical species are more or less isotropic, i.e., non-directional. These concepts are, therefore, less applicable, if forces with a strong directional character are involved, for instance H-bonds, covalent bonds, or other effects that cause pronounced asymmetry. An example of the latter can be found, for example, in the lead chalcogenide series: the binary compounds PbS, PbSe, and PbTe belong to the highly symmetric NaCl structure type, but PbO forms a relatively complicated layer-type structure with a severely asymmetric electron density around the Pb(II) cations (see Section 5.3.4). Another example are the two main allotropes of carbon, diamond and graphite, which cannot be sufficiently described by sphere packings or coordination polyhedra. Yet another example comprises certain classes of porous compounds, for instance, zeolites. The basic building blocks of zeolites can be described by coordination polyhedra: they are exclusively made of Al/SiO_4 tetrahedra. And, interestingly, all tetrahedra are exclusively corner-connected. But this description is, of course, far away from being sufficient to describe all the approx.

250 different zeolite types, all of which have characteristic channels and cages of different size in different mutual orientations.

In those and analogous cases, it is more appropriate to describe these compounds as a set of points (representing atoms or a group of atoms) that are connected to other points by lines (representing (chemical) bonds), forming a network (short form = net) of connected polygons. This type of representation of infinite, periodic structures was first introduced by A.F. Wells. His studies on this subject were summarized in his book *Three-Dimensional Nets and Polyhedra* published in 1977 [14]. A concise version has also been included as Chapter 3, *"Polyhedra and nets"*, in his classical textbook *Structural Inorganic Chemistry* [15].

Until approx. two decades ago, this network approach was mainly popular in the zeolite community. However, with the invention of metal-organic frameworks (MOFs), it has experienced a spectacular renaissance. The original approach of Wells has now been extended in the sense that in modern crystallographic topology, the nodes that are weaved together into a continuous network do represent not only single atoms but groups of atoms or polyhedral units that are connected to other units.

In Chapter 14, the two classes of porous substances – zeolites and MOFs – are introduced and the network approach for the classification of such structures is discussed in more detail.

3 Symmetry of crystals

The symmetry of crystals can be described on four levels of hierarchy (see Figure 3.1). On the first level, the *crystal system* is specified to which a crystal belongs, the second level specifies the *Bravais lattice*, or the type of centring, the third level is concerned with the symmetry of the outer shapes, i.e., the morphology of crystals, by specifying the *point group* of the crystal, and on the fourth level, the complete symmetry information of the crystal structure is given as the *space group*.

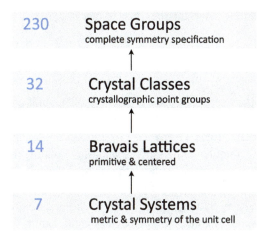

Figure 3.1: Four levels of the symmetry of crystals.

Let us begin with the first level, the crystal system. The answer to the question, what a crystal system is, is actually difficult, although a concise definition is available. The International Union of Crystallography (IUCr) defines a crystal system as follows:

> A crystal-class system, or crystal system for short, contains complete geometric crystal classes of space groups. All those geometric crystal classes belong to the same crystal system which intersect exactly the same set of Bravais classes. [16]

The problem with this definition is that almost nobody understands it. It refers to two other terms that are unknown at this point (geometric crystal class and Bravais class), and neither is it clear what a "complete" crystal class is nor what is meant by the term "intersect". In the given reference, the meaning of the latter term is specified as: "*A geometric crystal class and a Bravais class of space groups are said to intersect if there is at least one space group common to both.*" Does this help? The author thinks that this is not the case.

Crystal systems are a way to categorize crystals, i.e., a classification scheme, according to (i) the presence of symmetry elements that are *characteristic* for each of the crystal system and can be regarded as a *minimum* set of symmetry elements that have to be present in a crystal and (ii) the *maximum* point symmetry of the underlying translational lattice of the crystal. Furthermore, for the description of crystal structures, i.e., if the fractional coordinates of the atoms are specified, a system of coordinates is chosen that is adapted to the symmetry of the crystal. The symmetry of a crystal leads to certain requirements concerning the system of coordinates; in other words: the *lattice parameters* are subject to certain *restrictions*. All this still sounds very abstract. Therefore, we will approach the crystal systems on a slightly different path, by introducing the concept of the unit cell as the first level of hierarchy.

3.1 The concept of the unit cell

What all crystals have in common is that they are built up from a smallest unit and that the final crystal results from adjoining this unit in all three spatial directions in a way that no gaps remain, and no overlaps occur, i.e., by *pure translation* operations; this also means that all unit cells in a crystal have the same orientation. Building the crystal by adjoining unit cells that are rotated against each other is not allowed. The spatial directions do not necessarily have to correspond to the three perpendicular directions of a Cartesian coordinate system. Rather, a coordinate system is used that is adapted to the symmetry of the crystal. The outer shape of this smallest unit, the *unit cell*, is in all cases a particular type of hexahedron: a parallelepiped. A parallelepiped is a geometric body that is confined by six faces with three pairs of congruent faces that lie in parallel planes (see Figure 3.2).

3.1.1 The metric and the stoichiometry

The unit cell is, in its first instance, characterized by its so-called metric. The metric is a set of six parameters to unequivocally describe the unit cell: three for the lengths of the edges, a, b, and c, which are running per definition along the x-, y-, and z-axis of the coordinate system, respectively, and three for the angles between these edges, namely, the angle α, which defines the angle between edge b and c, the angle β that is enclosed by the edges a and c, and finally the angle γ, describing the aperture between edges a and b (Figure 3.3). As the edges have a certain *length* and a *direction* (along the axes of the system of coordinates), they can be regarded as *vectors* ($\vec{a}, \vec{b}, \vec{c}$) pointing from the origin to an equivalent point at the respective neighbouring unit cell. In this regard, it is also said that these vectors *span* the unit cell.

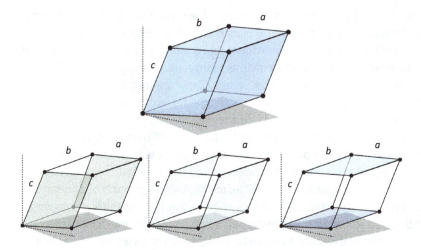

Figure 3.2: All unit cells are parallelepipeds, geometric bodies that are confined by six faces, with three pairs of congruent faces that lie in parallel planes.

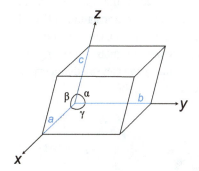

Figure 3.3: About the definition of the six lattice parameters.

As the only operation that is allowed to build the final crystal from unit cells is to replicate and adjoin them along all three spatial directions, the composition of the unit cell must correspond to the sum formula of the compound that has been crystallized.

3.1.2 Lattice points and motifs/bases

If we place an infinitesimal small point, i.e., a mathematical point at each corner of the unit cell of an infinite large bundle of unit cells, these points will form a *lattice*. A lattice is an arrangement of mathematical points in space (3D) or in the plane (2D), or on a line (1D), in which all points have the same surrounding. One consequence of this definition is that a lattice in its mathematical sense is infinitely extended. Real crystals, however, are not infinitely large and do have boundary surfaces. The surface-to-volume ratio of a crystal is a very important property that affects its chemical and physical behaviour as we will see later. But for now, we can approximate real crystals as quasi-infinitely extended as a crystal even with an edge length as small as 1 mm consists of approx. 10^{27} unit cells.

A lattice is characterized by its lattice vectors. These are translation vectors, which transfer the lattice points to the other ones along the three directions a, b, and c, and their magnitudes (or lengths) correspond to the periodicity along these directions, respectively. In that regard, a crystal has a lattice-*like* structure.[1] Of course, the lattice itself is not a physical object and is not part of the crystal structure; it only tells us that the building blocks of a crystal are arranged in a way that can be mathematically expressed by three lattice vectors. So, the lattice is the Aufbau or construction principle of a crystal.

Crystals are, of course, chemical entities and, therefore, the unit cell is not an empty body, but filled with atoms. Each (mathematical) lattice point *represents* a (real) chemical motif. This motif that is also called base can be a single atom or a very complicated molecule. It is important to realize that only in a very few cases atoms are located *at* the lattice points of the *underlying* lattice of a crystal structure, if we choose to place the lattice points at the corners of the unit cell. In fact, we are free to place the lattice points everywhere in the unit cell. As all constituents of a motif are subject of the same translation principle (expressed by the lattice vectors) they all build the *same* translation lattice, although parallel shifted with respect to each other; however, they are all congruent (superimposable). An example of applying the translation principle to a motif consisting of a schematic three-atomic molecule is shown in Figure 3.4. Finally, we can say that a crystal structure is a kind of a convolution function of a lattice with its motif.

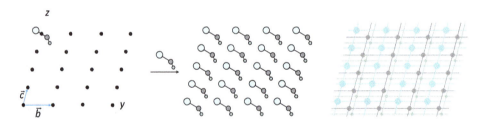

Figure 3.4: Application of the translation principle on the motif of a three-atomic molecule (left and middle); all constituents of a motif give congruent (superimposable) translation lattices (right).

---

1 The author is indeed a fan of precise wording. However, he has also noticed that in recent years there is an increasing tendency of nit picking in the scientific community. As an example, there are some people, who always raise their voice as soon as someone talks about two-dimensional materials and preach that only three-dimensional materials exist, because every object, and let it be a single atom, is already three-dimensional. Likewise, if they detect phrasings such "a crystal has a lattice structure", they immediately express their concern that in such wordings the necessary distinction between a lattice (a mathematical construct) and a (real) structure is dismissed.

3.1.3 The seven crystal systems

Interestingly, the number of principally different unit cell shapes – expressed by their metric – is very limited. All crystals can be classified according to not more than seven crystal systems. These seven crystal systems differ with respect to their lattice parameters; however, the classification scheme is not based upon the metric itself, but rather upon the symmetry of the crystal. The presence of certain symmetry elements that are characteristic for each crystal system leads to certain requirements on the metric. Furthermore, the crystal systems are also defined by their maximum point symmetry.

1. For crystals belonging to the *triclinic* crystal system there is either no symmetry present (except the symmetry element of identity; in crystallography it is expressed by the symbol 1, which is common to all objects) or maximal a centre of inversion, $\bar{1}$. This leads to no restrictions concerning the metric at all; the six lattice parameters can have any conceivable value.
2. If exactly one twofold rotational axis is present or one mirror plane or both a twofold axis and a mirror plane in the same direction, it requires that two of the angles are 90°; all other parameters are not restricted. Those crystals belong to the *monoclinic* crystal system.
3. Crystals which possess *three* twofold axes of rotation, or two perpendicular mirror planes and *one* further twofold axis of rotation require that all angles are 90°, but the length of the lattice vectors are not restricted. Such crystals belong to the *orthorhombic* crystal system.
4. The *tetragonal* crystal system is characterized by the presence of *one* fourfold axis of rotation or *one* fourfold rotoinversion axis. This leads to the restriction that all angles have to be 90° and that two of the lattice vectors must have the same length.
5. The presence of *one* threefold axis of rotation or *one* threefold rotoinversion axis can be realized by two different metrics, either by a rhombohedral axis system, in which all angles are equal as well as all length are equal, or by a hexagonal axis system, in which two angles have to be 90° and the remaining one 120°, and two of the lengths have to be identical. These are the characteristics of crystals belonging to the *trigonal* crystal system.
6. Crystals that possess *one* sixfold axis of rotation or *one* sixfold rotoinversion axis are classified into the *hexagonal* crystal system that requires that one angle is 120° and the other two equal to 90°, and that two of the vectors are of the same magnitude.
7. Finally, the *cubic* crystal is characterized by the presence of *four* (independent) threefold axes of rotation. This can only be achieved if all angles are 90° and if all lengths are identical.

In Table 3.1, the restrictions concerning the metric of the unit cell and the maximum point symmetry of the underlying translational lattice for the seven crystal systems are specified.

Table 3.1: The seven crystal systems, their restrictions concerning the metric, and their maximal point symmetry.

Crystal system	Restrictions concerning		Maximal point symmetry
	Length of axes	Angles of the cell	
Triclinic	None*	None*	$\bar{1}$
Monoclinic	None*	$\alpha = \gamma = 90°$	$2/m$
Orthorhombic	None*	$\alpha = \beta = \gamma = 90°$	mmm
Tetragonal	$a = b$	$\alpha = \beta = \gamma = 90°$	$4/mmm$
Trigonal	$a = b$	$\alpha = \beta = 90°, \gamma = 120°$ or	$\bar{3}m$
	$a = b = c$	$\alpha = \beta = \gamma$	
Hexagonal	$a = b$	$\alpha = \beta = 90°, \gamma = 120°$	$6/mmm$
Cubic	$a = b = c$	$\alpha = \beta = \gamma = 90°$	$m\bar{3}m$

*The respective parameters can have any conceivable value.

3.2 The 14 Bravais lattices – centring of unit cells

If we consider unit cells in which lattice points are only at the corners, then these unit cells or lattices are called *primitive*. The seven primitive unit cells or lattice types of the seven crystal systems are shown in Figure 3.5. The trigonal crystal system is special, because there are two types of trigonal crystals: if we want to describe all of them with primitive unit cells only, then we need two different unit cell shapes. Some of the crystals of the trigonal crystal system can be described with the same unit cell shape as in the hexagonal crystal system, namely, a hexagonal prism. However, there are also crystals of the trigonal crystal system, which must be described in their primitive lattice form as a *rhombohedron*. The restrictions concerning the metric are here: $a = b = c$ and $\alpha = \beta = \gamma$. Such a primitive rhombohedral unit cell, however, can always be converted into a cell with hexagonal axes/metric ($a = b$, $\alpha = \beta = 90°$ and $\gamma = 120°$), see below. This means that although there is a rhombohedral lattice, there is no rhombohedral crystal system.

A primitive unit cell comprises exactly one base or motif, and it is always the smallest possible unit cell. As there is usually not just one possibility to choose the lattice vectors for a given lattice, and because in certain cases the smallest possible unit cell is not the best choice, there are some rules concerning the choice of the unit cell:

1. The unit cell should be as *small* as possible. This means that the lattice vectors should be as *short* as possible.
2. At the same time, the unit cell should *reflect* the *symmetry* of the lattice. This means that the lattice vectors should run *parallel* to symmetry *axes* or *perpendicular* to *planes* of symmetry.
3. An additional rule is that the lattice vectors should be *orthogonal* (or *hexagonal*), i.e., they should enclose an angle of 90° (or 120°).

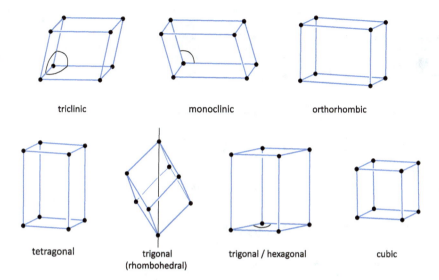

Figure 3.5: Overview of the seven primitive unit cells or lattice types of the seven crystal systems. In the primitive form, there are two different axis systems for crystals of the trigonal crystal system, a rhombohedral one with restrictions on the metric of $a = b = c$ and $\alpha = \beta = \gamma$ (typically but not necessarily, the angles are $\neq 90°$) and a hexagonal one with the metric $a = b$, $\alpha = \beta = 90°$ and $\gamma = 120°$.

The described relationships and the rules for choosing an appropriate unit cell are illustrated in Figure 3.6 using a two-dimensional example.

In principle, there are four types of centrings:
- Primitive, identical with no centring, symbol P.
- Base-centring, which means, that there is an additional lattice point at *one* of the faces, and dependent on which face it is, the symbol is A, B, or C. (Note: the face A is defined by the lattice vectors $\vec{b}$ and $\vec{c}$, the face B by $\vec{a}$ and $\vec{c}$, and the face C by $\vec{a}$ and $\vec{b}$).
- Body-centring with one additional lattice point right at the centre of the cell, symbol I (from the German word "Innen" = inside).
- Face-centring with additional lattice points at *all* faces; symbol F.

In addition, there is one more special type of centring that only exists for the trigonal crystal system, the rhombohedral centring, which is a special kind of body-centring, with two additional lattice points at the fractional coordinates (⅓, ⅔, ⅔) and (⅔, ⅓, ⅓); the symbol is R. This kind of centring is advantageous for those trigonal crystals whose smallest (primitive) unit cell is rhombohedral. In such rhombohedral cells, the threefold axis of rotation runs along the body-diagonal, something that should be avoided. The R centring leads to cells with hexagonal metric and the threefold axis of rotation is then running along the c-axis, in accordance with crystallographic convention.

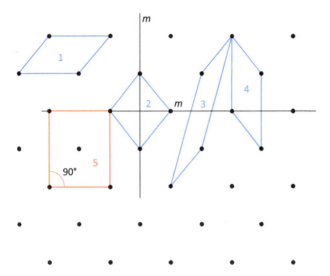

Figure 3.6: A section of a lattice in which all possible primitive unit cells are oblique; four of those possible unit cells are shown in blue (1, 2, 3, and 4). In these cases, it is preferable to choose a larger, a centred cell (shown in orange, 5) in order to achieve a higher number of right angles (or angles of 120°). In this way, the unit cell reflects clearly and in the best way the symmetry of the lattice, which has two perpendicular mirror planes (shown as black lines, *m*). In the centred cell 5, the lattice vectors are running perpendicular to the mirror planes.

In addition to the 7 primitive lattices, there are only further 7 centred lattices, giving in total 14 lattices that represent principally different translation lattices. This finding was the work of Auguste Bravais, 1811–1863, a French physicist, crystallographer, and polymath. He was able to show in 1848 that other types of lattice-like arrangements of points in 3D do not exist (the work was published only two years later [17]). The number 14 does not seem obvious at first glance. If we have seven crystal systems and at least four kinds of centrings, then the total number of different cells including centrings should be 28. There are two reasons that this number is reduced to 14: (i) some of the conceivable 28 lattice types can be converted into each other, meaning that they are identical; for instance a body-centred monoclinic lattice (type *I*) can always be converted to a one-sided face-centred monoclinic lattice (type *A*, *B*, or *C*); (ii) certain types of centring are forbidden in certain crystal systems for symmetry reasons; for instance, a one-sided face-centred lattice is not compatible with cubic symmetry.

In Figure 3.7, all 14 Bravais lattices together with their Pearson symbols are shown in a table-like overview; the crystal systems form the columns of this table and the type of centrings the rows.

The statement that the 14 Bravais lattices include 7 primitive and 7 centred lattices bears a potential misunderstanding that was, amongst others, picked up by Kettle and Norrby [18]. Of course, to every centred lattice there is a corresponding primitive one; it

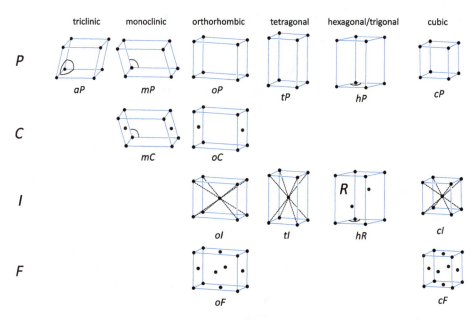

Figure 3.7: Overview of the 14 Bravais lattices, arranged according to the seven crystal systems. Below the unit cells the corresponding Pearson symbols are given, which consist of a small letter for the crystal system and a large letter for the type of centring. In order to avoid confusion between triclinic and tetragonal, the old term "anorthic" is chosen for the triclinic system, the corresponding symbol for triclinic-primitive is therefore, *aP*. The symbol *hP* is used for both primitive lattices of the trigonal and hexagonal system.

is just that the corresponding unit cells do not show the full symmetry of the underlying lattice. Two examples are depicted in Figure 3.8, the primitive unit cells of the body-centred and face-centred cubic lattice.

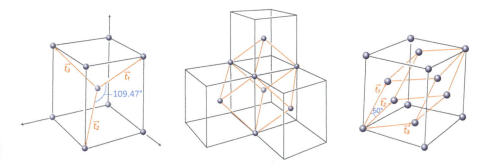

Figure 3.8: The shortest vectors ($\vec{t_1}, \vec{t_2}, \vec{t_3}$) between lattice points of the body-centred cubic lattice are running along the body-diagonals with a length of (0.5, 0.5, 0.5) and enclose an angle of 109.47° (left); the resulting primitive unit cell is a rhombohedron with that opening angle (middle); the shortest vectors between lattice points of a face-centred cubic lattice are running along the face-diagonals with a length of (0.5, 0.5, 0) and enclose an angle of 60° (right).

3.3 Crystallographic point groups (crystal classes) – the morphology of crystals

In this section, we will look at the symmetry of the outer shapes of crystals, i.e., the *morphology* of crystals. Although crystals exist in an incredible variety of shapes, from a symmetry point of view, they can be classified into only 32 principally different so-called *crystal classes*. All these 32 classes have different names and refer to a specific symmetry of crystal shapes. The specification of the symmetry is done by specifying the (crystallographic) point group. As the semantic difference between the terms crystal class (a geometric shape with a specific point symmetry) and crystallographic point group (the point group itself) is very subtle, the two terms are often used interchangeably by crystallographers and mineralogists.

The procedure for determining the crystal class is as follows:
1. We investigate the outer shape of a crystal and determine all point symmetry elements that can be found on this specific crystal. This is done systematically according to the so-called crystallographic viewing directions, which are different for each of the seven crystal systems (see Table 3.2).
2. We write down all symmetry elements that are clearly distinguishable from each other, i.e., we restrict ourselves to those which represent an independent (unique) symmetry element and are not transformed into each other by other symmetry elements or are generated automatically by the presence of other symmetry elements.
3. All crystals that have the same set of point symmetry elements belong to the same crystal class.

Note that the crystal class of a crystalline compound can also be unambiguously derived from its space group (see below). However, it is often the case that real crystal samples develop a crystal shape that has higher (hypermorphy, pseudo symmetry) or lower (hypomorphy) symmetry.

Table 3.2: Overview of the specified viewing directions for the seven crystal systems.

Crystal system	Crystallographic viewing directions
Triclinic	–
Monoclinic	b (c)*
Orthorhombic	a, b, c
Tetragonal	$c, a, [110]$
Trigonal	c, a
Hexagonal	$c, a, [210]$
Cubic	$c, [111], [110]$

*By convention, the direction in which symmetry elements appear in the monoclinic crystal system is the b-direction. However, in some areas, particularly in Eastern Europe, the c-direction is preferred.

3.3.1 Point symmetry elements

Because there are some special features concerning the external symmetry of crystals and because a different notation scheme for the specification of point symmetry is used (the international notation scheme of Hermann and Mauguin) compared to the point symmetry of molecules (here, the notation scheme after Schoenflies [19] is applied), it seems to be useful to shortly recap the different types of point symmetry that can be present in crystals.

Identity

All objects, even the most asymmetric ones, have at least one element of symmetry: the *identity*. It may seem somewhat contradictory that completely asymmetrical objects also have a symmetry element, but we can consider the identity also as a onefold axis of rotation. The respective operation is a rotation of 360° around an arbitrarily oriented axis that runs through the object. After this rotation the object is indistinguishable from its original appearance, which satisfies the prerequisite of being a symmetry operation (see Figure 3.9). Alternatively, we can consider the respective operation of the identity element as "do nothing".

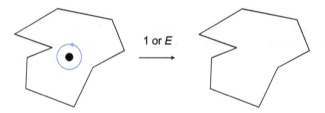

Figure 3.9: The symmetry element identity can also be regarded as a onefold axis of rotation; the corresponding symmetry operation is a rotation by 360° around that axis.

The Schoenflies symbol for identity is E (German for Einheitsoperator = unitary operator); however, in the international system, the onefold axis of rotation is specified, and the symbol is simply 1. This symmetry element is only explicitly given for a symmetry specification for an object if it is the only symmetry element.

Centre of inversion

Objects with a centre of inversion (also called symmetry centre) are constructed in such a way that there are always two corresponding parts, which have the same distance from this symmetry centre but are located in exactly opposite directions of space. In mathematical terms, for every point at the coordinates x, y, z, there must also be an identical point at the coordinates $-x, -y, -z$. The respective symmetry operation is a point reflection. Note that in two dimensions, a centre of inversion is identical with a twofold axis of rotation (see below).

The small letter *i* is used as a symbol for the inversion centre in the molecular world, while in crystallography, the symbol is just a 1 with a line above it: $\bar{1}$. It is pronounced as "one bar" or alternatively "bar one". It can be assumed that the original meaning is "minus one". The presence of a centre of inversion is only specified for an object, if it is – apart from the identity – the only symmetry element present; otherwise it is left out.

A regular octahedron and hexahedron (i.e., cube) are polyhedra that have a centre of inversion (see Figure 3.10), while a regular tetrahedron has no centre of inversion.

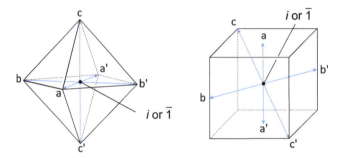

Figure 3.10: Polyhedra (octahedron, left, and cube, right) with a centre of inversion. Selected pairs of points (a, a', b, b', c, c') are shown that are transformed into each other by point mirroring at the inversion centre.

Mirror planes

An object that remains unchanged upon reflection (or is indistinguishable from its initial state after that reflection) has mirror symmetry (also called reflection or bilateral symmetry). The corresponding symmetry element is a mirror plane (or in 2D, a mirror line). Graphically, the mirror plane is symbolized in 2D drawings with a simple, continuous line and abbreviated with the letter *m* (from mirror). The skull of human beings and the ones of most animals have (almost perfect) mirror symmetry. An object can, of course, also have more than one mirror symmetry, for example, the letter H has a vertical and a horizontal mirror line (see Figure 3.11). An octahedron and a cube, for instance, have nine mirror planes each; in Figure 3.11 seven of the mirror planes of a cube are visualized.

Different from the molecular world, in which mirror planes are denoted with slightly different symbols if there is an additional rotational symmetry present (for instance, σ_v for mirror planes with normal vectors perpendicular to the axis of rotation and σ_h for mirror planes with a normal vector parallel to the axis of rotation), the orientation of the mirror plane is implicitly given with respect to the crystallographic viewing directions (see above). The resulting symbol for the two mirror lines

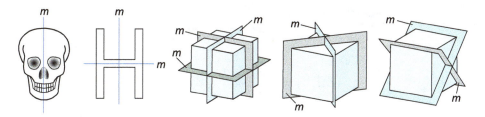

Figure 3.11: Examples of objects with mirror symmetry. Seven out of nine mirror planes of the cube are shown; it should be clear how the remaining two mirror planes are oriented.

of the letter H is simply *mm*. In Section 3.3.2, we will look at some examples, in order to explain in more detail which, and in which order the symmetry elements of crystal shapes are denoted.

Axis of rotation

Objects that are indistinguishable after a rotation of 360°/*n* (symmetry operation) have an *n*-fold axis of rotation (symmetry element; geometrically a line). We already introduced the "1-fold" axis of rotation; however, this is only equivalent to the symmetry element identity and not an element that specifies rotational symmetry. While ordinary objects can have rotational symmetry of any order, the rotational symmetry of crystals is limited to the presence of twofold, threefold, fourfold, and sixfold axes of rotation, which are shown together as point patterns with their graphical symbols synoptically in Figure 3.12. Note that a sixfold axis of rotation includes a twofold and a threefold rotational axis, and that a fourfold axis of rotation includes a twofold axis. Note further that these point ensembles also possess mirror symmetry. Axes of rotation are always specified *parallel* to the respective crystallographic viewing directions.

There are no axes of rotation of the order five and there are none with an order larger than six. Such rotational symmetries are incompatible with the periodicity of crystals, i.e., the periodicity of their underlying translational lattices. This is the reason why there are only 32 point groups in crystallography. A proof that only these rotational symmetries are allowed in crystals, sometimes also called the crystallographic restriction theorem, can be found in Ref. [20].

In solid-state chemistry, when we are working with crystal structures, for instance in crystal structure visualization software, we often look at point patterns, in which the points represent actually different kind of atoms. Often, the different types of atoms are visualized with (a) different colours and (b) with different sizes. We have to be aware of the fact that a symmetry relationship is only given, if the nature of the atoms is identical that are transformed into each by the symmetry operation. In this regard, it is a good exercise to consider how many mirror planes and how many axes of rotation are present in the patterns of Figure 3.13 (the answer is given in the caption of this figure).

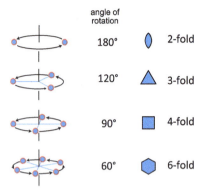

Figure 3.12: Overview of the axes of rotation in crystallography with point patterns of corresponding symmetry, the specification of the rotation angle in degree, and the graphical symbol of the respective symmetry element.

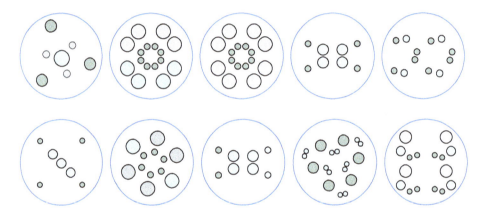

Figure 3.13: Point patterns with different types of rotational and mirror symmetry; Top from left to right: one set of three mirror planes, one threefold axis of rotation; one mirror plane; two sets of two mirror planes, fourfold axis of rotation; two mirror planes, one twofold axis of rotation; no mirror plane, no axis of rotation; Bottom from left to right: two mirror planes, one twofold axis of rotation; two mirror planes, one twofold axis of rotation; one mirror plane, no axis of rotation; one set of three mirror planes, one threefold axis of rotation; one mirror plane, no axis of rotation.

Rotoinversion axis

A rotoinversion axis is a symmetry element in which two operations are carried out subsequently; it is a *combined* symmetry operation: first, a rotation by 360°/n, where n is the order of the rotoinversion axis, and then a point mirroring at the centre (an inversion). Analogous to pure rotations, crystallographic relevant are only rotoinversions of the order 1, 2, 3, 4, and 6. The first two are not particularly interesting, since they can be described as an inversion centre ($\bar{1} = i$) or as a simple reflection ($\bar{2} = m$); they do not constitute symmetry elements on their own.

3.3 Crystallographic point groups (crystal classes) – the morphology of crystals

Depending on the order of the rotoinversion axis, there might be additional symmetry elements present. The following rules apply:
(1) Rotoinversion axes with odd order imply the simultaneous presence of a (simple) inversion centre.
(2) Rotoinversion axes of even order automatically contain an axis of rotation of half of the order: the rotoinversion axis $\bar{4}$ contains a twofold axis of rotation, and the rotoinversion $\bar{6}$ always automatically contains a threefold axis of rotation.
(3) If the order of the rotoinversion axis is even and not divisible by 4, then there is automatically a mirror plane perpendicular to this axis. As we only consider crystallographic rotoinversions, this principle holds for the two- and sixfold rotoinversion axis.

All crystallographic relevant rotoinversion axes are exemplified by respective point patterns with that symmetry in Figure 3.14.

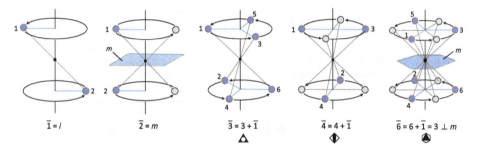

Figure 3.14: Crystallographic relevant rotoinversions illustrated by point patterns. A sixfold rotoinversion axis automatically contains a mirror plane perpendicular to the rotoinversion axis. Sites which are only temporarily realized (i.e., after the rotation part) are shown as grey circles.

While in crystallography this kind of symmetry is expressed with rotoinversion axes, the same type of symmetry is expressed in molecular chemistry by rotary-reflection axes. For every rotoinversion axis there is an equivalent rotary-reflection axis; however, usually the order has to be adjusted. The respective symmetry operation of an n-fold rotary-reflection axis (Schoenflies symbol = S_n; S stands for the German word "Spiegel" = mirror) is a combination of a rotation about the axis by 360°/n, followed by a reflection at a mirror plane that is oriented perpendicular to that axis, i.e., a horizontal mirror plane, symbolized in the Schoenflies notation as σ_h. If we compare the patterns that are produced by either a rotoinversion axis or a rotary-reflection axis, we see that the following equivalency exists: $\bar{1} \stackrel{\triangle}{=} S_2, \bar{2} \stackrel{\triangle}{=} S_1, \bar{3} \stackrel{\triangle}{=} S_6, \bar{4} \stackrel{\triangle}{=} S_4$, and $\bar{6} \stackrel{\triangle}{=} S_3$.

It can be shown relatively easily that rotary reflections with an odd order (S_1, S_3, S_5, ...) imply the additional presence of (i) a pure rotation and (ii) a horizontal mirror plane and that those of even order (S_2, S_4, S_6, ...) imply the presence of an inversion centre, if the order is not divisible by 4.

In Table 3.3, all point symmetry operations and elements for both the Schoenflies notation and the notation scheme after Hermann & Mauguin are summarized.

Table 3.3: Overview of all point symmetry elements and operations as well as the respective symbols according to Schoenflies and Hermann & Mauguin (H & M).

Symmetry element	Symmetry operation	Symbol after Schoenflies	Symbol according to H & M (international system)
Identity	Leave the object as it is (do nothing)	E	1
Centre of inversion	Point reflection at the centre	i	$\bar{1}$
Mirror plane	Reflection at a plane	σ (σ_v, σ_h, σ_d)	m
Axis of rotation	Rotation by 360°/n around that axis	C_n (n = 1, 2, 3, ..., ∞)	n (n = 1, 2, 3, 4, 6)
Rotoinversion axis	Rotation by 360°/n around that axis followed by a point reflection at the centre	–	$\bar{n}$ (n = 1, 2, 3, 4, 6)
Rotary-reflection axis	Rotation by 360°/n around that axis followed by a reflection at a mirror plane perpendicular to that axis	S_n (n = 1, 2, 3, ..., ∞)	–

3.3.2 The 32 crystal classes

Now that all point symmetry elements have been discussed, we can proceed with the symmetry characterization of crystals according to their outer shapes. We will go through each of the different crystal systems and discuss the occurrence and some of the morphological features of the crystal classes within the crystal systems.

Triclinic crystal system

In the triclinic crystal system, there are only two crystal classes, either characterized by identity (1) or by a centre of inversion ($\bar{1}$) as the only symmetry element. The first one is called the pedial crystal class. In this class all crystal faces are unique and are not symmetry related to each other. Such faces are called pedions (gr. for "foot face", i.e., basal face). This crystal class is very rare; only a very few crystals belong to this class. In the crystal class $\bar{1}$ there is only a centre of inversion, meaning that pairs of faces are related to each other through this centre of inversion. Such faces are called pinacoids (gr. for board, meaning pair of parallel faces); thus, this is the pinacoidal class, see also Figure 3.15. Chalcanthite ($CuSO_4 \cdot 5\ H_2O$) and many feldspar minerals, for instance albite ($NaAlSi_3O_8$) and kyanite (Al_2SiO_5), belong to this class.

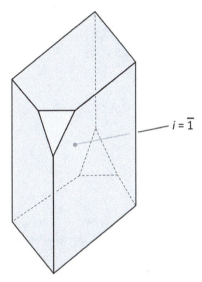

Figure 3.15: Example of a crystal shape with point symmetry $\bar{1}$.

Monoclinic crystal system

In the monoclinic crystal system, symmetry is present in only one direction; by convention, this should be the direction parallel to the b-axis. In the monoclinic-domatic class with point symmetry m, only a mirror plane (perpendicular to that direction) is present. Potassium tetrathionate ($K_2S_4O_6$) and the mineral clinohedrite ($CaZnSiO_4 \cdot H_2O$) belong to this crystal class.

If only a twofold axis of rotation (parallel to the b-direction) is present, the crystal class is called monoclinic-sphenoidal (from Greek sphenoeides = wedge-like). Axes of rotation that have no mirror plane perpendicular to that axis are called *polar* axes of rotation. Polar rotation axes can be recognized by the fact that the faces of the crystal develop differently in the respective direction and in the opposite direction (see Figure 3.16, middle). Prominent examples of crystals that belong to this class are tartaric acid and sucrose.

The remaining class in this crystal system is the monoclinic-prismatic class with point symmetry $2/m$. Note that the presence of both a mirror plane (perpendicular to the b-direction) and a twofold axis of rotation (parallel to the b-direction) automatically generates a centre of inversion ($\bar{1}$). However, this symmetry element is only specified in the point group symbol, if it is the only symmetry element of that geometric shape.

The monoclinic-prismatic class is by far the most frequently occurring crystal class and classical examples of crystals that belong to this class are gypsum ($CaSO_4 \cdot 2\,H_2O$), orthoclase ($KAlSi_3O_6$), and oxalic acid dihydrate ($C_2O_4 \cdot 2\,H_2O$).

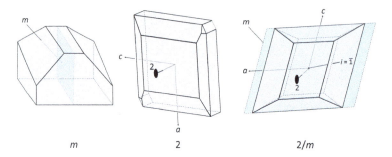

Figure 3.16: Examples of crystal shapes with point symmetry m (left), 2 (middle), and 2/m (right).

Orthorhombic crystal system

In the orthorhombic crystal system, symmetry elements are present in all three viewing directions (see Table 3.2). In the disphenoidal crystal class the point group symmetry is 222, i.e., there are only twofold axes of rotation along all directions (a, b, and c), but no other symmetry elements. The disphenoid faces that define this group consist of two faces on top of the crystal and two faces at the bottom of the crystal that are rotated to each other by 90°. Epsomite, i.e., $MgSO_4 \cdot 7\,H_2O$, is a classic example of a mineral with that morphology. A crystal shape with point symmetry 222 is depicted in Figure 3.17 (left).

The orthorhombic-pyramidal crystal class has two perpendicular mirror planes and a twofold axis of rotation along the third direction (by convention this should be the c-direction), leading to the point group $mm2$. In fact, this twofold axis of rotation is generated by the two perpendicular mirror planes. This is one of the most well-known symmetry rules: any pair of two perpendicular mirror planes generates automatically a twofold axis of rotation at their line of intersection. Therefore, one could argue that it would be sufficient to specify the crystal class as mm. The problem is that there are further symmetry relationships in the orthorhombic crystal system: both $2m$ and $m2$, that is, the combination of a twofold axis of rotation and a vertical mirror plane in a second viewing direction automatically generates a further mirror plane in the third direction. This means that $2m$ and $m2$ result equally in $mm2$. Furthermore, there is also the risk of confusion between $2m$ and $2/m$. In order to avoid such possible confusion, the IUCr has decided that in the orthorhombic system always three symbols or all three viewing directions have to be specified. Struvite ($NH_4MgPO_4 \cdot 6\,H_2O$) and hemimorphite ($Zn_4Si_2O_7(OH)_2 \cdot H_2O$) are minerals that often develop beautiful crystal shapes with the point symmetry $mm2$; see also Figure 3.17 (middle).

The remaining third crystal class in the orthorhombic crystal system is the orthorhombic-dipyramidal class with point symmetry mmm (short notation). As each two of the three mirror planes are oriented perpendicular to each other they automatically generate twofold axes of rotation at their lines of intersection, so that the full symbol is $2/m\ 2/m\ 2/m$. Amongst others, anglesite ($PbSO_4$), aragonite ($CaCO_3$), baryte ($BaSO_4$),

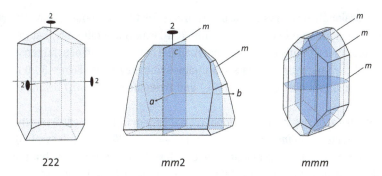

222   mm2   mmm

Figure 3.17: Examples of crystal shapes with point symmetry 222 (left), mm2 (middle), and mmm (right). Note that the right shape also possesses three twofold axes of rotation (not shown).

olivine $(Mg,Fe)SiO_4$, and topaz $(Al_2(F,OH)_2SiO_4)$ belong to this crystal class. A typical shape that possesses the point symmetry mmm is shown in Figure 3.17 (right).

Tetragonal crystal system

The characteristic feature of the tetragonal crystal system is the presence of either one fourfold axis of rotation or one fourfold rotoinversion axis (running parallel to the c-axis by convention). If they are the only symmetry elements, the respective crystal classes are called tetragonal-pyramidal (the reagent N-iodosuccinimide that is used in organic chemistry for iodination reactions belongs to that class) and tetragonal-disphenoidal (the mineral cahnite $(Ca_2B(OH)_4AsO_4)$ belongs to this class). Note that the fourfold rotoinversion implies simultaneously the presence of a simple twofold axis of rotation. This is reflected in the respective graphical symbol for that symmetry element, a filled ellipse in a hollow square standing on a corner. Figure 3.18 shows an example of two shapes that exhibit these symmetries. Note that both shapes have no centre of inversion.

In the first viewing direction, there can also be a mirror plane present (perpendicular to the c-axis), which leads to shapes with point symmetry 4/m (tetragonal-dipyramidal crystal class, see Figure 3.18). Scheelite $(CaWO_4)$, fergusonite $(YNbO_4)$, and the scapolite group (a group of tectoalumosilicates) are minerals belonging to this crystal class. Note that the mirror plane perpendicular to the fourfold axis of rotation generates a centre of inversion.

The second viewing direction in the tetragonal crystal system is the a-direction. The third viewing direction is not the b-direction because it is symmetry-equivalent to the a-direction, but the direction [110], i.e., parallel to the face diagonal. In both the second and third viewing direction there can be further twofold axes of rotation or mirror planes present.

Crystals belonging to the tetragonal-trapezohedral crystal class have the point symmetry 422. Please note that such a shape neither possesses a centre of symmetry nor any mirror plane (see Figure 3.18). Retgersite $(NiSO_4 \cdot 6 H_2O)$ belongs to this crystal class.

The tetragonal-scalenohedral crystal class is characterized by the point symmetry $\bar{4}2m$ (Figure 3.18). Chalcopyrite (CuFeS$_2$) and stannite (Cu$_2$FeSnS$_4$) are minerals that belong to this class. Note that a tennis ball[2] and a baseball too have this kind of symmetry.

The ditetragonal-pyramidal crystal class with point symmetry 4mm (Figure 3.18) is very rarely encountered. Diabolite (Pb$_2$CuCl$_2$(OH)$_4$) and fresnoite (Ba$_2$TiSi$_2$O$_8$) belong to this class.

The most symmetrical shape in the tetragonal crystal system (the holohedry) has the holohedral point symmetry 4/mmm – in short notation – or 4/m 2/m 2/m (full symbol), respectively, and is called the ditetragonal-dipyramidal class. Holohedral means that the point symmetry is identical with the point symmetry of the underlying lattice. Anatase (TiO$_2$) and zircon (ZrSiO$_4$) are the most prominent mineral members of this class, and they often appear as an overall stretched (pseudo)octahedron (not shown in Figure 3.18).

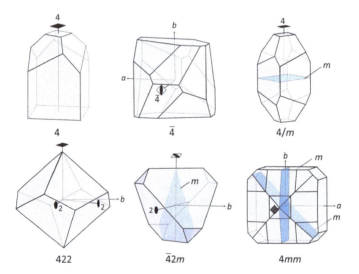

Figure 3.18: Examples of crystal shapes with point symmetry 4 (top, left), $\bar{4}$ (top, middle), 4/m (top, right), 422 (bottom, left), $\bar{4}2m$ (bottom, middle), and 4mm (bottom, right). Note that only the generating (unique) symmetry elements are drawn.

Trigonal crystal system

The characteristic feature of crystal shapes of the trigonal crystal system is that either one threefold axis of rotation or one threefold rotoinversion axis (both running parallel to the c-direction) is present. If these are the only symmetry elements present, they

---

[2] In geometry, there is an interesting theorem that is called the tennis ball theorem. The seams of a tennis ball divide the ball in two identical dumbbell-shaped pieces. The tennis ball theorem states that a smooth, closed curve on a surface dividing the sphere into two equal areas must have *at least four* points of inflection.

belong to the trigonal-pyramidal (3) and rhombohedral crystal class ($\bar{3}$), respectively (see Figure 3.19). It is interesting to note that further symmetry elements appear only in one further direction. In that direction a further twofold axis of rotation (trigonal-trapezohedron crystal class with point symmetry 32; the most common mineral in that class is quartz (SiO_2)), a further mirror plane (ditrigonal-pyramid crystal class, point symmetry 3m; a well-known mineral group that belong to this class is tourmaline), or both a twofold axis of rotation and a mirror plane can be present (ditrigonal-scalenohedral crystal class with point symmetry $\bar{3}2/m$, the minerals calcite ($CaCO_3$) and corundum (Al_2O_3) belong to this class); see also Figure 3.19.

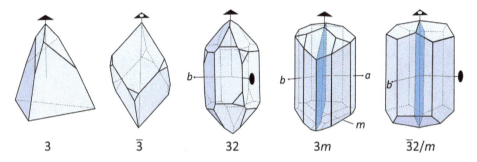

Figure 3.19: Examples of crystal shapes with point symmetry 3, $\bar{3}$, 32, 3m, and $\bar{3}2/m$.

Hexagonal crystal system

The hexagonal crystal system is characterized by the presence of either one sixfold axis of rotation or one sixfold rotoinversion axis (parallel to the c-direction). If these are the only symmetry elements, they constitute the hexagonal-pyramidal class with point symmetry 6 and the trigonal-dipyramidal class with point symmetry $\bar{6}$ (see also Figure 3.20). Note that the sixfold rotoinversion axis implies a simple threefold axis of rotation and a mirror plane perpendicular to that axis (3/m). Both crystal classes are very rare.

A sixfold axis of rotation and a mirror plane perpendicular to that axis leads to the hexagonal-dipyramidal crystal class with point symmetry 6/m. The apatite mineral group ($Ca_5(PO_4)_3(Cl,F,OH)$) belongs to this class.

In both the second (along the a-direction) and third viewing direction (along [210]) of the hexagonal crystal system, there can be further twofold axes of rotation or mirror planes present.

Crystals belonging to the hexagonal-trapezohedral class have the point symmetry 622. As in the analogous tetragonal class 422, such a shape neither possesses a centre of symmetry nor any mirror plane. The high-temperature modification of quartz (β-SiO_2) belongs to that crystal class, see also Figure 3.20.

The dihexagonal-pyramidal class is characterized by the point symmetry 6mm (Figure 3.20). Wurtzite, the hexagonal variety of ZnS, zincite (ZnO), and greenockite (CdS) belong to this crystal class.

The ditrigonal-dipyramidal crystal class has point symmetry $\bar{6}m2$. Benitoite (BaTiSi$_3$O$_9$) belongs to that crystal class. Note that shapes with a sixfold rotoinversion axis have a *trigonal* habit.

The most common crystal class in the hexagonal crystal system is the dihexagonal-dipyramidal class with point symmetry $6/m\,2/m\,2/m$ (full symbol) or $6/mmm$ (short symbol), respectively (see Figure 3.20). It constitutes the holohedry of the hexagonal crystal class. The mineral beryll (Be$_3$Al$_2$Si$_6$O$_{18}$), the elements magnesium and zinc for instance, and graphite belong to this crystal class.

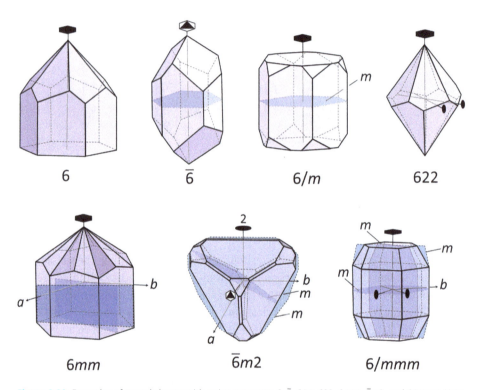

Figure 3.20: Examples of crystal shapes with point symmetry 6, $\bar{6}$, 6/m, 622, 6 mm, $\bar{6}m2$, and 6/mmm. Note that only the generating (unique) symmetry elements are drawn. Note further that shapes with a sixfold rotoinversion axis possess a *trigonal* crystal habit.

Cubic crystal system

The cubic crystal system is characterized by *four* threefold axes of rotation. In a cube, the shape with the highest point symmetry ($m\bar{3}m$, the hexakisoctahedral crystal class), these are the four body-diagonals that enclose an angle of $54°44'\,8'' = \tan^{-1}\sqrt{2}$ with an edge of the cube. However, there are further shapes with less symmetry that belong also to the cubic crystal system.

3.3 Crystallographic point groups (crystal classes) – the morphology of crystals

The least symmetrical shape in the cubic crystal system has point symmetry 23 and it is called the tetartoidal class. A tetartoid is a polyhedron consisting of 12 congruent but irregular pentagonal faces (see Figure 3.21) and is a special kind of pentagonal dodecahedron. Such a shape has only twofold and threefold axes of rotation but no centre of inversion or mirror planes, leading to enantiomorphy. Sodium chlorate ($NaClO_3$) and the mineral ullmannite (NiSbS) belong to this crystal class.

If there is an additional mirror symmetry perpendicular to the first viewing direction present, we arrive at the diploidal (or also sometimes called the disdodecahedral) crystal class with point symmetry $2/m\bar{3}$. The additional mirror symmetry generates a centre of inversion and converts the threefold axes of rotation into threefold rotoinversion axes. The mineral pyrite (FeS_2) sometimes crystallizes in that class, although the most abundant shape of pyrite crystals is a cube (a case of hypermorphy).

The gyroidal crystal class has point symmetry 432. Compared to the tetartoidal class, such shapes are characterized by additional fourfold axes of rotations, but again there is no centre of inversion nor are there any mirror planes, so that enantiomorphy is present. Only a very few substances crystallize in this gyroidal class; one example is $K_3Pr_2(NO_3)_9$ [21].

The hextetrahedral crystal class has point symmetry $\bar{4}3m$; that is also the point group of the symmetry of a regular tetrahedron. Note that in this crystal class no mirror symmetry perpendicular to the first viewing direction is present and that the threefold axes of rotation (along [111]) are polar, meaning that there is no centre of inversion. The hextetrahedral class is very common. For instance, the cubic variety of ZnS (sphalerite), the copper halides CuCl, CuBr, CuI, the semiconductors GaAs, InAs, and InSb as well as the minerals colusite ($Cu_{13}VAs_3S_{16}$), coloradoite (HgTe), and tetrahedrite ($Cu_{10}(Fe,Zn)_2Sb_4S_{13}$) belong to this crystal class.

The crystal class with the highest symmetry is the hexoctahedral class with point symmetry $m\bar{3}m$ (full symbol $4/m\bar{3}2/m$). It combines the rotational symmetry of the crystal class 432 with mirror symmetry perpendicular to the fourfold and twofold axes of rotation, turning the threefold axes of rotation into threefold rotoinversion axes and generating a centre of inversion. Many elements and compounds crystallize in this crystal class, for instance the metals Cu, Ag, Au, and Pt and the minerals halite (NaCl), galenite (PbS), fluorite (CaF_2), magnetite (Fe_3O_4), and spinel ($MgAl_2O_4$). Note that the regular octahedron also has the symmetry of this class.

Examples of shapes of all crystal classes of the cubic crystal system are gathered in Figure 3.21.

In Table 3.4 an overview of all 32 crystallographic point groups is given, both in long and short notation, and how they are distributed over the seven crystal systems. In addition, the viewing directions for each crystal system are given, the number of rotation axes of different order, the number of mirror planes, and whether this class has an inversion centre.

It is a good exercise to find *all* – not only the generating (unique) – symmetry axes and planes at the shapes that are depicted in Figures 3.16–3.21.

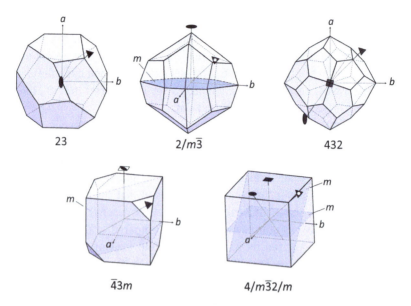

Figure 3.21: Examples of crystal shapes with point symmetry 23, $2/m\bar{3}$, 432, $\bar{4}3m$, and $4/m\bar{3}2/m$. Note that only the generating (unique) symmetry elements are drawn.

Table 3.4: Overview of all 32 crystallographic point groups.

No.	Crystal system	Name	Long symbol	Short symbol	Viewing directions	2	3	4	6	m	$\bar{1}$?
1	Triclinic	Triclinic-pedial*	1	1	None	–	–	–	–	–	No
2		Triclinic-pinacoidal	$\bar{1}$	$\bar{1}$		–	–	–	–	–	Yes
3	Monoclinic	Monoclinic-sphenoidal**	1 2 1	2	b	1	–	–	–	–	No
4		Monoclinic-domatic***	1 m 1	m		–	–	–	–	1	No
5		Monoclinic-prismatic	$1\frac{2}{m}1$	$\frac{2}{m}$		1	–	–	–	1	Yes
6	Orthorhombic	Orthorhombic-disphenoidal	2 2 2	2 2 2	a, b, c	3	–	–	–	–	No
7		Orthorhombic-pyramidal	m m 2	m m 2		1	–	–	–	2	No
8		Orthorhombic-dipyramidal	$\frac{2\ 2\ 2}{m\ m\ m}$	m m m		3	–	–	–	3	Yes

3.3 Crystallographic point groups (crystal classes) – the morphology of crystals

Table 3.4 (continued)

No.	Crystal system	Name	Long symbol	Short symbol	Viewing directions	2	3	4	6	m	$\bar{1}$	?
9	Tetragonal	Tetragonal-pyramidal	4	4	c, a, [110]	–	–	1	–	–		No
10		Tetragonal-disphenoidal	$\bar{4}$	$\bar{4}$		1	–	–	–	–		No
11		Tetragonal-dipyramidal	$\dfrac{4}{m}$	$\dfrac{4}{m}$		–	–	1	–	1		Yes
12		Tetragonal-trapezohedral	4 2 2	4 2 2		4	–	1	–	–		No
13		Tetragonal-pyramidal	4 m m	4 m m		–	–	1	–	4		No
14		Tetragonal-scalenohedral	$\bar{4}$ 2 m	$\bar{4}$ 2 m		3	–	–	–	2		No
15		Ditetragonal-dipyramidal	$\dfrac{4\,2\,2}{m\,m\,m}$	$\dfrac{4}{m}$ m m		4	–	1	–	5		Yes
16	Trigonal	Trigonal-pyramidal	3	3	c, a	–	1	–	–	–		No
17		Rhombohedral	$\bar{3}$	$\bar{3}$		–	1	–	–	–		Yes
18		Trigonal-trapezohedral	3 2 1	3 2		3	1	–	–	–		No
19		Ditrigonal-pyramidal	3 m 1	3 m		–	1	–	–	3		No
20		Ditrigonal-scalenohedral	$\bar{3}\,\dfrac{2}{m}\,1$	$\bar{3}$ m		3	1	–	–	3		Yes
21	Hexagonal	Hexagonal-pyramidal	6	6	c, a, [210]	–	–	–	1	–		No
22		Trigonal-dipyramidal	$\bar{6}$	$\bar{6}$		–	1	–	–	–		No
23		Hexagonal-dipyramidal	$\dfrac{6}{m}$	$\dfrac{6}{m}$		–	–	–	1	1		Yes
24		Hexagonal-trapezohedral	6 2 2	6 2 2		6	–	–	1	–		No
25		Dihexagonal-pyramidal	6 m m	6 m m		–	–	–	1	6		No
26		Ditrigonal-dipyramidal	$\bar{6}$ m 2	$\bar{6}$ m 2		3	1	–	–	4		No
27		Dihexagonal-dipyramidal	$\dfrac{6\,2\,2}{m\,m\,m}$	$\dfrac{6}{m}$ m m		6	–	–	1	7		Yes

Table 3.4 (continued)

No.	Crystal system	Name	Long symbol	Short symbol	Viewing directions	2	3	4	6	m	$\bar{1}$	?
28	Cubic	Tetartoidal	2 3	2 3	c, [111], [110]	3	4	–	–	–	–	No
29		Diploidal	$\frac{2}{m}\bar{3}$	$m\bar{3}$		3	4	–	–	–	3	Yes
30		Gyroidal	4 3 2	4 3 2		6	4	3	–	–	–	No
31		Hextetrahedral	$\bar{4}$ 3 m	$\bar{4}$ 3 m		3	4	–	–	6	–	No
32		Hexoctahedral	$\frac{4}{m}\bar{3}\frac{2}{m}$	$m\bar{3}m$		6	4	3	–	9	–	Yes

*gr. *pedion* = basal face; this is a crystal shape in which each type of face is represented only once.
**gr. *sphenoid* = wedge; denotes a pair of faces, a dihedron, in which the two faces can be transferred into each other by a twofold axis of rotation.
***gr. *doma* = house; denotes a dihedron in which the two faces can be transferred into each other by a mirror plane.
****gr. *scalenos* = limping, uneven; a scalenohedron denotes a polyhedron delimited by uneven triangles.

3.4 Further (micro-) translations – glide planes and screw axes

In order to be able to completely describe the symmetry of crystals, two further symmetry elements have to be introduced: glide planes (in 2D they are called glide lines) and screw axes (they necessarily require three-dimensional periodicity). Both symmetry elements are *coupled* symmetry elements, meaning that in both cases, the respective symmetry operations comprises two consecutive operations that have to be carried out. And both symmetry elements have a translational (or displacement) component, meaning that they do *not* belong to the *point* symmetry elements discussed above.

Altogether there are three types of symmetry elements in which translations are involved:
– translation vectors at which translations are carried out
– glide planes on which glide reflections (short: glides) are carried out, and
– screw axes along which screw rotations are performed.

Pure translations by whole units (from lattice point to lattice point along the three lattice vectors) were already introduced. And centrings must also be understood as translations, although in that case, lattice points (or motifs) are not shifted along the lattice vectors of the unit cell and the length of the translation vectors is smaller than that of an entire unit cell length.

The coupling of a reflection with a translation results in a glide reflection; the coupling of a rotation with a translation results in a screw rotation. Since the resulting patterns that are obtained by these symmetry operations must be compatible with the translation lattice, the translational component in restricted to certain values (see below).

3.4.1 Glide planes

Altogether, there are six different glide plane types that can occur in crystals: *a*, *b*, *c*, *n*, *d*, and *e*. For the first three (simpler) glide planes *a*, *b*, and *c*, the name coincides with the direction of the translation component: in the glide plane *a*, the translation takes place along *a* or the *x*-axis, and so on. And, for all of them, the translational component is exactly ½ of the unit cell length. With regard to the orientation of the mirror components of the glide planes to the coordinate system, they all have in common that these planes are either at least perpendicular to an axis or even parallel to a plane spanned by two vectors of the coordinate system. This means, fortunately, we do not have to deal with glide planes where the plane is completely oblique in the unit cell.

Often, the impact of glide planes on motifs is visualized with the help of a drawing of a 2D projection of the unit cell. We will do this here, too. For subsequent considerations, we uniformly select the (*a*,*b*) plane as the drawing plane. The *c*-axis should be perpendicular to this plane for simplicity (but this does not always have to be the case). For cases in which the glide plane is oriented parallel to the drawing plane, for simplicity, the concrete position along the *c*-axis should be $c = 0$, meaning that the glide plane and the drawing plane coincide. Later it will become clear that the position of the glide plane does not necessarily have to be at the height $c = 0$; often it is also at the positions $c = ½$ or $c = ¼$. If the glide plane is perpendicular to the drawing plane, it is graphically symbolized by a dashed line; if the glide plane lies parallel to the projection plane, an angle is placed at the top left and an additional arrowhead specifies the translation direction.

For each of the simple glide planes we only have to consider two scenarios, because the direction of the glide plane – expressed by the normal vector, which is perpendicular to this plane – cannot coincide with the direction of the translation component. All scenarios are visualized in Figure 3.22.

The remaining glides planes are the *diagonal* glide plane *n*, the *diamond-like* glide plane *d*, and the *double* glide plane *e*. The translation component for diagonal glide planes is $½(\vec{a}+\vec{c})$ if they are parallel to the c plane, $½(\vec{b}+\vec{c})$, if they are parallel to the a plane, and $½(\vec{a}+\vec{c})$, if they are parallel to the b plane. In the tetragonal and cubic crystal system there are also glide planes with a translation component of $½(\vec{a}+\vec{b}+\vec{c})$. In drawings, diagonal glides planes that are oriented perpendicular to the drawing plane are graphically symbolized by a dashed-dotted line.

The glide planes *d* are called diamond-like glide planes, because they are present, indeed, in the crystal structure of diamond. These are also diagonal glide planes, but the translation component is only ¼ instead of ½, i.e., either $¼(\vec{a}+\vec{b})$, or $¼(\vec{b}+\vec{c})$, or

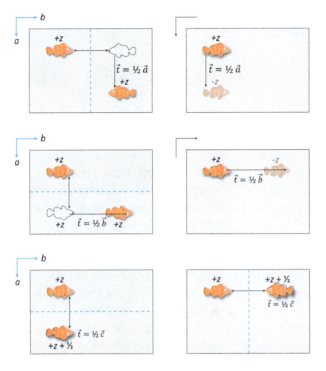

Figure 3.22: Illustrations of the impact of a glide plane a (top), b (middle), and c (bottom) on a fish motif. Positions that are only temporarily realized (i.e., after the reflection) are shown as dashed contour of the motif. Initially, the fish is located in all cases at the top left part of the unit cell somewhere above the drawing plane at the height +z along the c-axis.

¼($\vec{a} + \vec{c}$). Glide planes of type d can occur only in face- or body-centred structures. In the tetragonal and cubic crystal systems, the translation component can also be ¼($\vec{a} + \vec{b} + \vec{c}$). Diamond-like glide planes are drawn as an arrow-dashed line, if they are oriented perpendicular to the drawing plane.

Finally, the glide planes abbreviated with the letter e are reserved for instances in which actually two glide planes of the type a, b, or c that have the *same* orientation but different translation directions are present. Glide planes e that are oriented perpendicular to the drawing plane are symbolized by a double-dotted-dashed line.

The impact of the glide planes n, d, and e is illustrated by examples in Figure 3.23.

3.4.2 Screw axes

Screw rotations are a combination of pure rotations around an axis with a certain translation parallel to that axis. As rotations are involved, it is clear that for crystallographic objects the same restrictions apply as for pure rotations or rotoinversions, i.e., only two-,

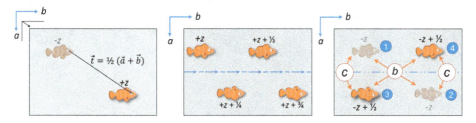

Figure 3.23: Left: example of a diagonal glide plane *n* that is oriented parallel to the drawing plane; Middle: example of a diamond-like glide plane *d* that is oriented perpendicular to the drawing plane; Right: example of a double glide plane *e*, here consisting of a glide plane *b* and *c* that are both oriented perpendicular to the drawing plane; here, there are two pairs of fishes, which are related by symmetry by the glide plane *b*, namely, each of the two fishes below and above the drawing plane (①+② and ③+④), while the relationship by symmetry of the pairs ①+③ and ②+④ is given by the glide plane *c*.

three-, four-, and sixfold screw axes are possible. The translational component is given with the help of an index m: A screw axis n_m, where m is an integer and always smaller than n, describes a corresponding operation in which an object must first be rotated by $360°/n$ about the axis and then has to be translated parallel to the axis by m/n units of an entire unit cell length. Furthermore, by definition, the rotation should always be carried out in the sense of a right-handed coordinate system. The rotational direction therefore results from the fact that the x-axis is rotated towards the y-axis. The translation is then always parallel to the z-axis, i.e., perpendicular to the (x,y) plane.

For pure rotations carrying out an n-fold (pure) rotation n times is equivalent to the identity operation. Analogous considerations for screw axes lead to the conclusion that carrying out an n_m screw rotation n times is equivalent to a lattice translation.

Applying the n_m scheme (with m always smaller than n) leads to the following screw axes:

2_1
$3_1, 3_2$
$4_1, 4_2, 4_3$
$6_1, 6_2, 6_3, 6_4,$ and 6_5

As the translational component of the 2_1 screw axis is ½ the impact of the respective symmetry operation on point patterns sometimes can be mixed up with that of a simple glide plane. Therefore, in the illustration in Figure 3.24 an asymmetric object is used instead of a point. However, for the sake of simplicity, indeed completely symmetrical points are used for the illustrations of the other screw axes.

If we now turn to threefold screw axes, an interesting property of screw axes become apparent, namely, that they can be *chiral*. A 3_1 screw axis produces a helical pattern that is usually called right-handed and is the mirror image of the left-handed helix created by a 3_2 screw axis (see Figure 3.25). Note that although the rotation is

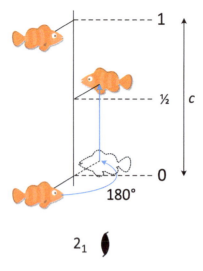

Figure 3.24: A section of a fish pattern having a 2_1 screw axis as a symmetry element. The associated operation consists of a rotation by 180° and a subsequent translation by ½ in the positive c- or z-direction. The graphical symbol consists of an ellipse with two small, curved hooks.

always carried out anticlockwise, a 3_2 screw axis produces a pattern that has an opposite handedness compared to the 3_1 screw axis. The corresponding point pattern of a 3_2 screw axis could also be reproduced by first rotating by −120° and then translation of ⅓ in the positive z-direction; however, because a negative rotation value is not permitted, this must be "compensated" for via the translational part, which is ⅔.

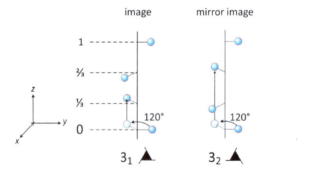

Figure 3.25: Point patterns generated by 3_1 and 3_2 screw axes, respectively, and the corresponding graphical symbols.

Among the fourfold screw axes, there is again an enantiomorphic pair (the right-handed 4_1 and the left-handed 4_3 screw axis) and the non-chiral 4_2 screw axis (Figure 3.26). Note that the screw axes 4_1 and 4_3 include a 2_1 screw axis and that the 4_2 screw axis includes a simple twofold axis of rotation.

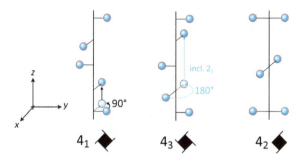

Figure 3.26: Illustrations of the screw axes of order 4 together with their graphical symbols. The two screw axes 4_1 and 4_3 form an enantiomorphic pair and behave like image and mirror image. Both contain a 2_1 screw axis, which is indicated at the 4_3 screw axis. The 4_2 screw axis is achiral and contains a simple twofold axis of rotation.

Turning to the remaining sixfold screw axes (illustrated in Figure 3.27), yet another new feature becomes visible. There are two enantiomorphic pairs (6_1 and 6_5 as well as 6_2 and 6_4), but while the point patterns that are generated by the 6_1 and 6_5 screw axes are *single* helices, the respective points of the 6_2 and 6_4 screw axes form intertwined *double* helices (see Figure 3.28). The 6_3 screw axis is (analogous to 4_2) nonchiral and includes a simple threefold axis of rotation. It is recommended as an exercise to verify that the following further relations are valid:

- the screw axes 6_1, 6_3, and 6_5 include a 2_1 screw axis
- the screw axes 6_2 and 6_4 include a simple twofold axis of rotation
- the screw axes 6_1 and 6_4 include a 3_1 screw axis
- the screw axes 6_2 and 6_5 include a 3_2 screw axis

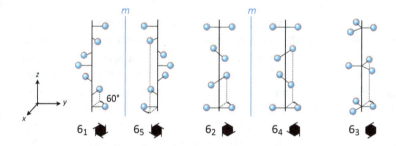

Figure 3.27: Overview of all screw axes of order 6 (together with their graphical symbols), each shown with a section of point patterns that are generated by these axes.

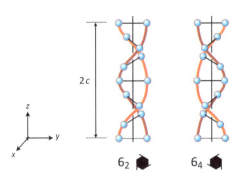

Figure 3.28: Illustration of the intertwined double helices that are generated by the screw axes 6_2 and 6_4.

3.5 The 230 space groups

From the combination of all point symmetry elements – inversion, rotation, reflection, mirroring, and rotoinversion – with the pure translation symmetry of all possible lattice types and the symmetry elements with translational components (glide planes and screw axes), we get exactly 230 different arrangements or types of periodic patterns in three dimensions. This was shown almost simultaneously and independently by Schoenflies [19] and Fedorov (1853–1919, a Russian mathematician, crystallographer and mineralogist) [22]. A proper definition of the term "space group" is that a space group is a *set of symmetry elements* – together with their respective symmetry operations – which *completely* describes the spatial arrangement of a 3D periodic pattern.

3.5.1 Nomenclature of space groups

The nomenclature of space groups is very similar to that of crystallographic point groups already discussed in Section 3.3. The name of a space group consists of a maximum of four symbols. In the so-called short notation, it can also be less. The space group symbol always begins with the Bravais type of the lattice, i.e., the indication of the type of centring (P, C, A, B, I, R, or F). The next three symbols specify symmetry elements with respect to the viewing directions, which vary from crystal system to crystal system (see Table 3.2).

Let us look at an example, namely $Pca2_1$, a space group that belongs to the orthorhombic crystal system. According to Table 3.2 the viewing directions in the orthorhombic crystal system are *a*, *b*, and *c*. This means that we have a primitive lattice (*P*), that there is a glide plane *c* perpendicular to the *a*-direction, a glide plane *a* perpendicular to the *b*-direction, and, finally, there is a 2_1 screw axis parallel to the *c*-direction. Please note that all letters in a space group symbol are set as italics, but all numbers are typeset normal.

What we should have in mind is that these specified symmetry elements in the space group symbol are not necessarily all symmetry elements that are present. However, they are sufficient, because the symmetry elements specified in this way are the so-called *generators*; from these, the complete set of symmetry elements can be derived. In cases of doubt, it is always recommended to have a look into the *International Tables for Crystallography, Volume A* [23], in which all space groups with their complete set of symmetry elements are listed (see below).

3.5.2 Deriving point groups from space groups

It is easy to derive the crystallographic point group from a given space group. The following steps are necessary:
- Leave out the Bravais type of the lattice.
- Convert symmetry elements with a translational component into their counterparts without them, i.e., glide planes become mirrors and screw axes pure axes of rotation.
- All point symmetry elements remain unchanged.

To give three examples: the point group of *Fddd* is *mmm*, *Pca2$_1$* becomes *mm2*, and *Ia$\bar{3}$d* is converted to *m$\bar{3}$m*.

3.5.3 International tables for crystallography

If you explore a crystal structure, it is always advisable to look simultaneously into the entry of the respective space group of that structure in the *International Tables for Crystallography, Volume A* [23]. Here, you will find information about:
- the long and short name of the space group
- the associated point group
- the crystal system to which the space group belongs
- a diagram with all symmetry elements and their locations within the unit cell
- a diagram of the general positions
- information about the choice of axes, the origin, and the asymmetric unit
- a list of all symmetry operations expressed as coordinate transformations
- a list of all principally different positions within the unit cell, i.e., Wyckoff positions, including their multiplicities and their site symmetry.

Diagram of symmetry elements and the general position diagram
Let us have a brief look at an example and explore some of the entries for the space group *P2$_1$/c* (belonging to the monoclinic crystal system), in its standard setting (unique axis = *b*, cell choice 1). The diagram of symmetry elements is shown in Figure 3.29 (left). It is drawn as projection along the *b*-axis, meaning we are looking at the (*a,c*) plane. Parallel to the projection plane is a glide plane, which is indicated by the oblique arrow

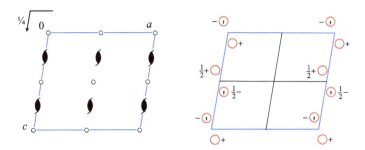

Figure 3.29: The symmetry element diagram (left) and the general position diagram (right) for the space group $P2_1/c$ in its standard setting viewed along the b-axis.

in the upper left corner. The height along the b-axis is ¼. In fact, ¼ always means ¼ *and* ¾, so there is a second glide plane at height $b = ¾$. There are also a number of 2_1 screw axes (ellipses with hooks) parallel to the b-direction, two of which are inside the cell and two more on the edges of the c-axis, at $c = ¼$ and ¾. Furthermore, there are inversion centres (drawn as small hollow circles) at all corners and all bisectors of the edges as well as in the centre of the cell.

If we now place an atom (symbolized with a red circle in Figure 3.29, right) at a general position (meaning that there is no point symmetry element) in the upper left part of the unit cell somewhere above the projection plane (indicated with a plus sign) and perform all the symmetry operations of the respective symmetry elements shown in Figure 3.29 (left), then we will end up with the complete scheme shown in Figure 3.29 (right). Some of the red circles are hollow, while others have a comma inside, meaning that these are mirror images of the ones without such a comma. Note that any reflection of a chiral motif turns it into its mirror image. The ½ is the short notation for "plus one half in the b-direction"; if an atom is transformed by a 2_1 screw axis from, say, initially "–" then the position along b is afterwards "–" plus one half, which is notated as "½ –".

Positions
This section inside the ITA for the space group $P2_1/c$ looks like this:

Coordinates

4   e   1   (1) x, y, z (2) $\bar{x}, y + 1/2, \bar{z} + 1/2$ (3) $\bar{x}, \bar{y}, \bar{z}$ (4) $x, \bar{y} + 1/2, z + 1/2$

2   d $\bar{1}$ ½, 0, ½   ½, ½, 0
2   c $\bar{1}$ 0, 0, ½   0, ½, 0
2   b $\bar{1}$ ½, 0, 0   ½, ½, ½
2   a $\bar{1}$ 0, 0, 0   0, ½, ½

In the first row, the general position (a site that is not located at a point symmetry element) is specified and then the different kinds of special positions are listed underneath. The first systematic elaboration of the point positions in the individual space groups was made by Ralph W. G. Wyckoff in 1922 [24], which is why the positions are also called Wyckoff positions. Wyckoff has labelled these positions with small Latin italicized letters (Wyckoff letters or Wyckoff symbols), starting with a small "a" for the most special position and then alphabetically ascending for the positions with lower symmetry – these letters are given in the second row. The order is, however, partly arbitrary. This is due to the fact that a special point position may occur multiple times, that is, may have the same site symmetry. Therefore, in these cases, you have to look up in the International Tables which site symmetry is exactly meant. In practice, for the specification of the position of an atom, often not only the Wyckoff letter but also the multiplicity (specified in the first column) is indicated, for instance, "$4e$". The multiplicity is the number of equivalent sites of a given position. They depend on the concrete site, i.e., if the position is general (x,y,z) or if the position coincides with the location of one or more point symmetry elements. This information is given in the third row and specifies the so-called site symmetry. The general position always has the site symmetry 1 (the identity symmetry element) and in the given example of the space group $P2_1/c$, there is only one other different kind of site symmetry, namely positions that lie at a centre of inversion ($\bar{1}$). From the fourth column on, the "Coordinates" are specified for the different positions; a bar above the letter is again to be interpreted as "minus". Surely you have noticed that the number of coordinates is identical to the given multiplicity in the first column, so the first row must be read as: if we have an atom that is located at a general position, i.e., the Wyckoff position $4e$, this general position (1) x, y, z (with site symmetry 1) has equivalent (symmetry-related) locations at (2) $\bar{x}, y+\frac{1}{2}, \bar{z}+\frac{1}{2}$, (3) $\bar{x}, \bar{y}, \bar{z}$, and (4) $x, \bar{y}+\frac{1}{2}, z+\frac{1}{2}$. Analogously, if we place an atom at the Wyckoff position $2a$, i.e., at the coordinates 0, 0, 0 there will be only one other symmetry-related position, namely 0, ½, ½.

In Chapters 4 and 5, we will make use of the Wyckoff positions when specifying crystal structures.

4 Densest sphere packings

In many crystal structures, the centres of some type of atoms are located in positions that correspond to the centres of spheres in *sphere packings*, and other types of atoms are in the interstitial sites (or voids) of such packings. These sphere packings, in particular *densest* (or closest) sphere packings, have relevance in such structures that are determined not by chemical but rather by geometrical features – meaning that they are arranged in a way that the space is filled to the maximum degree. Although there is an infinite number of densest sphere packings of varying kinds, two of them are of particular importance: the hexagonal closest packing (**hcp**) and the cubic closest packing (**ccp**). In 1611, Johannes Kepler, a German mathematician and astronomer, published his famous Kepler conjecture: it states that no arrangement of equally sized spheres filling space has a larger average density than that of the **hcp** and **ccp** arrangements. Interestingly, this conjecture was proven only a few years ago by Thomas C. Hales et al. [25–27].

Let us see how the **hcp** and **ccp** arrangements look like. We will start with one layer of densest packed (hard) spheres of equal size. The densest arrangement of spheres in a plane is a *hexagonal layer* of spheres. Each sphere is in contact with six neighbouring spheres within that layer. Between the spheres voids remain. The distance between one void and the *next-nearest* void is equal to the distance between the centres of two neighbouring spheres. The centres of the spheres are commonly denoted by A, while the voids are named B and C (see Figure 4.1). Now we add a second densest hexagonal layer on top of the first layer. In order to achieve a maximum density, the spheres should be placed at the indentations of the first layer, i.e., above the points B or C. Since we are dealing with hard spheres, it is clear that the spheres cannot be placed simultaneously over both positions; the distance is too close. Thus, we have two possibilities – we can place the second layer over the B or C positions; however, it is clear that identical arrangements result. Let us choose the position B for the moment (Figure 4.1, top-right). If we now add a third layer, there are again two possibilities. The centre of the spheres of this third layer can be either placed above the first layer A or above the positions marked with C (note the voids/indentations created by the second layer in Figure 4.1, top-right). This leads to two distinct *three-layer sequences*: *ABA* or *ABC* (Figure 4.1, bottom). In the first case the layers above and below the middle one are in the same positions (A) but in the second case the layers above and below the middle layer are in different positions (A and C). The two most simple densest sphere packings are obtained by repeating those stacking sequences, *ABAB* ... and *ABCABC* ..., and, in fact, these are the two most common stacking sequences that are found in nature.

The first one is called the hexagonal closest packing (**hcp**) and the second one, the cubic closest packing (**ccp**), and we will shortly see the reason for this nomenclature. Both packings have the same density, and the spheres fill 74.05% of the space. The perpendicular distance between two layers is $\sqrt{2/3} = 0.8165$. Each sphere is *in contact* with 12 neighbouring spheres (6 within the same layer, 3 above and 3 below the layer), i.e., the coordination number is 12. However, the coordination polyhedra are different: in the **ccp** arrangement, the centres of the neighbouring spheres build a *cuboctahedron*, whereas for the **hcp** arrangement, an *anticuboctahedron* is formed (see Figure 4.2).

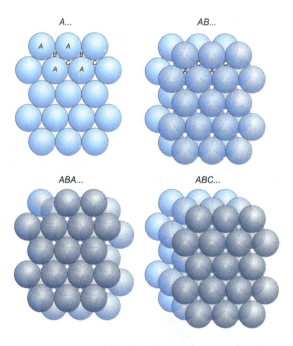

Figure 4.1: Sequential building of a three-layer slab from hexagonal densest layers, according to the **hcp** and **ccp** packing.

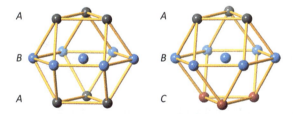

Figure 4.2: The coordination polyhedron of a sphere in the **hcp** (left, an anticuboctahedron) and **ccp** packing (right, a cuboctahedron); in both cases, the coordination number of the central sphere is 12. Note that the spheres are of the same type in the arrangements; they are coloured differently only for illustration purposes, emphasizing the different positions along the stacking direction.

Both kinds of densest sphere packings build a regular, strictly three-periodic array of spheres, meaning that they can be described in crystallographic terms.

4.1 The hexagonal closest packing

For the hexagonal unit cell of the **hcp** packing, it is most convenient not to choose one of the points A as the origin, because, if possible, the inversion centre should be at the origin (and thus also in the centre of the cell). The fact that the inversion centre cannot be located on the atoms is already evident from a symmetry examination of the coordination polyhedron: the anticuboctahedron has no inversion centre.

The hexagonal unit cell with the space group $P6_3/mmc$ (no. 194) comprises two spheres at the Wyckoff position 2c, with coordinates (⅓, ⅔, ¼) and (⅔, ⅓, ¾),[3] see Figure 4.3. Note that the planes normal to the c-direction that contain the centre of the spheres are mirror planes. If the spheres have unit diameter, then the a- (and b-) axis has the value 1, and the c-axis is then $2 \cdot \sqrt{2/3} = 1.633$.

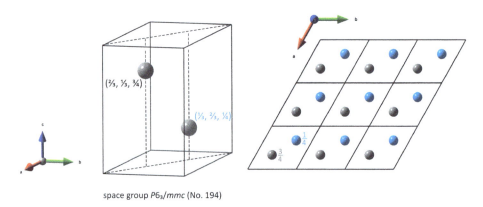

space group $P6_3/mmc$ (No. 194)

Figure 4.3: The unit cell of the **hcp** packing (left) and a projection of 3 × 3 unit cells along the c-axis (right). The spheres/atoms are drawn smaller than their actual size. Note that the spheres are of the same type and that they are coloured differently on the right panel for illustration purposes, only to emphasize their different heights along the c-axis.

In many textbooks about inorganic chemistry and solid-state chemistry and in numerous other sources on the internet, the illustrations of the unit cells of structures based on a **hcp** packing are incorrect (in the narrower sense) or at least misleading (in a wider sense). Let us dive a bit more deeply into this issue, which will be an excellent opportunity to foster our freshly gained knowledge concerning space groups and the *International Tables for Crystallography*.

---

[3] The Wyckoff position 2d with the positions (⅓, ⅔, ¾) and (⅔, ⅓, ¼) is equivalent.

A typical (but wrong) visual representation of the unit cell of an **hcp** arrangement that can be found in many textbooks and the illustration of the presence of a three-fold axis of rotation and a 6_3 screw axis are shown in Figure 4.4.

What do you notice? Well, the first thing is that while the unit cell depicted in Figure 4.3 is obviously centrosymmetric, the cells shown in Figure 4.4 are not. And this is clearly incorrect, because the space group of the **hcp** packing – $P6_3/mmc$ – has indeed a centre of inversion. And a comparison with a stripped-down version of the diagram of symmetry elements according to the International Tables reveals further discrepancies (Figure 4.5).

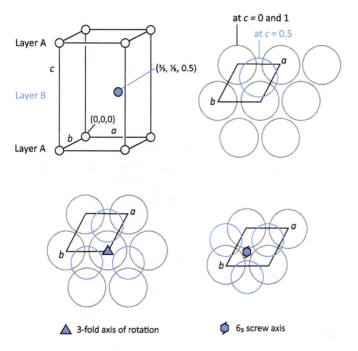

Figure 4.4: Two views of the hexagonal unit cell of an **hcp** arrangement according to many (incorrect) illustrations that can be found in textbooks and various sources on the internet (top), and visualization of the presence of the threefold axis of rotation and the 6_3 screw axis (bottom). Adapted and redrawn from Ref. [28].

The threefold axis of rotation (which is included in the sixfold rotoinversion axis) is not located at (0, 0, z), and vice versa, the 6_3 screw axis is not located at (⅓, ⅔, z) but at (0, 0, z). This means that in pictures like the one shown in Figure 4.4, the *relative* positions of the atoms are correct, but they are wrong with respect to the unit cell and to the locations of the symmetry elements. Applying the symmetry framework of the space group $P6_3/mmc$ to an atom at the origin (0,0,0) never gives a second coordinate (⅔, ⅓, 0.5), and vice versa. We can conclude that you have to be alarmed if you see **hcp** arrangements with atoms at the corners of the unit cell. Unfortunately, such

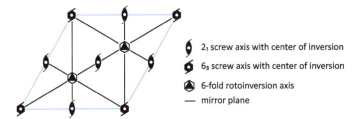

Figure 4.5: Locations of selected symmetry elements (see legend) in the unit cell of the space group $P6_3/mmc$.

wrong descriptions of the **hcp** arrangement do not occur only in textbooks but can also be found in research papers.

4.2 The cubic closest packing

In contrast to the **hcp** arrangement where a primitive unit cell is appropriate, a crystallographic meaningful description of the **ccp** arrangement is based on a face-centred cubic (**fcc**) unit cell (space group $Fm\bar{3}m$, no. 225), see Figure 4.6. The stacking direction of the densest layers is running parallel to the <111> directions, i.e., the body-diagonal of the unit cell. The spheres are placed at the Wyckoff positions $4a$ (0, 0, 0).

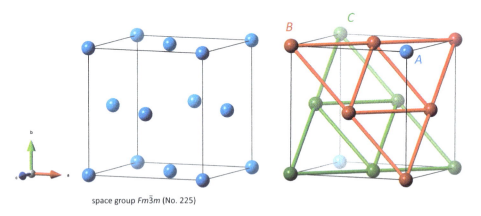

space group $Fm\bar{3}m$ (No. 225)

Figure 4.6: The unit cell of the **ccp** packing (left) and the stacking sequence of hexagonal densest layers emphasized by different colours and by bonds between the spheres within the same layer (right).

4.3 Voids in the hcp and ccp packing

In a densest sphere packing of n spheres, there are always n octahedral voids and $2n$ tetrahedral voids, no matter if it is the hexagonal or cubic densest sphere packing. It is important to realize where these voids are located and how the corresponding coordination polyhedra around these voids are linked together. It facilitates the understanding of structures that are based on densest sphere packings, in which these voids are occupied by other kind of atoms.

First of all, the common feature of both kinds of packings is how the tetrahedral and octahedral voids are generated. The tetrahedral voids are generated by three spheres in one layer (the base of the tetrahedron) and an additional sphere in the neighbouring layer, positioned at the indentation of these three spheres (at the tip of the tetrahedron), see Figure 4.7, left. The octahedral voids are generated by three spheres in one layer and three spheres in the neighbouring layer that are located again in the indentations of the layer below or above. Sometimes, it is not easy to see the octahedral shape of those six spheres if we look perpendicular to these stacked layers because the eye is more likely to recognize a trigonal antiprism at first. But this special trigonal antiprism is nothing else than a regular octahedron (see Figure 4.7, right).

The octahedral voids in the **hcp** packing are at the Wyckoff position $2a$ [(0, 0, 0) and (0, 0, ½)], i.e., at all corners of the unit cell and at all edge bisectors along the c-direction. The octahedra form face-sharing columns parallel to the c-direction, i.e., along the stacking direction of the successive layers. They are further edge-connected to the neighbouring face-sharing columns of octahedra (see Figure 4.8).

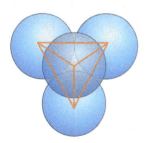 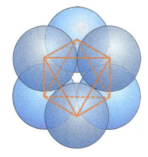

Figure 4.7: Tetrahedral (left) and octahedral (right) voids generated by four and six spheres in contact in densest sphere packings.

The tetrahedral voids in the **hcp** packing are located at the Wyckoff position $4f$ with coordinates (⅓, ⅔, z), (⅔, ⅓, z+½), (⅓, ⅔, –z), and (⅔, ⅓, –z+½) with $z = ⅛$ (0.125), see Figure 4.9, left. The tetrahedra can be divided into two sets, one with tetrahedra pointing upwards (in the positive c-direction, i.e., the stacking direction), and one with the tetrahedra pointing downwards. They will be called T+ and T– sites from now on. T+ and T– voids form pairs with a common face, while the T+/T– voids among themselves are only corner-

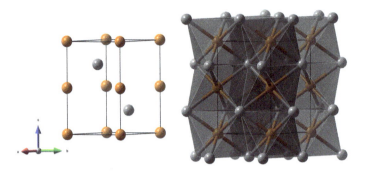

Figure 4.8: The unit cell of the **hcp** packing built by grey spheres and the locations of the centres of the octahedral voids shown as orange spheres (left). Slightly extended view of the unit cell of the **hcp** packing with the respective coordination polyhedra.

connected; T+ voids are further edge-connected to T− voids and vice versa, see also Figure 4.9, left.

The fact that face-sharing tetrahedral voids are present in the **hcp** arrangement is the reason for the fact that ionic/highly polar compounds with one type of atom building the **hcp** packing and another type of atom filling *all* T sites do not exist, because the atoms at the centre of the tetrahedra are very close together.

Now let us take a look at the octahedral and tetrahedral voids in the **ccp** arrangement. The octahedral voids in the **ccp** arrangement are located at the Wyckoff position 4b, (½, ½, ½), i.e., they are at the centre and at all edge-bisectors of the unit cell, see Figure 4.10 (left). The octahedral sites fall – as the packing itself – on a face-centred cubic lattice, displaced by (½, ½, ½) with respect to the lattice built by the **ccp**. In contrast to the **hcp** arrangement, the octahedra have only common edges but not common faces.

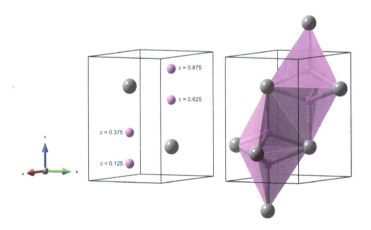

Figure 4.9: The unit cell of the **hcp** packing, together with the locations of the centres of the tetrahedral voids, shown as pink spheres (left), and visualization of the respective coordination polyhedra (right).

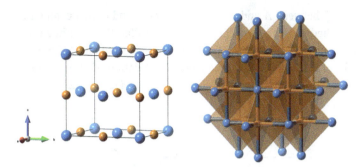

Figure 4.10: The unit cell of the **ccp** packing, together with the locations of the centres of the octahedral voids, shown as orange spheres (left), and visualization of the respective coordination polyhedra (right).

The tetrahedral voids of the **ccp** arrangement are at the Wyckoff position 8c, (¼, ¼, ¼), (¼, ¼, ¾), (¼, ¾, ¾), (¾, ¾, ¾), (¾, ¾, ¼), (¾, ¼, ¾), (¼, ¾, ¼), and (¾, ¼, ¼). The centres of the tetrahedral voids fall on a primitive cubic lattice, as highlighted in Figure 4.11, left. As in the **hcp** arrangement, they can be divided into the two sets pointing upward (T+) and downward (T−) sites; however, the direction here are the <111> directions, the body diagonals. Opposed to the **hcp** arrangement, in the **ccp** arrangement, the tetrahedra share only edges but not faces. The tetrahedra of one set are corner-connected to the tetrahedra of the same type and edge-connected to the tetrahedra of the other set.

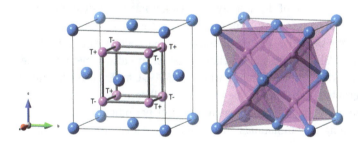

Figure 4.11: The unit cell of the **ccp** packing, together with the locations of the centres of the tetrahedral voids, shown as pink spheres (left), and visualization of the respective coordination polyhedra (right).

4.4 Locations of interstitial sites along the packing direction

The last thing we want to look at in our discussion of sphere packings is the location of the voids along the packing direction of densest layers. With reference to Figures 4.1 and 4.7, one can see that the centre of an octahedral void between two densest packed layers (AB) is at the C position and that it lies exactly midway between the two layers. Because the letter C is already reserved for a third densest layer above that site, it is common practice to name the octahedral interstitial sites with Greek letters: γ is located between A and B, β between A and C, and α between B and C. In a competing notation system, the tetrahedral site layers, T+ and T−, are called β and α; however, we will call them just T+ and T− sites/layers. These layers are located along the packing direction at a height of 0.25 and 0.75, respectively. The location of all interstitial site layers between a two-layer slab of densest layers is shown in Figure 4.12.

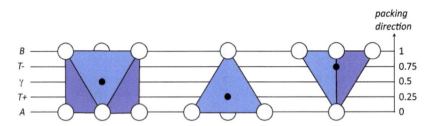

Figure 4.12: The location of the centre of an octahedral (left), upward-pointing T+ (middle) and downward-pointing T− (right) site between two densest packed layers (AB).

4.5 Stacking variants/polytypes

As stated at the beginning of this chapter, there is an infinite number of different densest sphere packings. The **hcp** and **ccp** arrangement are just the two most simple ones. However, the stacking sequence of densest layers does not have to strictly follow the AB ... (**hcp**) or ABC ... (**ccp**) pattern. Any other stacking sequence that satisfies the condition that two layers of the same type are never stacked above each other also leads to a densest sphere packing. All these possible variants together are also sometimes collectively called Barlow packings, after William Barlow (1845–1934, an English amateur geologist and crystallographer) [29]. Most of the metals crystallize according to the **hcp** or **ccp** arrangement. However, more complicated, but still periodic stacking variants occur, for instance, among the lanthanoids and actinoids. A widely used nomenclature system for stacking variants was introduced by Jagodzinski (1916–2012, a German physicist, mineralogist, and crystallographer) in 1949 [30]: A layer surrounded by two different layers is given the letter c (for cubic) and a layer that is surrounded by two identical layers is given the letter h (for hexagonal). This means that any stacking sequence expressed by the letters A, B, and C can be converted to a sequence of letters h and c, and vice versa. In the case of the indication of the sequences by h and c – completely and exactly as is expressed in the notation AB ... and ABC ... by the ellipses that they repeat infinitely – only the periodically occurring repeat unit is indicated. In Figure 4.13, a few examples are shown from which the principle should become clear.

The packing with the sequence hc is also known as the **double-hcp** arrangement and comprises four layers. As this is the packing sequence of lanthanum (and some

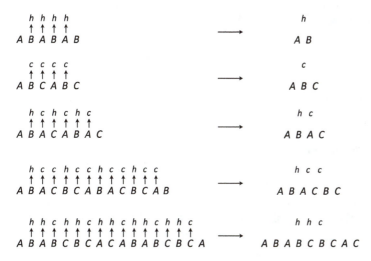

Figure 4.13: Examples of stacking sequences, expressed by the letters A, B, and C, their conversion into a sequence of Jagodzinski symbols h and c, and their reduction to their basic repeat unit.

other lanthanoids and actinoids), it is sometimes also called the lanthanum packing or lanthanum structure type (*Strukturbericht* type A3′).

A disadvantage of the notation system of Jagodzinski is that it is not easy to immediately recognize how many layers are in the repeat unit. Note, for instance, that the repeat unit of *hcc* comprises six layers, whereas *hhc* comprises nine layers. Therefore, sometimes, yet another nomenclature is used that specifies the number of layers (n) in the repeat unit and the lattice type: nC stands for cubic (the only type to consider in the cubic system is $n = 3$), nH stands for hexagonal, nR for rhombohedral, and nT for trigonal (but not rhombohedral) structures. The Jagodzinski symbol h is thus denoted as $2H$, hc as $4H$, and hhc as $9R$. One drawback of this specification method is that if the number of layers is large, there is generally more than one stacking sequence that leads to the same symbol. As mentioned before, some metals show a more complicated stacking sequence of densest layers than the two prototypical **hcp** and **ccp** arrangements. Compounds that exist in two or more different combinations of layer-like structural units that differ only in the stacking sequence (and orientation of the layers with respect to the crystallographic axes) of the individual layers, but not in their principal structure and composition are called *polytypes*. A compound that is particularly prone to the occurrence in very many polytypes is silicon carbide (SiC), also known as carborundum. The compound exists in more than 200 different polytypes. The $3C$, $4H$, $6H$, and $15R$ structures are the most common polytypes, but the existence of a $1200R$ (!) polytype with a lattice constant of $c = 3,015$ Å is also documented [31].

5 Some important structure types

5.1 Metal structures – sphere packings in action

Metals are characterized by the fact that their valence electrons are delocalized over the whole crystal. The attractive forces between the atoms of the metal are *non-directional*. For this reason, the structures of metals are highly influenced by geometrical factors, i.e., by the question of how to fill the space as effectively as possible, or in other words, how to achieve a dense or densest packing of the atoms. In addition to these space-filling considerations, there are other secondary factors that determine the final structure of a metal such as the electron configuration or valence electron concentration. Nonetheless, the majority of the metals of the periodic table (see Figure 5.1) crystallizes in the **ccp**, **hcp**, or **double-hcp** arrangement, i.e., in one of the different densest packing arrangements. A second set of metals crystallize in the somewhat less dense body-centred cubic (**bcc**) packing and only a few remaining metals of the periodic table crystallize in other special structures; we will look at a few of them.

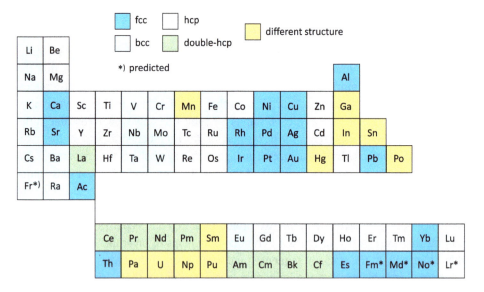

Figure 5.1: Overview of the structure of metals along the periodic table.

5.1.1 Cu (ccp or fcc) (*Strukturbericht* type A1)

The copper structure has been designated as the prototypical structure of all elements crystallizing in the face-centred cubic (**fcc**) packing, the cubic closest packing (**ccp**). The stacking sequence of layers of densest-packed spheres is *ABCABC* ... and the octahedral and tetrahedral voids remain unoccupied. As already explained in Section 4.2, these layers of the densest spheres are stacked along the [111] direction (the body diagonal) and the spheres fill 74.05% of the space. In Figure 5.2, a perspective view on the unit cell as well as the most important crystallographic structure descriptors are given. Each copper atom has 12 nearest neighbours, forming a cuboctahedron. These cuboctahedra are stacked along all crystallographic directions, sharing their square faces (Figure 5.3).

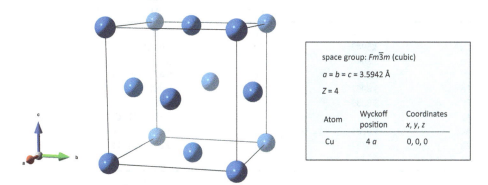

Figure 5.2: Perspective view on the crystal structure of Cu (left), the prototypical **ccp (fcc)** structure type, and the most important crystallographic descriptors (right).

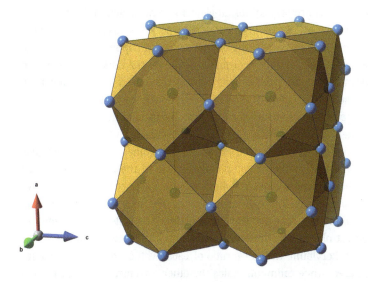

Figure 5.3: Stacked cuboctahedra in the **ccp (fcc)** structure.

An overview of elements that crystallize in the **ccp** structure is given in Table 5.1. Note that not only metallic elements do crystallize in this structure type but also some of the noble gases.

Table 5.1: Elements that crystallize in the **ccp** structure type (space group $Fm\bar{3}m$), together with their unit cell dimensions; if no temperature is given, the dimensions correspond to room temperature.

Element	a (Å)	Element	a (Å)
Ni	3.5240	Pb	4.9502
Cu	3.5942	Th	5.0843
Rh	3.8044	Ce	5.1604
Ir	3.8389	Ar	5.42 (−189 °C)
Pd	3.8907	Yb	5.481
Pt	3.9239	Ca	5.582
Al	4.0495	Kr	5.68 (−191 °C)
Au	4.0783	Sr	6.0849
Ag	4.0857	Xe	6.24 (−185 °C)
Ne	4.52 (−268 °C)		

5.1.2 Mg (hcp) (*Strukturbericht* type A3)

The structure of magnesium is prototypical for all elements crystallizing in the **hcp** structure, with a stacking sequence *ABAB* ... of layers of densest-packed spheres. Magnesium is, so-to-speak, the hexagonal counterpart to the **ccp** structure of copper. The fraction of space-filling of the packing (0.7405) and the coordination number (12) is identical. However, the coordination environment of the nearest neighbours around each Mg atom builds an anticuboctahedron instead of a cuboctahedron as in the case of Cu, compare Figure 4.2.

If the atoms are supposed to be ideally hard spheres, the ratio of the lattice constants c/a in an **hcp** arrangement is equal to $\sqrt{8}/\sqrt{3} = 1.63299$. Magnesium, with an experimentally determined c/a ratio of 1.6234, is indeed the metal that comes closest to this ideal value, and is the reason why it is the prototypical structure of the **hcp** structure type rather than the other "**hcp** arranged" metals, which have to be described as *slightly distorted variants* of the **hcp** structure. These distortions along the c-axis lead to structures in which the interactions within the same plane are either larger or smaller than the interactions between the densest planes. The divalent metals with an s^2 electron configuration are of particular interest: Beryllium, with a c/a ratio of 1.5681, is the extreme example for compressed **hcp** structures, while cadmium builds the other extreme, with a c/a ratio of 1.8856 (Table 5.2). This structure already has a distinctly layered character, and the coordination number would be more appropriately attributed as 6 + 6, (six nearest neighbours within one hexagonal densest plane and six (3 + 3) next-nearest neighbours in the plane above and below. These large differences between elements with the same outer electron configuration but different core electron configurations (Be: [s^2] s^2, Mg: [s^2p^6] s^2, Zn and Cd: [s^2d^{10}] s^2) are also reflected in some of their largely varying physical properties, for instance their melting points (Table 5.2). The reason for this trend can only be understood based on band structure calculations [32]. It is found that the c/a variation is closely connected to (i) the different extent of s-p mixing that can be developed by those elements and (ii) a delicate interplay between the total energy contributions of the band energy and the electrostatic Madelung energy.

Table 5.2: Elements that crystallize in the **hcp** packing, with their unit cell dimensions, c/a ratios, and melting points.

Element	a (Å)	c (Å)	c/a ratio	Melting point (°C)
Be	2.2858	3.5843	1.5681	1,287
Mg	3.2904	5.2103	1.6234	650
Zn	2.6649	4.9468	1.8563	420
Cd	2.9788	5.6167	1.8856	321
Sc	3.3080	5.2653	1.5917	1,541
Y	3.6451	5.7305	1.5721	1,526
Ti	2.9506	4.6788	1.5857	1,668
Zr	3.2312	5.1477	1.5931	1,857
Hf	3.1946	5.0511	1.5811	2,233
Tc	2.735	4.388	1.6044	2,157
Re	2.760	4.458	1.6152	3,186
Co	2.507	4.069	1.6231	1,495
Ru	2.7058	4.2816	1.5824	2,334
Os	2.7353	4.3191	1.5790	3,130

The unit cell and the crystallographic descriptors of Mg are depicted in Figure 5.4. The positions of the two Mg atoms of the unit cell are at fractional coordinates (⅓, ⅔, 0.25) and (⅔, ⅓, 0.75), corresponding to the Wyckoff position 2c with site symmetry $\bar{6}m2$. The space group is $P6_3/mmc$.

5.1.3 Lanthanoids and actinoids – stacking variants

As already briefly mentioned in the previous chapter, some of the lanthanoids and actinoids also crystallize in the form of densest packed layers, though they show more complicated stacking variants. A decisive influence of the f electrons is observed: the more f electrons, the higher the proportion of h layers (layers that are surrounded by two identical layers above and below the stacking direction). However, it is not the case that the f electrons are the driving force for the formation of h layers. It is rather

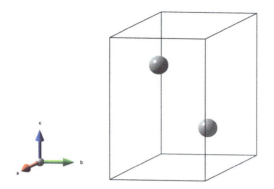
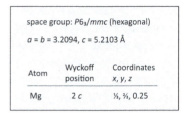

Figure 5.4: Unit cell of the prototypical **hcp** structure of Mg (left) and the most important crystallographic descriptors (right).

the other way around: With the f shell being increasingly filled, the nuclear charge also increases and because the f electrons are very diffuse, the higher nuclear charge acts more strongly on these f electrons than on the $5d$ and $6s$ shell electrons, meaning that the influence of the f electrons decrease with increasing atomic number. To put it another way, the more the f electrons gain influence, the larger is the tendency to form c layers (layers surrounded by two different layers above and below the stacking direction). The influence of the f electrons can also be seen if we look at the stacking variants that the lanthanoids form at higher pressures. With increasing pressure, the proportion of c layers increases (see Table 5.3). This is due to the fact that the external pressure compresses the outer ($5d$ and $6s$) electron shells to a larger extent than the f shell and therefore the f electrons gain influence.

α-Lanthanum itself, with no f electrons, builds the prototypical element with a **double-hcp** structure (*Strukturbericht* type A3'), see also Section 4.5.

Table 5.3: Phases (stacking variants) formed by the lanthanoids as a function of pressure.

Elements	Ambient pressure	Moderate pressure	High pressure
La, Ce, Pr, Nd	hc	c	
Sm	hhc	hc	c
Gd, Tb, Dy, Ho, Tm, Er	h	hhc	hc

Europium (**bcc**) and ytterbium (**ccp**) are exceptions in the lanthanoid series, and they also show a deviation of the usual orbital filling scheme: In europium, the f shell is (in advance) half-filled ($4f^7\ 6s^2$ instead of $4f^6\ 5d^1\ 6s^2$) and in ytterbium, the f shell is (in advance) completely filled ($4f^{14}\ 6s^2$ instead of $4f^{13}\ 5d^1\ 6s^2$).

5.1.4 W (bcc) (*Strukturbericht* type A2)

The structure of tungsten is prototypical for all elements crystallizing in the body-centred cubic (**bcc**) structure (Figure 5.5). Note that in contrast to the **ccp** and **hcp** packing, this structure is not a densest or closest packing. Accordingly, the space-filling fraction is only 0.6802 and the coordination number is reduced from 12 to 8. However, the six next-nearest neighbouring atoms are only 15% further away, meaning that the coordination number can also be described as 8 + 6. In tungsten, the distance between two nearest and two next-nearest neighbours is 2.7412 Å and 3.1652 Å, respectively. If we apply the formula to calculate the effective coordination number given in Section 2.3.2, we get a value of ECoN = 12.53, a value that is higher than the coordination number of the **hcp** or **ccp** arrangement.

In Table 5.4, the elements of the periodic table that crystallize in the **bcc** arrangement are listed together with their unit cell dimensions.

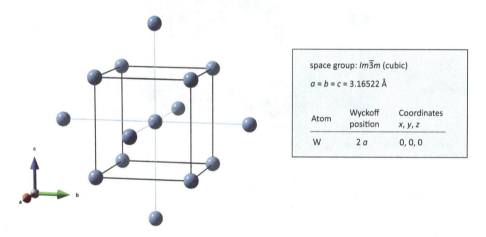

Figure 5.5: Slightly extended unit cell of the prototypical **bcc** structure of W (left) and the most important crystallographic descriptors (right).

Table 5.4: Elements that crystallize in the **bcc** packing, together with their unit cell dimensions.

Element	a (Å)	Element	a (Å)
Fe	2.8664	Li	3.5092
Cr	2.8846	Na	4.2906
Nb	3.006	Eu	4.578
V	3.028	Ba	5.019
Mo	3.1469	K	5.32
W	3.1562	Rb	5.70
Ta	3.3026	Cs	6.14 (−10 °C)

5.1.5 α-Polonium

α-Polonium, a chalcogen, is a very rare and highly radioactive metal that is chemically similar to Se and Te. It is the only metallic element that crystallizes (at ambient conditions) in a primitive cubic structure, with a lattice parameter of $a = 3.359$ Å (space group $Pm\bar{3}m$, no. 221), with Po at the Wyckoff position 1a at (0,0,0). A 2 × 2 × 2 supercell is shown in Figure 5.6; the coordination number is six.

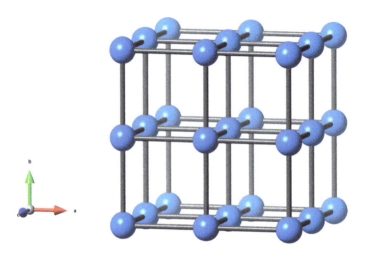

Figure 5.6: Primitive cubic structure of α-Po.

5.1.6 Special metal structures

Tin
Tin occurs in two allotropic forms. Below 13.2 °C, α-Sn is present; it is a brittle non-metal that forms covalent bonds and crystallizes in the diamond structure (coordination number four). The "white", metallic modification, β-Sn, crystallizes in the tetragonal crystal system (space group $I4_1/amd$, no. 141). It is the prototypical structure of the *Strukturbericht* type A5. The Sn atoms are coordinated by four neighbouring atoms in the form of a compressed tetrahedron (Sn-Sn = 3.022 Å), shown as thick sticks in Figure 5.7, and by two further neighbours that are located in the direction of the tetragonal *c*-axis above and below the atoms, respectively (Sn-Sn = 3.181 Å), shown as dotted blue lines in Figure 5.7. This results in a total coordination number of 4 + 2 = 6. A further compression along the *c*-axis would result in six equivalent neighbours in form of an octahedron, an arrangement that corresponds to a primitive cubic lattice. Thus, there is a relationship with the structure of α-Po.

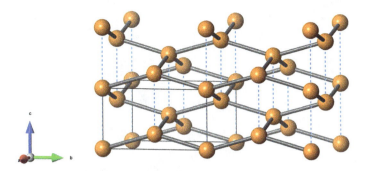

Figure 5.7: Crystal structure of β-Sn.

Tellurium

Screw axes are relatively common in molecular crystals or extended solid-state structures consisting of several elements. However, their presence in pure chemical elements is very rare. One of the few chemical elements with a screw axis is tellurium. It has a 3_1 screw axis and an interesting crystal structure. It crystallizes in the space group $P3_121$, which belongs to the trigonal crystal system. The lattice parameters are $a = b = 4.456$ Å, $c = 5.921$ Å, $α = β = 90°$, $γ = 120°$. The Te atoms build helices coiling around the edges of the unit cell in the crystallographic c-direction (see Figure 5.8, left); the Te-Te distance is 2.8345 Å. If we do not consider only the nearest neighbouring of the tellurium atoms that build this 3_1 screw axis but also to the nearest neighbouring atoms within the (a,b) plane, hexagonal layers become apparent (note that they are not densest layers!). They are stacked in an analogous manner as the Te atoms, meaning that these layers also build 3_1 helices (see Figure 5.8, right).

Although not so easy to recognize, the structure is also related to the structure of α-Po; every Te atom has six neighbours in the form of a distorted octahedron (two in the distance of 2.835 Å and further four in the distance of 3.4192 Å. This means that the deviation from a primitive cubic arrangement is not negligible, but it is also not that pronounced that it would not be possible to see the somehow crinkled cube-like arrangement, compare Figure 5.9.

Manganese

Four allotropes of manganese are known. They occur at successively higher temperatures and are labelled α (stable at room temperature), β (from 700 °C on), γ (1,097 °C), and δ (1,133 °C). α-Mn has a remarkably complex crystal structure, with a total of 58 atoms per body-centred unit cell (space group $I\bar{4}3m$, no. 217, $a = 8.911$ Å), comprising four crystallographically distinct Mn atoms. The crystallographic specification is given in Table 5.5.

The overall picture of the unit cell with 58 atoms is probably not very instructive (Figure 5.10), although the different Mn species are coloured differently. It is helpful to look at the coordination polyhedra of each of the Mn species that have coordination

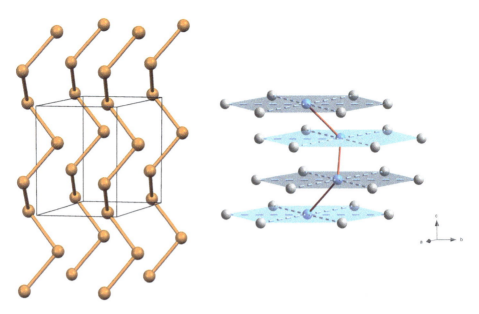

Figure 5.8: Two views of the crystal structure of tellurium. The tellurium atoms form helical chains that coil around the edges of the unit cell parallel to the c-direction, according to a 3_1 screw axis (left). Emphasis of the hexagonal layers in the (a,b) plane that likewise form a helix-like arrangement along the c-direction. One of the 3_1 screws is highlighted by blue tellurium atoms connected by red bonds (right) (rearranged and reprinted from [33] with kind permission from SpringerNature © 2020).

numbers between 12 and 16. The Mn1 atoms (at the centre and the corners of the unit cell) are surrounded by 16 neighbours (4 Mn2 and 12 Mn4) in form of a so-called 16-vertex Frank-Kasper (FK) polyhedron (compare Section 13.5.1, specifically Figures 13.10 and 13.12). Note that the Mn1 atoms have no Mn3 atoms as nearest neighbours. The Mn2 atoms are also surrounded by 16 neighbours in form of a 16-vertex FK polyhedron (one Mn1 atom, six Mn3, and nine Mn4 atoms). The Mn3 atoms are surrounded by a 13-vertex polyhedron (two Mn2 atoms, six Mn3 atoms, and five Mn4 atoms). And finally, the Mn4 atoms have 12 neighbouring atoms in the form of an icosahedron (one Mn1, three Mn2, five Mn3, and three Mn4). All coordination environments of the different Mn species are shown in Figure 5.11 (top row). A clearer picture of the structure of α-Mn is obtained if we also investigate the second coordination environment of the Mn1 atoms. These are simply cuboctahedra, all consisting of Mn3 atoms at a distance of 4.51 Å. Altogether, we have a body-centred packing of Mn1 atoms that are surrounded by 16-vertex FK polyhedra, and these are surrounded, in turn, by cuboctahedra (see Figure 5.11, bottom panels). So, the motif consists of 29 atoms (1 + 16 + 12) and because the structure is body-centred, we have 2 × 29 = 58 atoms per unit cell.

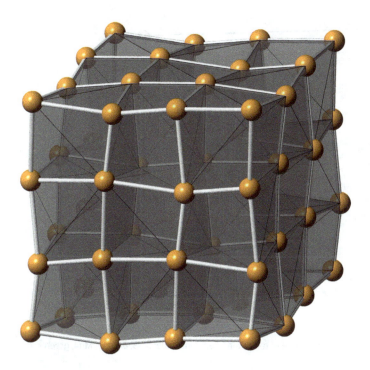

Figure 5.9: Alternative view of the structure of Te where the Te atoms are placed at the corners of crinkled cubes; view approximately along [210].

Table 5.5: Crystallographic data for α-Mn.

Atom	x	y	z	Wyckoff position	Site symmetry
Mn1	0.00000	0.00000	0.00000	2a	$\bar{4}3m$
Mn2	0.31787	0.31787	0.31787	8c	.3m
Mn3	0.35706	0.35706	0.03457	24g	..m
Mn4	0.08958	0.08958	0.28194	24g	..m

5.2 (Ionic) compounds based on densest sphere packings

The shape of monoatomic ions can be, in the first instance, approximated as spheres. Based on the principles derived in Chapter 4, a number of (mainly) ionic structures can also be described as being based on sphere packings in which the larger ion (usually the anion) builds the sphere packing and the smaller ion (usually the cation) completely or partially fills either the octahedral or tetrahedral voids (interstitial sites) or both of them. However, there are several problems with these kinds of descriptions: (i) the size of an ion is not a fixed value; it is dependent on the compound

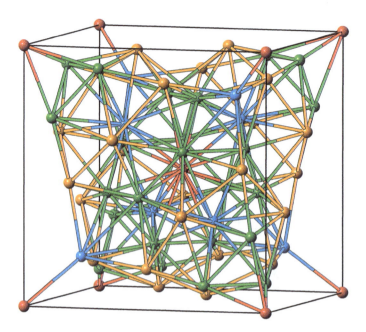

Figure 5.10: Unit cell of the crystal structure of α-Mn. Mn1, red; Mn2, blue; Mn3, orange; Mn4, green.

that the ion is incorporated in, (ii) it is unclear how to reliably measure or calculate an ion radius because any definition of an ion radius contains some degree of arbitrariness, (iii) only very few compounds are completely ionic, most of them have a certain fraction of covalent bonding, (iv) why should anions with like charges pack in a densest manner, and (v) there are several exemptions that violate the predictions that are made (if a cation fits well into a T or O site) based on existing values of ion radii. Having said this, the concept described above is – nevertheless – useful to memorize the structural principles of certain structure types. The only presumption that has to be dropped is that the ions of the larger ion type *are in contact with each other*, i.e., they probably do *not* build *densest packing* of spheres. However, what we know about these compounds from X-ray crystallography is that the *midpoints* of the ions are at sites that coincide exactly with the fractional coordinates of the **hcp** or **ccp** ensemble and their octahedral and/or tetrahedral interstitial sites. This means that the compounds we discuss here can probably be described as "expanded" versions of the **hcp/ccp** packing in which the ions with like charges are not in contact. A term for such arrangements that was coined by Michael O'Keeffe (*1934, a British-American chemist) is "eutactic" (from Greek = well-arranged). The reader should keep this in mind even if we continue to speak about compounds that can be described as based on **hcp** or **ccp** packings for simplification purposes. Some compounds can also be described as eutactic sphere packings even if they have metallic or covalent bonds. This

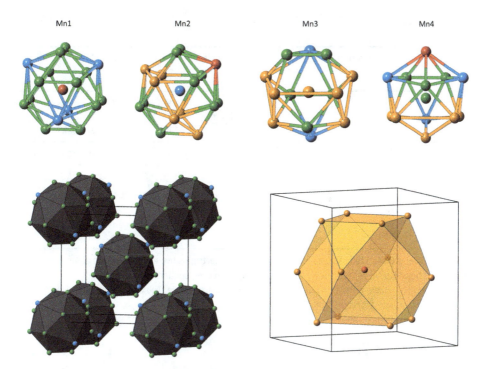

Figure 5.11: Coordination polyhedra of the four crystallographically distinct Mn species (top row); the body-centred packing of the 16-vertex FK polyhedra (bottom, left); exemplarily one cuboctahedron that encloses the FK polyhedron is shown at the bottom-right. Mn1, red; Mn2, blue; Mn3, orange; Mn4, green.

is, for instance, the case for TiO (metallic) and TiC (covalent), both belong to the structure type of NaCl.

In Table 5.6, an overview of some compounds that can be described as based on densest sphere packings (**hcp/ccp**) is given, together with the fractions of occupied tetrahedral and octahedral voids.

Table 5.6: Overview of some compounds based on densest sphere packings.

Packing	Composition	Interstitial sites			Example
		T+	T–	O	
ccp	A	–	–	–	Cu, Au, structure type A1
	AX	–	–	1	NaCl
	AX	1	–	–	ZnS (sphalerite)
	ABX_2	1	–	–	$CuFeS_2$ (chalcopyrite)
	AX_2	–	–	½	$CdCl_2$
	A_2X, AX_2	1	1	–	K_2O, CaF_2

Table 5.6 (continued)

Packing	Composition	Interstitial sites			Example
		T+	T−	O	
	AX_3	–	–	⅓	$CrCl_3/AlCl_3/RhBr_3$ *
	A_3X	1	1	1	Li_3Bi **
	AB_2X_4	⅛	⅛	½	$MgAl_2O_4$ (spinel)
hcp	A	–	–	–	Mg, structure type A3
	AX	–	–	1	NiAs
	AX	1	–	–	ZnS (wurtzite)
	AX_2	–	–	½	CdI_2
	AX_3	–	–	⅓	BiI_3
	A_2X_3	–	–	⅔	Al_2O_3
	AB_2X_4	⅛	⅛	½	Mg_2SiO_4 (olivine)

The letters A and B are used as placeholders for the more electropositive elements of the compounds and should not be confused with the designation of the layers in layer sequences.
*In these structures, the anions do not build an exact **ccp**-like structure. The overall symmetry is reduced to the monoclinic space group $C2/m$.
**Here, only intermetallic phases are known, which are treated separately in Chapter 11.

5.3 Compounds based on a cubic closest packing

5.3.1 NaCl (*Strukturbericht* type B1)

In the NaCl structure type, the chloride anions build an eutactic **ccp** packing in which all octahedral interstitial sites are occupied by sodium cations; the tetrahedral interstitial sites remain unoccupied. These sites are at the centre of the cell and at all edge bisectors of the cube (see Figure 5.12). This means that the sodium ions also build an **fcc** lattice that is shifted by (0.5, 0, 0) along the crystallographic directions. Therefore, we have the choice to place either the Na or Cl ion at the origin (0,0,0); often, the Na ion is placed at the origin, which is a little bit counterintuitive. The chloride anions are surrounded by six sodium ions in form of an octahedron, and vice versa; the coordination number is therefore 6:6. The coordination polyhedra share 12 common edges, but they do not share common faces; note that the triangular faces of the octahedron also correspond to the triangular faces of the (empty) tetrahedral voids. This is difficult to illustrate, but an attempt is made in Figure 5.12 (bottom right).

The number of structures that crystallize in the NaCl structure type is very high and these compounds actually comprise not only ionic but also some covalent structures or intermetallic phases. In detail, compounds that are analogous to NaCl are:

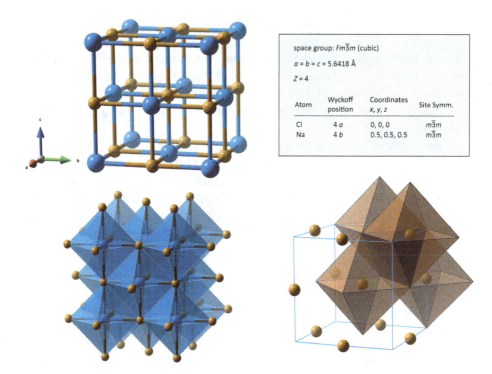

Figure 5.12: Crystallographic parameters and different illustrations of the NaCl structure. Top left: Ball-and-stick model; Cl, blue; Na, orange. Bottom left: ClNa$_6$ coordination octahedra (blue). Bottom-right: positions of the sodium ions (at the centre of the cell and at all edge-bisectors) and a group of four NaCl$_6$ octahedra (orange) around the vicinity of an empty tetrahedral void (dark blue).

- alkali metal halides, *except* CsCl, CsBr, and CsI;
- high-temperature modifications of the ammonium halides NH$_4$Cl, NH$_4$Br, and NH$_4$I, note, however, that the corresponding low-temperature phases belong to the CsCl structure type (see below);
- alkali metal hydrides (LiH, NaH, KH, RbH, and CsH)
- alkaline earth metal chalcogenides, *except* Be salts and MgTe;
- the transition metal monoxides TiO, VO, MnO, FeO, and NiO;
- nitrides und carbides of the group 4 and 5 metals: TiN, TiC, VC, and CrN;
- LaN, LaP, LaAs, LaSb, LaBi, LaS, LaSe, and LaTe (as well as analogous compounds of other rare-earth metals and actinoids);
- EuO, UO, PuO, and AmO;
- CdO, AgF, AgCl, AgBr, and the high-pressure modification of AgI;
- high-pressure modifications of InAs, ZnO, CdS, CdSe, CdTe, and InTe;
- high-temperature modification of CuI.

Disordered and distorted variants, derivatives, and relationships to other structures

Some non-binary compounds can also be described as based on the NaCl structure type. This is the case for some ternary M(I)-M(III) chalcogenides, in which the two metal ions are completely disordered, i.e., randomly distributed over the octahedral interstitial sites. This type includes, e.g., the compounds γ-LiTlO$_2$, α-LiFeO$_2$, α-NaTlO$_2$, LiBiS$_2$, LiYS$_2$, and NaLaS$_2$.

If the two metal cations are not statistically distributed but form an ordered arrangement in which the two metal types occupy alternating octahedral interstitial site layers (γ, α, β) between the **ccp** stacking sequence (along [111]) of densest layers (ABC...), then we arrive at a sequence A-γ(cation 1)-B-α(cation 2)-C-β(cation 1)-A-γ'(cation 2)-B-α'(cation 1)-C-β'(cation 2), and so on (see Figure 5.13). The highest possible symmetry of this structure is that of the trigonal space group $R\bar{3}m$ with one formula unit per unit cell.

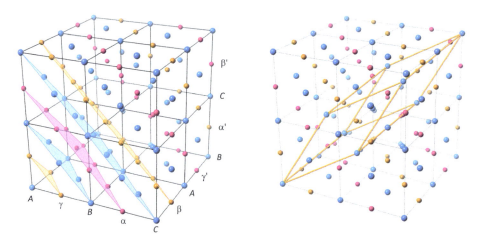

Figure 5.13: Relationship between the original **ccp** NaCl structure and the rhombohedral structure of compounds in which two cation types (orange and pink) occupy alternating octahedral interstitial site layers. A set of parallel planes with Miller index (111) are highlighted in the left picture; the primitive new rhombohedral unit cell is highlighted in the right picture.

Dependent on the nature and size of the cations, it can be expected that in such ordered structures some distortions occur, in contrast to structures with randomly distributed cations. This may involve compression or elongation along the single threefold axis while otherwise maintaining the ideal symmetry $R\bar{3}m$ or may comprise additional displacements so that not all ions of one type are symmetry-equivalent, leading to the space group $P\bar{3}m1$. We can compare the periodicity along the densest layers if the new rhombohedral cell is transformed to a rhombohedrally-centred cell with hexagonal axes. Compared with an ideal **hcp** stacking sequence (ABAB...) in which the c/a ratio is 1.633, this stacking sequence has a c/a ratio of 4.89898 (3/2 × 1.6333 × 2, see Figure 5.14).

Values smaller/larger than 4.89898 are indicative of a contraction/expansion along the threefold axis of rotation (the body-diagonal in the rhombohedral cell).

Compounds that crystallize in these ordered arrangements are AgBiS$_2$ (c/a = 4.683, $P\bar{3}m1$), AgBiSe$_2$ (c/a = 4.715, $P\bar{3}m1$), LiNiO$_2$ (c/a = 4.925, $R\bar{3}m$, see Figure 5.15), TlBiTe$_2$ (c/a = 5.272, $R\bar{3}m$), TlSbTe$_2$ (c/a = 5.272, $R\bar{3}m$), and α-NaFeO$_2$ (c/a = 5.326, $R\bar{3}m$).

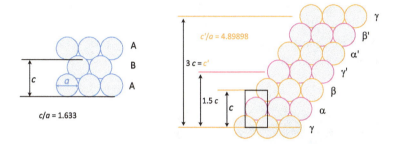

Figure 5.14: Comparison of the c/a ratio of the unit cell of an ideal **hcp** arrangement (left) with that of an ideal arrangement with a cubic-like stacking sequence ABCABC... in which the octahedral interstitial site layers γ-α-β are occupied by two different cations in an alternating fashion (right).

Yet another type of ordering of the cation arrangement is present in the ternary metal oxide γ-LiFeO$_2$. Here, in an alternating fashion two layers along the cube edges are occupied with Li, followed by two layers of Fe, and so on (see Figure 5.15). The

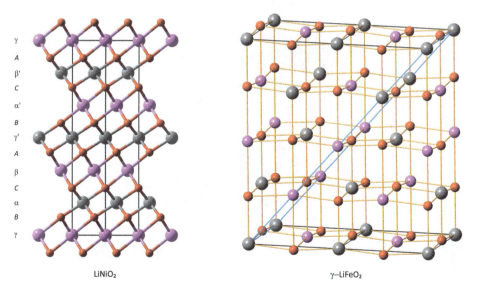

Figure 5.15: The structure and stacking sequence of LiNiO$_2$ (left, view along [110]) and the structure of γ-LiFeO$_2$ (right, two unit cells along the a-direction are shown); oxygen, red; Li, pink; Ni and Fe, grey.

oxygen atoms are slightly displaced, compared with an ideal **ccp** arrangement and the overall symmetry is reduced to the tetragonal space group $I4_1/amd$.

A structure type that can be described as NaCl-related is calcite ($CaCO_3$). This type is obtained when all Na cations are replaced by Ca cations, and all Cl ions by carbonate anions. The trigonal-planar carbonate ions are oriented perfectly perpendicular to the stacking direction c in the rhombohedrally-centred cell with hexagonal axes and they show an alternating orientation (rotation by 180° in the (a,b)-plane) from layer to layer (see Figure 5.16). Examples of other compounds that also crystallize in the calcite structure type are the carbonates $MgCO_3$, $CoCO_3$, $ZnCO_3$, $MnCO_3$, $FeCO_3$, and $CdCO_3$, the nitrates $LiNO_3$ and $NaNO_3$ and the borates $ScBO_3$, $InBO_3$, and YBO_3.

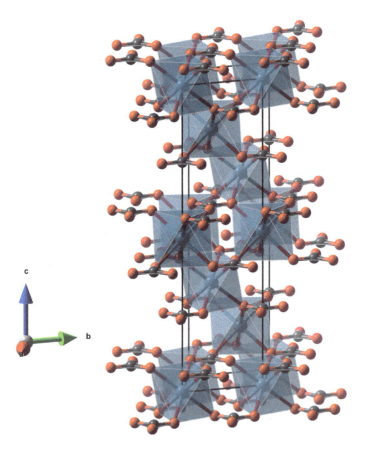

Figure 5.16: The crystal structure of calcite ($CaCO_3$). Oxygen, red; carbon, grey; CaO_6 octahedra, blue-grey.

5.3.2 CdCl$_2$ (*Strukturbericht* type C19)

In the CdCl$_2$ structure type, the chloride anions build an eutactic **ccp** packing and the cadmium cations occupy all octahedral interstitial sites of every second such layer of sites, meaning that every other layer remains completely unoccupied. Cadmium chloride thus has a distinctly layered character. The overall stacking sequence is A-γ-B-☐-C-β-A-☐-B-α-C-☐ (Figure 5.17). The Cd ions are surrounded by six Cl ions in the form of an octahedron and the Cl ions form the tip of a trigonal pyramid with three Cd^{2+} as the base (CN = 6:3). The octahedra are connected to adjacent octahedra by six common edges, forming a layer in which each halide atom belongs to three octahedra simultaneously (Niggli formula: MX$_{6/3}$), see Figure 5.17 (bottom right).

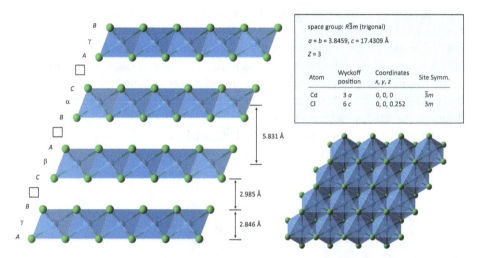

Figure 5.17: Crystal structure of CdCl$_2$, (Cl: green, CdCl$_6$ octahedra: blue), together with the stacking sequence and selected (inter)layer distances (left, view along [110], c-axis pointing upwards), view perpendicular to one layer consisting of edge-sharing CdCl$_6$ octahedra (bottom right) as well as the most important crystallographic parameters (top right).

Other compounds that crystallize in the CdCl$_2$ type of structure are:
- MgCl$_2$, MnCl$_2$, FeCl$_2$, CoCl$_2$, and NiCl$_2$,
- NiBr$_2$, ZnBr$_2$, and CdBr$_2$,
- NiI$_2$.

If the cations build the **ccp** packing of densest layers and the anions occupy the octahedral voids, we arrive at the respective *anti*-CdCl$_2$ type. This ion arrangement is realized in Cs$_2$O, the only compound that has been assigned to this structure type.

5.3.3 CrCl$_3$, AlCl$_3$, and RhBr$_3$

If we follow the previous logic and further reduce the occupation of the octahedral voids in a packing of cubic closest-packed anions, we arrive at the structures with stoichiometry MX$_3$. In principle, two alternatives are possible: Every octahedral layer is occupied by ⅓ or every second layer is occupied by ⅔ and the layers in between remain completely empty. While there is no example for the first possibility, several compounds realize the second type of arrangements. However, as newer single-crystal studies revealed, the anions do not build an ideal cubic-like *ABC* stacking sequence; the anions are not all symmetry-equivalent anymore, and the overall symmetry is reduced to the monoclinic space group $C2/m$. In fact, the stacking sequence is *ABCDEFG*, meaning that every 8th layer is exactly on top of each other. Nevertheless, these structures build similar layered-type structures as CdCl$_2$, with the difference that the octahedra of one layer has not six but only three common edges with neighbouring octahedra (all anions belong to two octahedra simultaneously, Niggli formula MX$_{6/2}$), leaving holes in the sheets that also have an octahedral shape (see Figure 5.18).

In the literature, this structure type is also known as AlCl$_3$ or CrCl$_3$ type; however, in the ICSD, the prototypical compound RhBr$_3$ is chosen as reference. Compounds that belong to the RhBr$_3$ structure type are:
- AlCl$_3$, YCl$_3$, CrCl$_3$, MoCl$_3$, β-TcCl$_3$, RhCl$_3$, RuCl$_3$, and IrCl$_3$,
- InBr$_3$, BiBr$_3$, IrBr$_3$, GdCl$_3$, CfBr$_3$, and RhBr$_3$,
- VI$_3$, CrI$_3$.

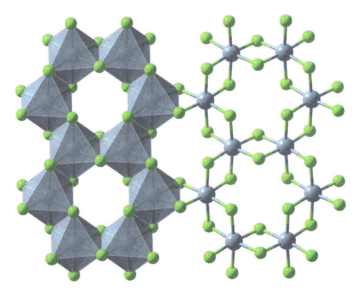

Figure 5.18: View perpendicular to the sheets of edge-connected octahedra in the structure of AlCl$_3$.

5.3.4 CaF$_2$ (*Strukturbericht* type C1)

After having discussed eutactic **ccp**-based structures in which the octahedral interstitial sites are (completely or partly) occupied, we now move on to those in which the tetrahedral interstitial sites are occupied.

The fluorite (CaF$_2$) structure type is characterized by an eutactic **ccp** packing of the larger atom/ion type in which *all* tetrahedral interstitial sites are occupied by the smaller atom/ion type; the octahedral voids remain unoccupied (see Figure 5.19). Looking only at the fluoride anions, they are arranged according to a cubic-primitive lattice. In the picture of the coordination polyhedra, in CaF$_2$ the Ca ions are surrounded by eight fluorine ions in form of a cube (the centre of every second cube remains empty), and the fluorine anions – as they are located in the tetrahedral voids – are surrounded by Ca^{2+} ions in form of a tetrahedron (CN = 8:4). The cubes share all 12 edges (but no faces) with neighbouring cubes and the tetrahedra shares all six edges (but no faces) with neighbouring tetrahedra.

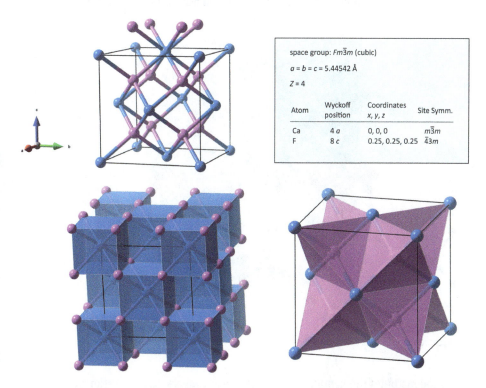

space group: $Fm\bar{3}m$ (cubic)

$a = b = c = 5.44542$ Å

$Z = 4$

Atom	Wyckoff position	Coordinates x, y, z	Site Symm.
Ca	4 a	0, 0, 0	$m\bar{3}m$
F	8 c	0.25, 0.25, 0.25	$\bar{4}3m$

Figure 5.19: Crystallographic parameters of the crystal structure of CaF$_2$ (fluorite) and three different views of the structure; top-left: ball-and-stick model of CaF$_2$ (Ca, blue; F, pink), in which the unit cell is slightly extended in one direction to emphasize the cubic environment of the Ca ions; bottom-left and -right: visualization of the Ca (cubes) and F (tetrahedra) coordination polyhedra, respectively.

Compounds that crystallize in the fluorite structure type are:
- fluorides and oxides of elements with a large atomic radius, *e.g.*, SrF_2, BaF_2, CdF_2, HgF_2, $\beta\text{-}PbF_2$, YbF_2, CeO_2, ThO_2, UO_2, NpO_2, PuO_2, CmO_2, and AmO_2,
- some hydrides, *e.g.*, VH_2, LaH_2, YH_2, CmH_2, DyH_2, SmH_2, PrH_2, and PuH_2.

Some non-binary compounds also crystallize in the fluorite structure in which either the larger or smaller atoms/ions are statistically distributed over their sites. This is the case for:
- PrOF, CfOF, PuOF, $NaHoF_4$, $NaErF_4$, $NaYF_4$, $NaYbF_4$, $KLaF_4$, $KCeF_4$, $RbGdF_4$, $CaUO_4$, $TlBiF_4$, $BiUO_4$, SmHO, HoHO, $TiVH_4$, and $NbVH_4$.

Ternary systems in which the anions are not statistically distributed are built by the rare earth oxyfluorides YOF, NdOF, EuOF, GdOF, TbOF, DyOF, HoOF, and ErOF, all of which crystallize in a rhombohedral structure (space group $R\bar{3}m$). The anion ordering is accompanied by a small distortion from the ideal cubic fluorite lattice and the YO_4F_4 coordination polyhedra are slightly distorted accordingly (see Figure 5.20).

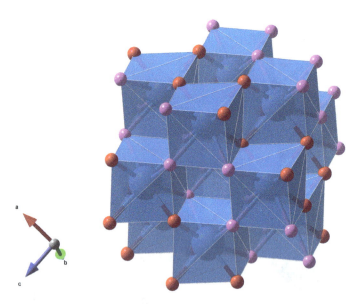

Figure 5.20: Rhombohedral crystal structure (space group $R\bar{3}m$) of YOF. Due to the anion ordering, the YO_4F_4 coordination cubes are slightly distorted (Y, blue; O, red; F, pink).

If the anions build the **ccp**-like packing of densest layers and the cations occupy the tetrahedral voids, we arrive at the respective *anti*-CaF_2 type, sometimes also referred to as the Li_2O structure type. Compounds that crystallize in this type include:
- many of the alkali metal chalcogenides like Li_2O, Li_2S, Li_2Se, Li_2Te, Na_2O, Na_2S, Na_2Se, Na_2Te, K_2O, K_2S, K_2Se, K_2Te, Rb_2O, and Rb_2S,

- the carbide of beryllium, Be_2C,
- the intermetallic compounds Mg_2Si, Mg_2Ge, Mg_2Sn, Mg_2Pb, In_2Au, In_2Pt, and Ga_2Au,
- phosphides of the platinum metals Ir and Rh, i.e., Ir_2P and Rh_2P.

A ternary compound in which two metals are randomly distributed over the tetrahedral sites is LiMgN.

Relationships to other structure types

In the CaF_2 structure type, all tetrahedral interstitial sites are occupied. If one-half of the atoms from the tetrahedral sites are removed, compounds with the stoichiometry MX are obtained. Depending on which of the four remaining tetrahedral interstitial sites are occupied, one arrives at different structure types. Starting from these new structure types, one can again remove half of the remaining tetrahedral positions and arrive at structures in which only one-quarter of all tetrahedral interstitial sites are occupied. This leads to compounds with the stoichiometry MX_2 again, in which the metal and non-metal atoms are interchanged. Let us follow this genealogy of structures stepwise, beginning with CaF_2.

When all tetrahedral sites that are in the same T layer are removed, the structure type of (red) α-PbO (the mineral is known as litharge; *Strukturbericht* type B10) is obtained, in which layers of edge-sharing OPb_4 (slightly distorted) tetrahedra are present, while the Pb atoms form the tip of a square pyramid whose base is formed by four oxygen atoms (CN = 4:4); the Pb atoms lie alternately above and below the centres of these squares (Figure 5.21). Red PbO crystallizes in the tetragonal space group $P4/nmm$ (no. 129) with two formula units per unit cell. The only two other compounds of that structure type are the blue-black SnO and the high-pressure modification of CuBr. The only compound that crystallizes in the *anti*-PbO structure type is LiOH.

The layered structure of α-PbO has long been associated with the stereochemically active electron lone pair of Pb(II), with a $6s^2$ electron configuration, leading to a highly asymmetric electron density around the Pb ions. The four surrounding oxygen atoms of Pb(II) are on one side and the lone pair points to the opposite direction (between the layers). This directional character of this stereochemically lone pair was usually explained by a hybridization of the $6s$ states and the unfilled $6p_z$ states, thereby lowering the internal electronic energy by a second-order Jahn-Teller mechanism. However, as other Pb(II) compounds do not show this behaviour, the structure of PbO is in need of an explanation; PbS, for instance, crystallizes in the highly symmetrical 6-coordinated NaCl structure type. Such an explanation was given by an analysis of DFT-based band structure calculations by Walsh and Watson [34] that were experimentally supported soon after [35]. According to the band structure calculations, the decisive interactions are anion-dependent and of indirect nature. Most of the $6s$ states of the Pb(II) ions are at the *bottom* of the valence band. The mixing of pure metal $6s$ and $6p$ states is rather small and cannot account for the large distortion.

The main states at the *top* of the valence band are generated by mixing Pb(II) 6s states with anion *p* states with an antibonding character, but these antibonding states alone do not result in an asymmetric electron distribution around the Pb(II) ions. Only *subsequent* interactions of these states with the unoccupied Pb(II) $6p_z$ states are the driving forces for the distortion of the crystal structure. These subsequent interactions lead to an increased electron density on one side of the Pb(II) ion and a decreased electron density on the other side (the side of the four oxygen atoms), thereby reducing the antibonding interaction with the 2p oxygen states (see Figure 5.22). This is not observed in PbS because of the much weaker interaction between the Pb 6s and S 3p states due to the higher energy of S 3p orbitals, compared to O 2p.

The structure of BiOCl can be described as a derivate of the red PbO structure type: here, the Pb atoms are substituted with Bi atoms, the oxygen atoms are unchanged, and four additional chlorine atoms build the base of another square-pyramid with the Pb atoms that are at the opposite site of the PbO_4 square pyramids and that are rotated by 90° (resulting in a slightly-distorted square antiprism). This is the structure type of PbFCl (*Strukturbericht* type $E0_1$, again space group $P4/nmm$), see Figure 5.23.

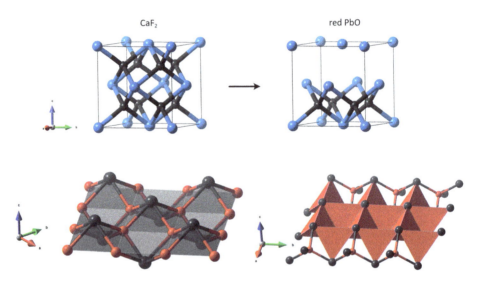

Figure 5.21: Derivation of the PbO structure type from the CaF_2 structure type by removing all tetrahedral sites of every second tetrahedral layer (top), and the structure of red PbO in which the Pb atoms are at the tip of the square pyramids built by O atoms (bottom left); the O atoms are surrounded by four Pb atoms in the form of a slightly distorted tetrahedra.

Starting again from the CaF_2 structure type, another structure type is obtained if every second T site within all T layers is removed, or, in other words, if only T sites remain occupied that are of the same type, i.e., T+ or T–. This is the case for the structure type of zincblende, ZnS, (the mineral is known as sphalerite, *Strukturbericht* type B3). Usually, this structure type is described in such a way that the zinc ions build the **ccp**

Figure 5.22: Schematic orbital interaction scheme in PbO. Left: highly populated antibonding states through Pb 6s and O 2p interactions; right: subsequent interactions with the unfilled Pb $6p_z$ orbital that leads to an increased population of the Pb $6s$-$6p_z$ hybrid states and a depopulation of the antibonding Pb 6s-O 2p states (redrawn and adapted from [34]).

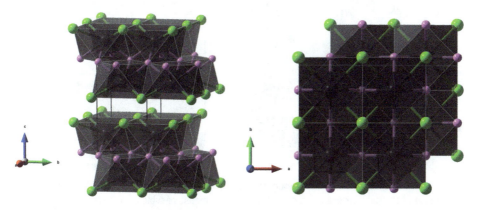

Figure 5.23: Two different views of the structure of PbClF (Pb, black; Cl, green; F, pink). The Pb atoms are surrounded by four F and four Cl atoms in the form of a square antiprism.

packing and the sulphur atoms occupy the tetrahedral sites, but the opposite description is, of course, also correct and completely equivalent. Note that all ZnS_4 tetrahedra are oriented in the same way and that they are corner-connected (Figure 5.24). Note further that sphalerite is, in contrast to red PbO, not a layered structure.
Further compounds that crystallize in the ZnS (sphalerite) structure type are:
- CuCl, CuBr, and CuI,
- BeS, ZnSe, ZnTe, and CdTe,
- β-BN, AlP, AlSb, GaP, GaAs, GaSb, and InSb,
- β-SiC.

If the two atomic species of ZnS (sphalerite) are substituted by atoms of the same element, one ends up with the diamond structure (*Strukturbericht* type A4). Diamond crystallizes in the cubic space group $F\bar{d}3m$ (no. 227, with $a = b = c = 3.56712$ Å), with a total of eight carbon atoms per cell at the Wyckoff position 8a (0, 0, 0). The relationship to the ZnS (sphalerite) structure is shown in Figure 5.25.

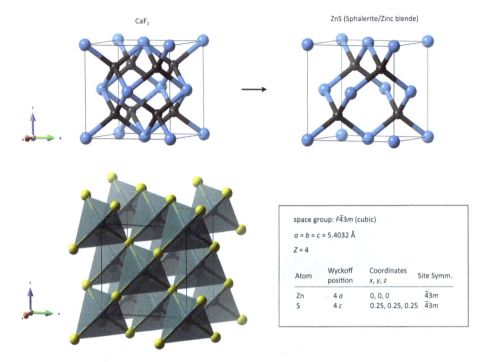

Figure 5.24: Derivation of the ZnS (sphalerite) structure type from the CaF$_2$ structure type by removing every second T site (top); structure and crystallographic parameters of cubic ZnS (bottom).

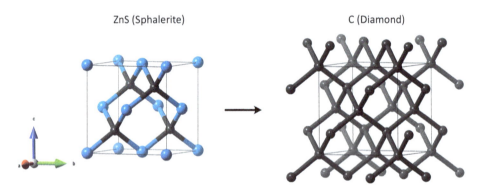

Figure 5.25: Relationship of the structures of ZnS (sphalerite) and the diamond allotrope of carbon.

Some ternary compounds also belong to the ZnS (sphalerite) structure type. In the following compounds, the two metal atoms are statistically distributed over the Zn sites: $CuSi_2P_3$, $CuGe_2P_3$, and Cu_2MX_3 (M = Ge, Sn and X = S, Se, and Te).

If the two metal atom types are ordered, another structure type is obtained, namely that of $CuFeS_2$ (chalcopyrite, see Figure 5.26). Here, the tetrahedral sites are alternatingly occupied by the two metal atoms along the fourfold axis of rotation (along the *c*-direction) and the symmetry is reduced to the tetragonal space group $I\bar{4}2d$. Many compounds of the type M(I)M'(III)X$_2$ (M(I) = Li, Ag, Cu; M'(III) = Al, In, Ga, Tl, Fe; X = S, Se, Te) and M(II)M'(IV)X$_2$ (M(II) = Be, Mg, Zn, Cd; M'(IV) = Si, Ge, Sn; X = P, As, Sb) belong to this structure type.

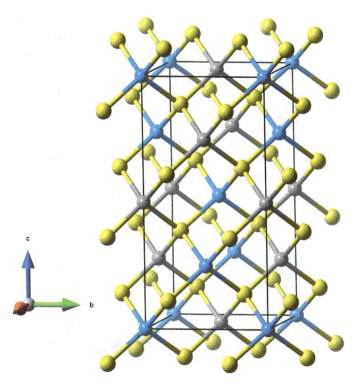

Figure 5.26: Crystal structure of CuFeS$_2$ (chalcopyrite). Cu, blue; Fe, grey; S, yellow. Note that the *c*-axis is doubled compared to that of CaF$_2$.

Once again, starting from the CaF$_2$ structure, yet another structure type is obtained when two atoms of the same type are removed in each tetrahedral layer (T+ or T–), but alternating from T site layer to T site layer. This leads to the structure type of PtS (*Strukturbericht* type B17). The Pt atoms are surrounded by four S atoms in a rectangular-planar manner (the S-Pt-S angles are 82.73° and 97.27°, respectively). These rectangles are edge-linked in one direction and form parallel bands. They are linked on both sides via common sulphur atoms with corresponding bands perpendicular to the former. The S atoms are surrounded by four Pt atoms in the form of a distorted tetrahedron (CN = 4:4). These tetrahedra are edge-connected in the *c*-direction and further linked by common Pt atoms in the *a*- and *b*-direction (see Figure 5.27). PtS crystallizes in the tetragonal space group $P4_2/mmc$ (no. 131) with two formula units per cell.

Only four compounds belong to the PtS structure type. These are:
- PtS, PdS, PtO, and PdO.

CuO and AgO (precisely Ag(I)Ag(III)O$_2$) build monoclinic-distorted variants of the PtS type.

All relevant structure types with a **ccp** arrangement of one type of atoms and a T site occupancy of ½ have been discussed. Three further structure types can be derived in which the occupancy is once again halved, i.e., structures in which only ¼ of the T sites are occupied. Two of them can be derived from the PbO structure type. If one-half of the oxygen atoms is removed in such a way that in one cell the T sites of one type (T+ or T−) and in the neighbouring cell along one direction the other type (T− or T+) are removed, and the cations and anions are exchanged (see Figure 5.28), the structure type of red HgI$_2$ (*Strukturbericht* type C13) is obtained, crystallizing in the tetragonal space group $P4_2/nmc$ (no. 137) with two formula units per unit cell. This is a layered compound of corner-shared HgI$_4$ tetrahedra in which the tetrahedra of neighbouring layers are pointing in opposite directions (see Figure 5.28, bottom-right). Only one other compound belongs to that structure type, namely γ-ZnCl$_2$.

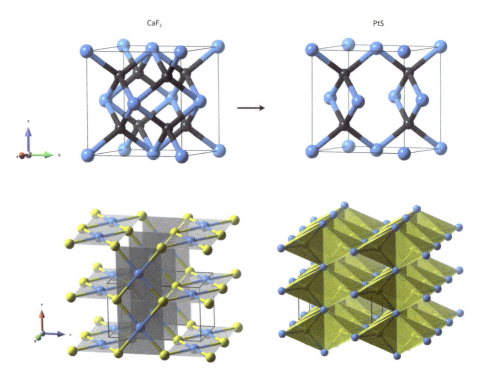

Figure 5.27: Derivation of the PtS structure type from CaF$_2$ (top) and the structure of PtS, shown as coordination polyhedra of the Pt (bottom-left) and S atoms (bottom-right).

The second structure type that can be derived from the PbO structure type is that of SiS_2 (*Strukturbericht* type C42) Once again, one-half of the oxygen atoms is removed, but this time in a manner that in one cell a T− and a T+ site remain occupied and in the neighbouring cell along one direction the T+ and T− site remain occupied. Changing the cation positions with anions, and vice versa, leads to chains of edge-sharing tetrahedra running along the *c*-axis, as realized in SiS_2 (see Figure 5.28), which crystallize in the orthorhombic space group *Ibam* (no. 72) with four formula units per unit cell.

The third structure type in which only ¼ of the T sites are occupied can be derived from the ZnS (sphalerite) structure type if again cations and anions are exchanged and one-half of the T sites are removed in such a way that a space net of corner-shared tetrahedral sites results. The prototypical structure of this type is α-$ZnCl_2$. In this compound, which crystallizes in the tetragonal space group $I\bar{4}2d$ (no. 122), the chloride ions are cubic closest packed and the $ZnCl_4$ tetrahedra form helices along the *c*-direction (see Figure 5.29).

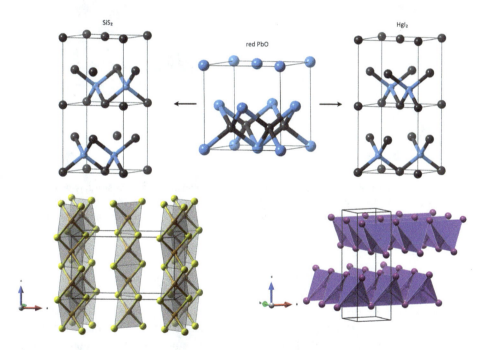

Figure 5.28: Derivation of the SiS_2 and HgI_2 structure type from red PbO (top), and structures of SiS_2 (bottom-left) and HgI_2 (bottom-right) in a coordination polyhedron representation.

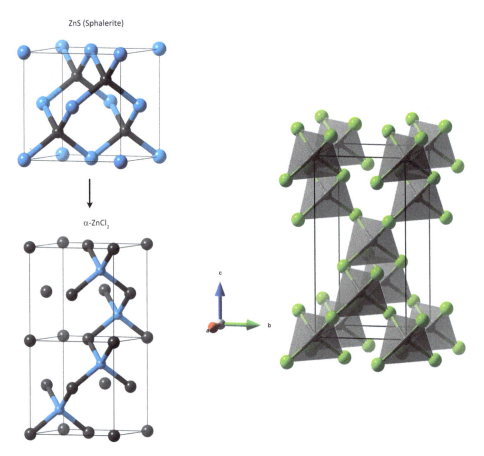

Figure 5.29: Derivation of α-ZnCl$_2$ structure type from ZnS (sphalerite) (left) and the structure of α-ZnCl$_2$ in which the ZnCl$_4$ tetrahedra are emphasized (right; Zn: grey, Cl: green). Note that the origin is shifted by [0.25, 0.25, 0.125] in the new tetragonal setting (with a doubling of the cell along the c-direction) with respect to the original cubic setting of ZnS (sphalerite).

5.3.5 MgAl$_2$O$_4$ (spinel) (*Strukturbericht* type H1$_1$)

Up to now, we have discussed compounds with a **ccp** arrangement of one type of ions in which either the octahedral or tetrahedral voids are partially or completely occupied by other types of ions. With the spinel type, we now come to an important type of structure in which both the octahedral and the tetrahedral voids are (partially) occupied. In the prototypical structure of MgAl$_2$O$_4$, the oxide ions form a (slightly distorted) cubic dense sphere packing (stacking sequence *ABC*...), in which ⅛ of the tetrahedral voids are occupied by Mg ions and ½ of the octahedral voids are occupied by Al ions. MgAl$_2$O$_4$ crystallizes in the cubic space group $Fd\bar{3}m$ (no. 227) with eight formula units per unit cell.

The structure of the spinel is insufficiently accessible if we consider only the corresponding coordination polyhedra (i.e., MgO$_4$ tetrahedra and AlO$_6$ octahedra) of the metal cations (see Figure 5.30). Instead, it is useful to identify Al$_4$O$_4$ cube-like units as one of the fundamental building blocks, where the corners of this cube are alternately occupied by Al and O atoms. These cubes are corner-connected via shared Al atoms to four neighbouring Al$_4$O$_4$ cubes in a tetrahedral fashion, i.e., each Al atom belongs to two such Al$_4$O$_4$ units. Starting from such a cube placed right in the centre of the cell (0.5, 0.5, 0.5, Wyckoff position 8b, origin setting 1), the first four tetrahedrally arranged neighbouring cubes are then at the centres of every second octant of the cubic unit cell (¼, ¾, ¼), and the next-nearest neighbouring cubes are then placed at all edge bisectors of the unit cell (Figure 5.31, top-left); in other words: if we connect the centres of the cubes, a diamond net (**dia**) is obtained.

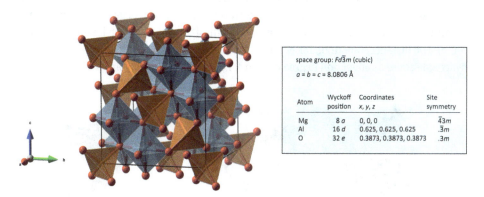

Figure 5.30: Crystal structure and crystallographic parameters of MgAl$_2$O$_4$. MgO$_4$ tetrahedra, orange; AlO$_6$ octahedra, blue-grey; oxygen, red.

The Mg atoms, i.e., the centres of the MgO$_4$ tetrahedra are at the equivalent Wyckoff position 8a (0, 0, 0), meaning that they occupy the remaining complementary positions around the central Al$_4$O$_4$ cube, and are further placed at all corners and at all face-centres of the unit cell (Figure 5.31, bottom-left). Each MgO$_4$ tetrahedron is connected to four Al$_4$O$_4$ cubes via common oxygen atoms (Figure 5.31, top-right). Considering only the Mg atoms, they likewise form a diamond net, meaning that the spinel structure can also be regarded as a twofold interpenetrating diamond net (**dia-c2**), see Figure 5.31 (bottom-right).

At first glance, the distribution of cations among the voids in the spinel structure appears to be a clear-cut affair in which the cations with the smaller ionic radius occupy the tetrahedral voids (T) and those with larger radius occupy the octahedral voids (O): $A^T B_2{}^O O_4$. In fact, these relationships are much more complicated and already not fulfilled in the prototype Mg^{T}Al$_2{}^O$O$_4$: $r(Mg^{2+})$ = 72 pm, $r(Al^{3+})$ = 54 pm. This once again shows that the rationalization of structures based only on the sizes of the species that are involved is insufficient. Another aspect to be considered is the electrostatic part of the lattice energy and thus the Madelung constant for the different configurations. However, even this does not provide a sufficient explanation for the observed ion distribution for all spinel structures and shows that we are not dealing with purely ionic compounds here. Finally, in the case of transition metal involvement, crystal field stabilization energies must be considered, which in turn depend on the geometry of the surrounding ligands (modifying the splitting of the d orbitals), the number of d electrons, the spin pairing energy, and if anions other than oxygen are involved, the ligand strength.

Also, the fact that there are numerous so-called *inverse spinel structures* in which one-half of the B cations occupy tetrahedral sites and the A ions occupy the octahedral sites, $B^T(AB)^O O_4$, indicates that the preference of an ion species for one of the available interstitial sites is not particularly strong. Between normal and inverse spinels there are also arbitrary intermediate stages that can be characterized by the inversion degree λ, defined as the fraction of B cations at tetrahedral sites, with the limiting cases $\lambda = 0$ for normal and $\lambda = 0.5$ for inverse spinels, respectively. The inversion degree can also be temperature-dependent, and it depends on the synthesis route (high- or low-temperature route). For instance, one finds for $NiMn_2O_4$ synthesized by the so-called low-temperature hydroxide route an inversion degree of $\lambda = 0.4$, i.e., $(Ni_{0.2}Mn_{0.8})^T(Ni_{0.8}Mn_{1.2})^O O_4$.

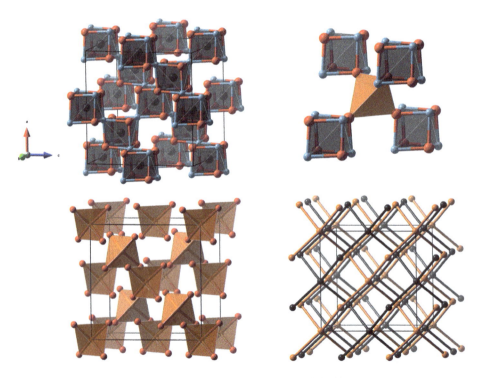

Figure 5.31: Structural features of $MgAl_2O_4$ emphasizing the Al_4O_4 cube-like building units at Wyckoff position 8a (top-left), the connection of four cubes via a MgO_4 tetrahedron (top-right), the spatial arrangement of the MgO_4 tetrahedra (bottom-left), and the resulting twofold interpenetrated diamond net (**dia-c2**), obtained by connecting the centres of the Al_4O_4 cubes (black) and the Mg atoms (orange), respectively (bottom-right); oxygen, red.

Compounds that crystallize in the normal spinel structure are:
- $A(II)^T B(III)_2{}^O O_4$ with:
  - A = Mg, Cr, Mn, Fe, Co, Ni, Cu, Zn, Cd, and Sn, and
  - B = Al, Ga, In, Ti, V, Cr, Mn, Fe, Co, Rh, and Ni,
  - these also comprise some binary compounds with the metal ion in a mixed valence state, e.g., $Mn^T Mn_2{}^O O_4$ (= $Mn_3 O_4$), and $Co^T Co_2{}^O O_4$ (= $Co_3 O_4$),
- $A(II)^T B(III)_2{}^O X_4$ with X = S, Se, and Te:
  - $CoCo_2 S_4$ (= $Co_3 S_4$), $HgCr_2 S_4$, $CdCr_2 Se_4$, $ZnCr_2 Se_4$, and $CuCr_2 Te_4$,
- $A(IV)^T B(II)_2{}^O X_4$ with X = O and S:
  - A = Ge, Sn,
  - B = Mg, Fe, Co, Ni for X = O, and Cu for X = S;
- $A(VI)^T B(I)^O O_4$, for instance $MoNa_2 O_4$, and $WNa_2 O_4$, and
- $ZnK_2(CN)_4$.

Inverse spinel structures are realized, for instance, in the compounds:
- $A(II)B(III)_2 X_4$ (i.e., $B(III)^T[A(II)B(III)]^O X_4$) with X = O and S:
  - $FeFe_2 O_4$ (= $Fe_3 O_4$), $CoFe_2 O_4$, $NiFe_2 O_4$, $CuFe_2 O_4$, $MgFe_2 O_4$, $MgIn_2 O_4$, $CoIn_2 S_4$, and $CrAl_2 S_4$,
- $A(IV)B(II)_2 O_4$:
  - $SnMg_2 O_4$, $SnMn_2 O_4$, $SnCo_2 O_4$, and $SnZn_2 O_4$,
  - $TiMg_2 O_4$, $TiMn_2 O_4$, $TiFe_2 O_4$, $TiCo_2 O_4$, and $TiZn_2 O_4$,
  - $VMg_2 O_4$, $VCo_2 O_4$, and $VZn_2 O_4$, and
- $NiLi_2 F_4$.

Furthermore, variants exist in which the ions are substituted with lower- and higher-valent species, for instance, $(LiAl_3)^T Al_2{}^O O_8$ or $LiMn_2 O_4$ (= $Li(I)^T[Mn(III)Mn(IV)]^O O_4$). The latter compound is used as a component in cathodes for lithium-ion batteries; however, the market share is comparably low: more than 90% of the lithium-ion batteries are based on the layer compound $LiCoO_2$.

5.3.6 CaTiO₃ (perovskite, *Strukturbericht* type E2₁)

The structure type of perovskite occasionally gives rise to confusion because the structure of the eponymous mineral, perovskite ($CaTiO_3$), is a distorted variant of the actual (undistorted) structure type. Opinions also differ as to which compounds should be added to the perovskite family at all [36, 37].

Let us start with the ideal undistorted cubic variant (the aristotype) that satisfies the formula ABX_3 and is realized, for example, by $SrTiO_3$. Here, the Sr and O atoms *together* build a **ccp** arrangement (stacking sequence *ABC*...), and the Ti atoms occupy ⅛ of the octahedral sites, specifically those that are exclusively enclosed by oxygen atoms. As Sr is part of the sphere packing, it is surrounded by 12 oxygen atoms in form of a cuboctahedron. SrTiO$_3$ crystallizes in the space group $Pm\bar{3}m$ with one formula unit per unit cell. In the perovskite structure, there are two equivalent possibilities with respect to the distribution of atoms on the unit cell or which type of atom is placed at the origin, respectively: either Ti or Sr can be placed at the origin (Wyckoff position 1a, 0,0,0) and the respective other metal species is then in the centre of the cell (Wyckoff position 1b, ½, ½, ½), see Figure 5.32. In each case, TiO$_6$ octahedra are present, which are corner-linked in all three directions to neighbouring TiO$_6$ octahedra (Niggli formula: TiO$_{6/2}$). Between each eight of these octahedra a Sr cation is located.

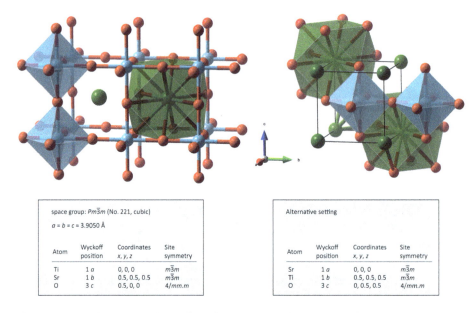

space group: $Pm\bar{3}m$ (No. 221, cubic)			
$a = b = c = 3.9050$ Å			
Atom	Wyckoff position	Coordinates x, y, z	Site symmetry
Ti	1 a	0, 0, 0	$m\bar{3}m$
Sr	1 b	0.5, 0.5, 0.5	$m\bar{3}m$
O	3 c	0.5, 0, 0	4/mm.m

Alternative setting			
Atom	Wyckoff position	Coordinates x, y, z	Site symmetry
Sr	1 a	0, 0, 0	$m\bar{3}m$
Ti	1 b	0.5, 0.5, 0.5	$m\bar{3}m$
O	3 c	0, 0.5, 0.5	4/mm.m

Figure 5.32: Crystal structure and crystallographic parameters of the undistorted perovskite structure of SrTiO$_3$ in two different settings; Ti at the origin (left), Sr at the origin (right). In the left picture, a 1 × 2 × 1 super cell is shown. Ti, light-blue; Sr, green; O, red.

An evaluation of whether an undistorted or distorted perovskite structure is realized can be obtained from the Goldschmidt tolerance factor [38], which is based on the ionic radii of the species involved and is calculated as follows:

$$t = \frac{r(A) + r(X)}{\sqrt{2} \cdot [r(B) + r(X)]}$$

Despite its simplicity and the fact that perovskites are not pure ionic compounds, the tolerance factor has reasonable predictive power, especially for oxides (X = O) and also for some fluorides (X = F): the ideal-undistorted cubic structure is observed in a range of t from 0.89 to 1, distorted structures at values between 0.8 and 0.89, and at values smaller than 0.8 the ilmenite type is formed rather than the perovskite type. If t > 1, meaning that large A and small B cations are present, a hexagonal instead of a cubic packing of the AX_3 layers is preferred. In the case of chlorides and sulphides (X = Cl, S), the limiting values of the tolerance factor for a certain type tends to decrease: cubic and distorted phases are formed for t in the range 0.8–0.9, and hexagonal perovskites are formed if t is larger than 0.9.

Distorted ABX_3 perovskites

In the majority of simple synthetic and naturally occurring perovskites, the main mode of distortion is tilting (rotation) of the BX_6 octahedra. This tilting occurs if the size of the A cation is too small for the twelvefold coordination site between the eight surrounding BX_6 octahedra. This means that the corner-connected BX_6 octahedra can be regarded as a flexible scaffold – the flexibility is provided by the bendable B-O-B connections, literally acting as a hinge. In this way, the scaffold adjusts the size of the void to the size of the A cation, which is also accompanied by a decrease in the coordination number of the A cation. Usually, the BX_6 octahedra are assumed to be rigid, but not necessarily ideal. The tilting of the BX_6 octahedra are accompanied by a symmetry reduction from the space group $Pm\bar{3}m$ to that of a certain hettotype. For instance, the mineral perovskite ($CaTiO_3$) adopts the orthorhombic space group $Pbnm$ with the Ca cations in an eightfold coordination, known as the $GdFeO_3$ structure type.

The distortions of perovskites are most commonly described by a system of nomenclature that was introduced by Glazer in 1972 [39]. In this scheme, the tilting of the BX_6 octahedra is described in terms of rotations around three orthogonal Cartesian axes, coinciding with the three axes of the parent cubic unit cell. In the general case of unequal rotation angles around the x-, y-, and z-axis, the rotation is denoted as a, b, and c (degrees). If the tilt is exactly the same about two axes, then the character is repeated. For instance, if the amount of rotation around the x- and y-axis is the same but different from the amount around the z-axis, this is denoted as aac. Furthermore, with the help of a superscript (0, + or –), it is specified if all the successive layers of octahedra perpendicular to a specific rotation axis have the same sense of tilt (+), i.e., an in-phase tilt, or are rotated in an opposite sense, i.e., have an anti-phase tilt (–). If there is no rotation perpendicular to a specific axis, the superscript is zero (0). Thus, the ideal undistorted cubic perovskite type has the tilt symbol $a^0a^0a^0$, the symbol $a^+b^+c^+$ indicates three unequal angles of rotation about the x-, y-, and z-axis with consecutive octahedra along the same axis rotating in the same sense, and for equal angles of rotation the notation would become $a^+a^+a^+$.

While Glazer initially identified 23 such tilt systems, which correspond to a particular space group, Howard and Stokes [40, 41] were able to show by group theoretical considerations that some of the 23 systems were combinations of one or more of the 15 basic tilting modes they identified. They are given in Table 5.7, together with the original tilting system number of Glazer and the corresponding space group. Note that the given space groups are those of maximum symmetry, meaning that they reflect only the tilting of the octahedra. If other distortions are present, such as Jahn-

Teller distortions, a space group with lower symmetry is obtained. This is, for instance, the case for $KCuF_3$ – a $a^0a^0a^0$ tilt system – in which the elongation of the CuF_6 octahedra lowers the symmetry from cubic $Pm\bar{3}m$ (no. 221) to tetragonal $I4/mcm$. Further, note that for perovskite phases that have a repeat unit of more than two octahedra in each axial direction, a more complex notation system is needed.

Table 5.7: Overview of the tilt systems for BX_6 octahedra in ABX_3 perovskites.

Phase	Tilt class	Tilt symbol	Tilt system no. according to Glazer	Space group (no.)
Zero-tilt	0 0 0	$a^0a^0a^0$	23	$Pm\bar{3}m$ (221)
One-tilt	0 0 −	$a^0a^0c^-$	22	$I4/mcm$ (140)
	0 0 +	$a^0a^0c^+$	21	$P4/mbm$ (127)
Two-tilt	0 − −	$a^0b^-b^-$	20	$Imma$ (74)
		$a^0b^-c^-$	19	$C2/m$ (12)
	0 + −	$a^0b^+c^-$	17	$Cmcm$ (63)
	0 + +	$a^0b^+b^+$	16	$I4/mmm$ (139)
Three-tilt	− − −	$a^-a^-a^-$	14	$R\bar{3}c$ (167)
		$a^-b^-b^-$	13	$C2/c$ (15)
		$a^-b^-c^-$	12	$P\bar{1}$ (2)
	+ − −	$a^+b^-b^-$	10	$Pnma$ (62)
		$a^+b^-c^-$	8	$P2_1/m$ (11)
	+ + −	$a^+a^+c^-$	5	$P4_2/nmc$ (137)
	+ + +	$a^+a^+a^+$	3	$Im\bar{3}$ (204)
		$a^+b^+c^+$	1	$Immm$ (71)

In Figure 5.33, three of the 15 systems are exemplified, showing the BX_6 octahedra in the three perpendicular directions of the parent cubic system, namely those of the zero-tilt phase $a^0a^0a^0$ of $SrTiO_3$, those of the one-tilt in-phase system of $CsSnI_3$ ($a^0a^0c^+$) (present between 350 and 425 K), and those of the mineral perovskite, $CaTiO_3$, which is a three-tilt system $a^+b^-b^-$, in which one tilt is in-phase and two tilts are out-of-phase but of equal amount.

Stacking variants

In the cubic perovskite, the stacking sequence of the densest AX_3 layers is $ABC...$ or, expressed by the Jagodzinski symbol, c, meaning that only corned-shared BX_6 octahedra are present. Although these layers become undulated in the distorted/tilted perovskite variants, still the octahedra are exclusively corner-connected. However, within the super-rich family of perovskite structures, also members with other stacking variants of densest layers exist, with a varying degree of h layers. At such h layers the octahedra share common faces. The amount of h layers depends on the nature and (relative) size of the involved metal ions but can also be dependent on the synthesis

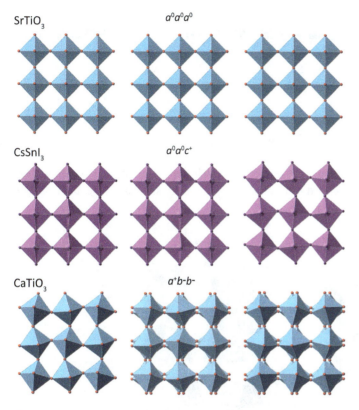

Figure 5.33: Three examples of perovskite tilt systems according to the nomenclature of Glazer; the BX_6 octahedra of the compounds $SrTiO_3$ (top), $CsSnI_3$ (middle), and $CaTiO_3$ (bottom) are shown along the three perpendicular cubic or pseudo cubic axes.

method. For instance, it is possible to obtain hexagonal $BaTiO_3$ (space group $P6_3/mmc$, no. 194) with a layer sequence of *hcc*, the slightly oxygen-deficient phase $BaMnO_{3-x}$ (x ~ 0.04) is present as the 4*H* polytype, with an *hc* layer sequence, and $BaNiO_3$ consists of exclusively *h* layers, see Figure 5.34.

The perovskite structure family consists of a huge number of members; so only a few representative compounds are listed in the following:
- A(I)B(V)O_3 with
  - $KNbO_3$ and $KTaO_3$
- A(II)B(IV)O_3 with
  - $CaTiO_3$, $SrTiO_3$, $BaTiO_3$, $BaSnO_3$, $CdSnO_3$, $CaIrO_3$, $PbTiO_3$, $PbZrO_3$, $SrCoO_3$, $SrMoO_3$, $SrRuO_3$, and $(Fe,Mg)SiO_3$,
- A(III)B(III)O_3 with
  - $LaAlO_3$, $LaTiO_3$, $LaMnO_3$, $LaCoO_3$, $NdAlO_3$, $ErCoO_3$, $HoCrO_3$, $YbMnO_3$, $BiFeO_3$, $GdFeO_3$, $LaCoO_3$, and $BiInO_3$,

- A(I)B(II)F$_3$ with
  - NaMgF$_3$, NaFeF$_3$, KMgF$_3$, KCrF$_3$, KMnF$_3$, KFeF$_3$, KCoF$_3$, KNiF$_3$, KCuF$_3$, KZnF$_3$, CsPbI$_3$, and AgMgF$_3$,
- A(I)NbO$_2$F with A = Li, Na, K,
- KMnCl$_3$, CsCdCl$_3$, CsHgCl$_3$, CsAuCl$_3$, and CsGaCl$_3$,
- CsCdBr$_3$a and CsHgBr$_3$,
- CsSnI$_3$,
- CaTiS$_3$, SrTiS$_3$, SrZrS$_3$, BaTiS$_3$, and BaZrS$_3$.

Furthermore, there are a few compounds in which the cation and anion sites are exchanged; they crystallize in the so-called *anti*-perovskite structure type, for instance, ISAg$_3$, CuNMn$_3$, PbNCa$_3$, and AuOCs$_3$.

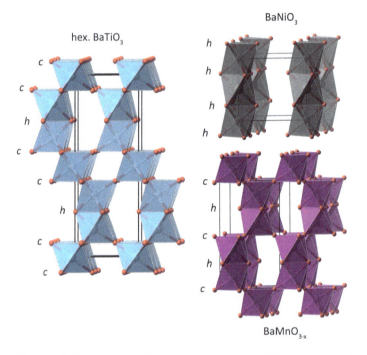

Figure 5.34: Three examples of perovskite structures with different amounts of *h* layers; hexagonal BaTiO$_3$ with an *hcc* stacking sequence of BaO$_3$ densest layers (left, Ba atoms not shown), BaMnO$_{3-x}$ (x ~ 0.04) with an *hc* stacking sequence (bottom-right), and BaNiO$_3$ consisting exclusively of *h* layers (top-right); view approximately along [100].

Several members of the perovskite structure family are technically very important. Barium titanate, BaTiO$_3$, and lead zirconate titanate, Pb(Zr$_x$Ti$_{1-x}$)O$_3$, abbreviated PZT, are important dielectric ceramics; these compounds, including their phase transitions, are discussed in Section 12.4.2. Hybrid organic-inorganic perovskites are another emerging class of compounds that could potentially find many applications. They are presented in the following section.

Hybrid organic-inorganic perovskites

The A site in the perovskite structure type is relatively large so that a number of complex ions, including organic ones, like ammonium, NH_4^+, methyl ammonium, $(CH_3NH_3)^+$, (abbreviated in literature as MA), tetramethylammonium, $[(CH_3)_4N]^+$ (TMA) or formamidinium $(H_2N=CHNH_2)^+$ (FA), can occupy this position. The most important ones are those with Pb or Sn at the B site and X = Cl, Br, I. Those organic–inorganic hybrid perovskites have sparked enormous interest among researchers in the past few years. These materials have been employed in a wide array of applications, such as in solar cells, photodetectors, light-emitting diodes (LEDs), transistors, and lasers, owing to their excellent optoelectronic performances, such as adjustable bandgaps, high absorption coefficients, and long charge-carrier diffusion lengths. In particular, the power conversion of such hybrid perovskites-based solar cells is noteworthy as their photo efficiency exceed 20%. From a structural point of view, the complex cation is typically disordered at room temperature, forming the cubic perovskite structure. For the prominent member of $MAPbCl_3$ – already characterized for the first time in 1987 – phase transitions occur upon cooling, first to a tetragonal phase at 177 K and then to an orthorhombic phase below 170 K. In this orthorhombic phase, the MA ions are ordered, the distorted $PbCl_6$ octahedra are tilted, and the Pb ions are off-centred within the octahedra (see Figure 5.35).

It is also possible to exchange the X site of the perovskite with ditopic multiatomic species such as cyanide (CN^-), azide (N_3^-), dicyanometallate (for instance, $[Ag(CN)_2]^-$), formate ($HCOO^-$), and borohydride (BH_4^-). This gives rise to the possibility to tune the size of the A site of the perovskite and to include even larger complex organic moieties at that site, for instance, tetrapropylammonium or triethylenediammonium (diprotonated diazabicyclo[2.2.2]octane ($DABCO^{2+}$)) cations, which in turn further alter the electronic properties of such hybrid organic-inorganic perovskites.

The literature concerning hybrid organic-inorganic perovskites is vast. As an entry point, the interested reader is referred to the monography in Ref. [42].

Double perovskites $A_2BB'X_6$

Ordered members of the perovskite supergroup can be obtained if either or both of the A and B site cations are substituted with another species and if they are at crystallographic distinct sites. If these cations are ordered at only one site, the compounds are termed double perovskites, whereas if ordering occurs at both sites, they are referred to as complex or quadruple perovskites. The most common of the B site ordered perovskites have the general formula $A_2BB'X_6$, where B and B' are different cations in octahedral coordination, located at crystallographically-distinct sites. The A

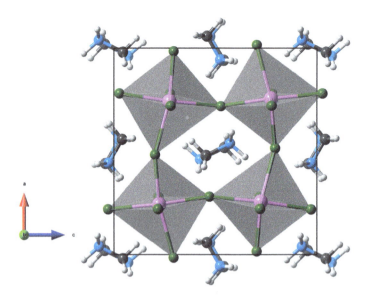

Figure 5.35: Crystal structure of the ordered orthorhombic phase of $CH_3NH_3PbCl_3$ (space group *Pnma*, no. 62); the methylammonium ions are rotated by 180° in successive layers along the *b*-direction. Pb, pink; Cl, green; C, dark-grey; N, blue; H, white.

site ordered double perovskites AA'BX$_6$ and quadruple perovskites AA'BB'X$_6$ have not yet been found as minerals but are well-known as synthetic phases.

Compounds with equal proportions of B and B' cations are termed 1:1 B-site ordered perovskites and as the BX$_6$ and B'X$_6$ octahedra are arranged in a chessboard-like pattern (as the Na and Cl ions in NaCl), they are also called rock salt-type double perovskites. Ideally, they exhibit long-range order with no mixing of the cations over the two available crystallographic sites; however, site mixing is well-known in synthetic double perovskites. If the B/B'X$_6$ octahedra are not tilted, the compounds adopt the space group $Fm\bar{3}m$ (no. 225), with a doubled dimension of the unit cell compared to the primitive cell of the single perovskite ABX$_3$. With octahedron tilting, eleven space groups of reduced symmetry are possible. In Figure 5.36 the structure of a tetragonal double perovskite compound, Sr$_2$FeMoO$_6$ (space group $I4/m$, no. 87), is shown.

In double (and quadruple) perovskites, the basic structural features (corner-shared B/B'X$_6$ octahedra) are maintained but they allow for a wider range of B cation species to be incorporated, which enables access to an expansive range of alternative compositions and magnetic and photo-physical properties. Halide double perovskites are an emerging subject of current research [43].

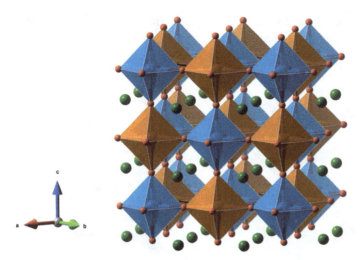

Figure 5.36: Crystal structure of the ordered double rock salt-type perovskite compound Sr_2FeMoO_6. Sr, green; Fe, blue; Mo, orange; O, red.

Relationship with other structure types

ReO_3 (*Strukturbericht* type $D0_9$)

The prototypical compound ReO_3 is closely related to the ideal cubic perovskite structure of $SrTiO_3$. Instead of the corner-shared TiO_6 octahedra in $SrTiO_3$, ReO_6 corner-sharing octahedra are present that build a primitive cubic array; the oxygen atoms are at the unit cell edge bisectors and are linearly coordinated by two Re atoms. The only difference is that the centre of the cell, surrounded by eight ReO_6 octahedra, is unoccupied, see Figure 5.37, left. Therefore, as a kind of didactic mnemonic, one could formulate: "ReO_3 = $SrTiO_3$ − Sr". ReO_3 crystallizes in the cubic space group $Pm\bar{3}m$ (no. 221) with one formula per unit cell. ReO_3 is a rather unusual oxide compound with respect to two aspects: (i) it is the only stable trioxide compound of the group 7 elements (Mn, Tc, Re), (ii) it has a very high electrical conductivity (almost as high as that of Cu) and – typically for a metallic behaviour – the conductivity decreases with increasing temperature. The formation of the conductivity band results from the overlap between the Re 5d (t_{2g}) and O 2p ($p_{x/y}$) orbitals. The metallic-like properties of ReO_3 are also reflected in the metallic lustre of the (deep red) crystals it forms.

Compounds that crystallize in the ReO_3 structure type are:
- ReO_3, UO_3, AlF_3, ScF_3, $TiOF_2$, NbF_3, TaF_3, TaO_2F, MoF_3, Na_3N, and Cu_3N, whereas the latter two represent compounds with an *anti*-ReO_3 structure.

Some hydroxides of trivalent metals, *e.g.*, $In(OH)_3$ and $Sc(OH)_3$, form structures that are closely related to the structure of ReO_3. As in ReO_3, corner-linked metal-oxygen octahedra are present, but in contrast to those in ReO_3, they are strongly tilted with respect to each other. In terms of Glazer's perovskite tilting notation, it would be a $a^+a^+a^+$ system. Furthermore, the hydroxide ions that are coordinated to the trivalent metal ions form hydrogen-bonded square 4-rings (see Figure 5.37, right).

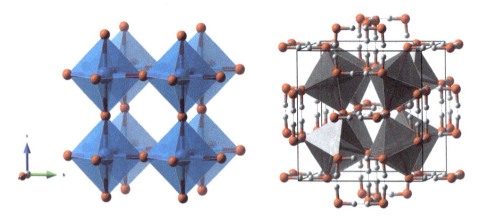

Figure 5.37: Crystal structure of ReO$_3$ (left) and In(OH)$_3$ (right). Re, blue; In, grey; O, red; H, white. Note that the hydrogen atoms in In(OH)$_3$ are disordered over two symmetry-equivalent positions with an occupancy of 0.5.

Tungsten bronze

The starting point of tungsten bronzes is tungsten trioxide (WO$_3$) in which mainly strongly electropositive metals (mainly alkali metals) are incorporated in non-stoichiometric amounts, wherein a corresponding amount of the W(VI) is reduced to W(V). The name bronze is not related to the structure of the alloy between copper and tin but mainly due to the similar colour of some of the tungsten bronze phases. While WO$_3$ is structurally similar to ReO$_3$ – corner-linked WO$_6$ octahedra are present – WO$_3$ has completely different properties compared to ReO$_3$. The octahedra are significantly tilted towards each other and a triclinic (< 17 °C) or a monoclinic phase (> 17 °C) is formed around room temperature. And while ReO$_3$ has rather metallic properties, WO$_3$ is a semiconductor (band gaps between 2.6 and 3.1 eV have been reported). If, now, alkali metals are incorporated into the WO$_3$ host structure, it remains identical up to a certain degree of doping, and then phases with higher symmetry are adopted. For sodium, the host structure of WO$_3$ remains unchanged up to a doping level of ~ 0.1, and above that level, a metallic tetragonal tungsten bronze (TTB) is formed. The sodium atoms are located inside the square and (irregular) pentagonal channels running along the c-axis (see Figure 5.38, left). For potassium also hexagonal tungsten bronzes (HTB) are known and the K atoms reside in hexagonal channels running along the c-axis (see Figure 5.38, right).

K$_2$NiF$_4$ and Ruddlesden-Popper phases

The structural type of K$_2$NiF$_4$ is the prototype consisting of alternating (incomplete) NaCl-like (AX) and (incomplete) perovskite-like (ABX$_3$) layers of different thickness, as shown

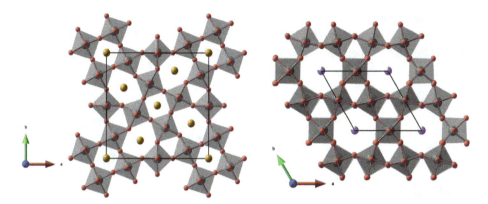

Figure 5.38: Crystal structure of a tetragonal sodium ($Na_{0.48}WO_3$, left) and a hexagonal potassium tungsten bronze ($K_{0.26}WO_3$, right). Na, yellow; K, purple; O, red.

in Figure 5.39 (left). To illustrate that both structure types are present, it might be helpful to rewrite the sum formula as $KNiF_3 \cdot KF$. K_2NiF_4 crystallizes in the body-centred tetragonal space group $I4/mmm$. The Ni atoms are octahedrally surrounded by six fluorine ions. Because the perovskite- and NaCl-blocks are incomplete and interrupted, the coordination number of K is reduced from 12 to 9 as shown in the outer most left of Figure 5.39.

Examples for compounds that crystallize in the K_2NiF_4 structure type are:
- K_2CuF_4,
- A_2MgF_4, A_2FeF_4, A_2CoF_4, and A_2NiF_4, with A = K, Rb, and Tl,
- A_2BCl_4, with A = Rb and Cs, and B = Cr, Mn, and Cd,
- Sr_2BO_4, with B = Ti, Zr, Hf, Mo, Mn, Tc, Ru, Rh, and Ir,
- Ba_2SnO_4 and Ba_2PbO_4,
- RE_2BO_4, with RE = La, Ce, Pr, and Nd, and B = Ni and Cu.

The K_2NiF_4 structure type is the simplest type of a larger family of the so called Ruddlesden-Popper (RP) phases, with the general formula of $A_{n+1}B_nX_{3n+1}$ that can be rewritten as $A_{n-1}A'_2B_nX_{3n+1}$, where A and A' are alkali, alkaline earth, or rare-earth metals, B is a transition metal, and X is usually O, F, or Cl. The A cations are part of complete perovskite blocks and have a coordination number of 12, the A' cations are at the boundary between the perovskite and rock-salt-like blocks, and have a coordination number of 9; the B cations are transition metals that are octahedrally, square-pyramidal, or square-planar coordinated. Theoretically, compounds with any number of n in the series $n = 1,2,3, ..., \infty$ ($n = \infty$ is identical with the pure perovskite phase ABX_3), with increasing numbers of perovskite layers are conceivable, but usually only the first 3 are experimentally accessible as phase-pure compounds. RP phases are named after S.N. Ruddlesden and P. Popper, who first synthesized and described a series with $n = 1, 2$, and 3 for the Sr-Ti-O system in 1957/1958 [44, 45].

Similar to the parent perovskite phases, RP phases can possess interesting optical, magnetic, and electric properties. For instance, the oxocuprate $La_{1.85}Ba_{0.15}CuO_4$ paved the way for high-temperature superconductors (cf. Section 11.2.3).

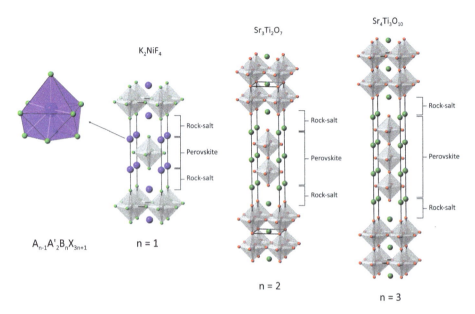

Figure 5.39: Crystal structure of K_2NiF_4 (left), $Sr_3Ti_2O_7$ (middle), and $Sr_4Ti_3O_{10}$ (right), which represent Ruddlesden-Popper phases of the general formula $A_{n-1}A'_2B_nX_{3n-1}$ with n = 1, 2, and 3, respectively. In these examples for n = 2 and 3, A = A' = Sr. F, light-green; O, red; Sr, dark-green; K, purple; B-cation, grey.

5.4 Compounds based on a hexagonal closest packing

5.4.1 NiAs (*Strukturbericht* type B8)

In the prototypical structure of nickel arsenide (the mineral is known as nickeline or niccolite), the As atoms form a hexagonal closest packing (stacking sequence *ABAB*...) and the Ni atoms occupy all octahedral interstitial sites. In contrast to the cubic counterpart of NaCl, the atoms in NiAs have different coordination environments. While the Ni atoms are surrounded by six As atoms in form of an octahedron, the As atoms are surrounded trigonal-prismatically by six Ni atoms (CN = 6:6), see Figure 5.40. Note that the Ni atoms (occupying the y sites) are stacked in a primitive fashion (meaning that the atoms lie directly on top of each other) along the *c*-direction, and that every second trigonal prism is empty. Three trigonal prisms in the same plane meet at a common edge. The $NiAs_6$ octahedra have two common faces with neighbouring octahedra (along the *c*-direction) and are further edge-connected (along *a* and *b*).

While the NaCl structure type is preferred for highly polar/ionic compounds (this is favourable due to electrostatic reasons, since each atom has only atoms of the other element in its vicinity), the NiAs structure type is dominant for element combinations that have a rather low electronegativity difference. Note that face-sharing octahedra are present that would be highly unfavourable for ionic species. In fact, the Ni-Ni distance (2.516 Å) is only slightly larger than the Ni-As distance (2.439 Å). There are clear signs that metal-metal interactions are present in compounds of the NiAs structure type: They have a metallic lustre, they show electrical conductivity, the metal does not have to be present in stoichiometric amounts, and the c/a-axis ratio is dependent on the electron configuration of the metal component. Furthermore, for many compounds the c/a ratio is smaller than the ideal value (1.633) of the **hcp** packing. Usually, the rough trend observed is that the more electron-rich the metal, the smaller the c/a ratio (see Table 5.8). This means that a considerable shrinkage in the c-direction takes place, i.e., in the direction in which the metal atoms are closest to each other. Note, however, that also an expansion that takes place in the (a,b) plane leads likewise to a decrease of the c/a ratio.

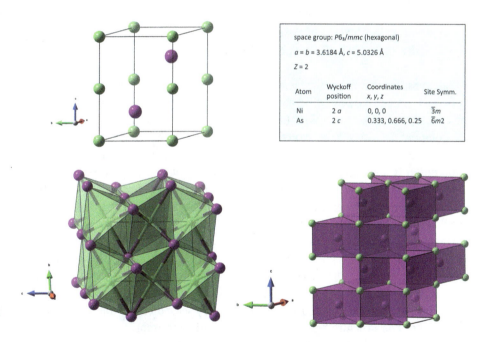

Figure 5.40: Unit cell and crystallographic parameters of NiAs (top), and visualization of the coordination polyhedra of Ni (bottom-left) and As (bottom-right). Ni, green; As, purple.

Table 5.8: c/a-axis ratio of some compounds of transition metal antimonides, selenides, and tellurides that crystallize in the NiAs structure type. NiAs itself has a c/a ratio of 1.391.

compound	TiSb	VSb	CrSb	MnSb	FeSb	CoSb	NiSb
c/a ratio	1.549	1.276	1.327	1.398	1.264	1.328	1.305
compound	TiSe	VSe	CrSe	MnSe	FeSe	CoSe	NiSe
c/a ratio	1.677	1.631	1.642	1.628	1.402	1.460	1.462
compound	TiTe	VTe	CrTe	MnTe	FeTe	CoTe	NiTe
c/a ratio	1.667	1.540	1.566	1.630	1.487	1.383	1.351

The NiAs structure type is realized in element combinations of metals from group 4 to 10 and, in particular, the non-metals/metalloids P and S and their higher homologues, *e.g.*:
- TiP and VP,
- TiAs, MnAs, CoAs, and NiAs,
- TiSb, VSb, CrSb, MnSb, FeSb, CoSb, NiSb, IrSb, PdSb, and PtSb,
- MnBi, NiBi, RhBi, InBi, and PtBi,
- TiS, VS, CrS, FeS, CoS, and NiS,
- TiSe, VSe, CrSe, MnSe, FeSe, CoSe, and NiSe,
- TiTe, VTe, CrTe, MnTe, FeTe, CoTe, NiTe, and PdTe.

It is also possible that either the metal or the non-metal/metalloid site in the NiAs structure type is occupied by two different elements (not necessarily in equal amounts), as is the case, for instance, for (Co,Ni)As (langistite) and Pt(Sb, Bi) (stumpflite).

Compounds with a NiAs structure often show a certain phase width, meaning that a certain fraction (up to approx. 20%) of the metal atoms may be missing. The composition is then $M_{(1-x)}X$ (x = 0–0.2). The vacancies may be distributed statistically or in an ordered fashion. In the latter case, they are superstructures of the NiAs type, which can be found, for example, in polytypes of sub-iron(II) sulphides (the mineral is known as pyrrhotite) such as Fe_7S_8, Fe_9S_{10}, $Fe_{10}S_{11}$, $Fe_{11}S_{12}$, which crystallize either in the monoclinic or hexagonal crystal system. The stoichiometric (FeS) variant is known as the mineral troilite. While troilite is non-magnetic, the magnetic properties of the non-stoichiometric $Fe_{(1-x)}S$ phases are dependent on the Fe content. Fe-rich phases are antiferromagnetic, while more Fe-deficient phases are ferrimagnetic.

If the metal atoms are removed only from every second octahedral interstitial site layer of the NiAs structure type, there can be a continuous transition from the NiAs to the CdI_2 structure type (see next section). This is, for instance, known for $Co_{(1-x)}Te$. CoTe (x = 0) crystallizes in the NiAs type, while $CoTe_2$ (x = 0.5) belongs to the CdI_2 type.

MnP can be considered as a distorted variant of the NiAs structure in which also the distance of the metal atoms in the (a,b) plane is decreased. In NiAs, the Ni atoms have only two nearest neighbours (in the c-direction) within a distance of approx. 2.52 Å, while the next-nearest neighbours are at a distance of 3.618 Å. In MnP, each

Mn atom has *four* nearest neighbours within a distance of approx. 2.82 Å. The Mn atoms build zigzag chains in one direction, which are further connected to the neighbouring zigzag chains via Mn-Mn links. Also, the P atoms in MnP come significantly closer together (2.66 Å) compared to the As atoms in NiAs (3.27 Å) and form infinite zig-zag chains, too (see Figure 5.41). Therefore, another possible interpretation of the MnP structure is to consider the P atoms as a polyanionic Zintl substructure (P_∞^-). MnP crystallizes in the orthorhombic space group *Pnma* (no. 62) (a = 5.2580 Å, b = 3.1720 Å, c = 5.9180 Å) with four formula units per unit cell. MnP has not only an interesting structure but magnetic properties, too. It is one of the typical itinerant *helimagnets*. The helimagnetic phase is present below 47 K. Above 47 K, two different ferromagnetic phases exist [46]. Helimagnetism is a form of magnetic ordering where spins of neighbouring magnetic moments arrange themselves in a helical pattern. It results from the competition between ferromagnetic and antiferromagnetic exchange interactions (see also Chapter 9).

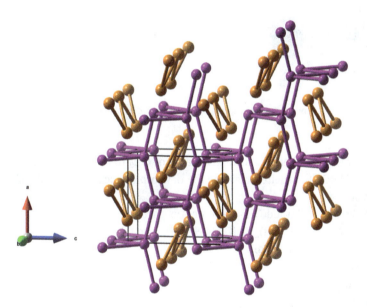

Figure 5.41: Crystal structure of MnP (Mn, purple; P, orange).

5.4.2 CdI$_2$ (*Strukturbericht* type C6)

In the prototypical structure of cadmium iodide, the I atoms form a hexagonal closest packing (stacking sequence *ABAB*...) and the Cd atoms occupy all octahedral interstitial sites of every second octahedral layer, meaning that every other second such layer is completely empty, leading to an overall stacking sequence *A*-γ-*B*-☐ (see

Figure 5.42). Analogous to the cubic counterpart $CdCl_2$, the Cd ions are surrounded by six Cl ions in form of an octahedron and the I ions form the tip of a trigonal pyramid with three Cd^{2+} as base (CN = 6:3). The octahedra are connected to adjacent octahedra by six common edges, forming a layer in which each halide atom belongs to three octahedra simultaneously ($MX_{6/3}$), compare also Figure 5.17. Cadmium iodide is a layered-type compound; only relatively weak van-der-Waals forces are present between the individual layers.

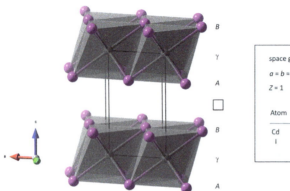

Figure 5.42: Crystal structure together with the stacking sequence (left) and crystallographic parameters (right) of CdI_2.

Meanwhile, a considerable number of stacking variants of CdI_2 have been found, so that $Cd(OH)_2$ (or even $Mg(OH)_2$, brucite) might be the new prototypical compounds of the CdI_2 structure type.

Compounds that belong to $CdI_2/Cd(OH)_2$ structure type are:
- $Mg(OH)_2$, $Ca(OH)_2$, $Mn(OH)_2$, $Fe(OH)_2$, $Co(OH)_2$, $Ni(OH)_2$, and $Cd(OH)_2$,
- $MgBr_2$, $TiBr_2$, VBr_2, $MnBr_2$, $FeBr_2$, $CoBr_2$, and $NiBr_2$,
- MgI_2, CaI_2, PbI_2, TiI_2, VI_2, MnI_2, FeI_2, and CoI_2,
- TiS_2, ZrS_2, PtS_2, and SnS_2,
- $TiSe_2$, $ZrSe_2$, and $PtSe_2$,
- $TiTe_2$, $ZrTe_2$, and $PtTe_2$.

Also some ternary compounds crystallize in the CdI_2 structure type, e.g., BiTeBr, in which the Te and Br atoms are statistically distributed. Interestingly and in contrast to BiTeBr, in BiTeI, the metalloid/non-metal atoms are both ordered, reducing the symmetry to the space group $P3m1$ (no. 156).

In $CrBr_2$, CrI_2, and $CuBr_2$, a huge Jahn-Teller distortion is present, leading to isolated, infinite edge-sharing chains of square-planar coordinated Cr/Cu atoms running along the b-axis of the monoclinic crystal structure (space group $C2/m$), see Figure 5.43.

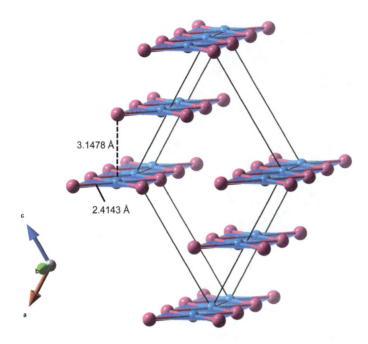

Figure 5.43: Crystal structure of CuBr$_2$ (Cu, blue; Br, pink).

If the metal atoms build the **hcp** packing and the non-metal atoms occupy the octahedral voids, the respective *anti*-CdI$_2$/Cd(OH)$_2$ type is present. Compounds that crystallize in this type are Ag$_2$F and Ag$_2$O.

5.4.3 β-V$_2$N (*Strukturbericht* type L'3$_2$)

There are further structure types with a hexagonal densest packing of one type of atoms, in which one-half of the octahedral interstitial sites are occupied; however their distribution is different with respect to the fraction of occupancy among these octahedral layers. Another possibility to arrive at a stoichiometry of MX$_2$ or M$_2$X (*anti*-type) is that the octahedral layers are alternately occupied by ⅔ and ⅓. The stacking sequence is then A-γ$_{2/3}$-B-γ'$_{1/3}$. The prototypical compound of that type is V$_2$N (space group $P\bar{3}1m$, no. 162). The octahedra of the γ layer are connected analogously to the ones in the layers of AlCl$_3$ (compare Figure 5.18). The octahedra in the γ'-layer are isolated within that layer but have three common corners with the octahedra of the γ layer; they lie above and beneath the holes of that layer (see Figure 5.44). V$_2$N, Cr$_2$N, and Fe$_2$N are compounds that crystallize in this structure.

There are also several ternary/quaternary compounds in which two/three different elements occupy the octahedral interstitial site layers in an ordered manner or

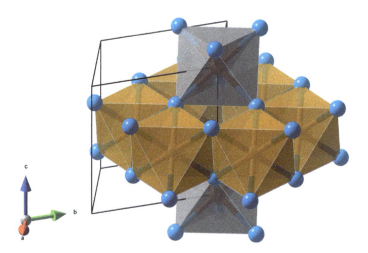

Figure 5.44: Crystal structure of V_2N; octahedra of the γ layer are shown in orange, octahedra of the γ' layer in grey; blue, V atoms; orange, N atoms.

those in which two different elements build the **hcp** packing. In the ICSD the reference compound of that type is $PbSb_2O_6$. Other compounds that are isostructural are:
- Li_2ZrF_6, Li_2HfF_6, Li_2NbF_6, Li_2NbOF_5, and Li_2PbF_6
- $CaSb_2O_6$, $CaAs_2O_6$, $SrSb_2O_6$, $SrRu_2O_6$, $BaSb_2O_6$, $CdSb_2O_6$, and $PbSb_2O_6$,
- $LaTiSbO_6$,
- $MnAs_2O_6$, $NiAs_2O_6$, $CoAs_2O_6$, $PdAs_2O_6$, $CdAs_2S_6$, $HgAs_2O_6$, and $SrAs_2O_6$,
- UV_2O_6 and UCr_2O_6.

5.4.4 $CaCl_2$ (*Strukturbericht* type C35), TiO_2 (rutile) (*Strukturbericht* type C4), and FeS_2 (marcasite) (*Strukturbericht* type C18)

In the $CaCl_2$ structure type, the chlorine atoms build an almost ideal **hcp** packing and the Ca atoms occupy one-half of the octahedral interstitial sites in every octahedral layer. This is just another possibility to arrive at a stoichiometry of MX_2. Note that, in contrast, in CdI_2, every second octahedral layer remained unoccupied, while every other such layer is completely occupied.

The octahedra build infinite edge-shared strands along the c-direction, and these strands are further connected by corners to neighbouring strands. Note that the octahedra have alternating orientations along the stacking sequence (A-$\gamma_{1/2}$-B-$\gamma'_{1/2}$) and that there is already a small deviation from an ideal **hcp** packing: the densest layers are slightly undulated. The $CaCl_6$ octahedra are slightly compressed along the axial ligands (Ca-Cl_{ax} = 2.71 Å; Ca-Cl_{eq} = 2.77 Å). $CaCl_2$ crystallizes in the orthorhombic space group *Pnnm* (no. 58), see Figure 5.45. This undulation is even more pronounced in the TiO_2 (rutile) structure type, see below.

Compounds that crystallize in the CaCl$_2$ structure type are:
- MgF$_2$, NiF$_2$, CaCl$_2$, CaBr$_2$, YBr$_2$, PtO$_2$, and RuO$_2$.

The reference compound of the respective *anti*-type of CaCl$_2$ is Fe$_2$C; some borides, carbides, and nitrides belong to this type, i.e., Pd$_2$B, Fe$_2$C, Co$_2$C, Rh$_2$C, and Co$_2$N.

The structure type of TiO$_2$ (rutile) is closely related to the structure of CaCl$_2$; however, the oxide ions in TiO$_2$ do *not* build an ideal **hcp** packing but the dense packed layers are considerably undulated. Apart from that, the distribution of the metal atoms among the octahedral interstitial sites and the connection scheme of the occupied octahedra is identical with those in the CaCl$_2$ structure type so that it might be justified to mention the rutile structure within this Section 5.4 (structures based on a **hcp** packing). While the CaCl$_6$ octahedra are slightly compressed in the direction of the axial ligands, the TiO$_6$ octahedra are elongated (Ti-O$_{ax}$ = 1.986 Å, Ti-O$_{eq}$ = 1.945 Å) in that direction. The oxide ions are in a rather unusual cation environment; they are surrounded by three Ti atoms in the form of an isosceles, almost equilateral triangle (CN = 6:3). If only the Ti atoms are considered, they form a tetragonally body-centred packing.

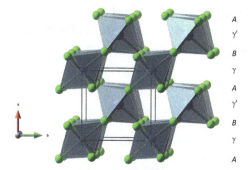

space group: *Pnnm* (orthorhombic)

a = 6.240, b = 6.430, c = 4.200 Å

Z = 2

Atom	Wyckoff position	Coordinates x, y, z	Site Symm.
Ca	2 a	0, 0, 0	..2/m
Cl	4 g	0.275, 0.325, 0	..m

Figure 5.45: Crystal structure and stacking sequence (left) as well as the most important crystallographic parameters of CaCl$_2$ (right); CaCl$_6$ octahedra, pale blue; Cl atoms, green.

What is different with respect to the octahedra in CaCl$_2$ is that the octahedra of the neighbouring octahedra strands have a different mutual orientation: They are perfectly perpendicular (90°) to each other in TiO$_2$, while the axial directions of the neighbouring octahedra form an angle of 75°/105° in CaCl$_2$. This twisting of the octahedra strands is even more pronounced in marcasite, a polymorph of FeS$_2$ (likewise, as CaCl$_2$, it crystallizes in the space group *Pnnm*). The driving force for this is that the sulphur atoms have moved towards each other to form S$_2^{2-}$ dumbbells. The longitudinal axes of the octahedra of the neighbouring octahedra strands enclose an angle of 47°/133° (compare Figure 5.46).

It is useful to review the fundamental differences between an ideal **hcp** packing and the tetragonal distorted variant that is present in rutile. In the **hcp** packing, the

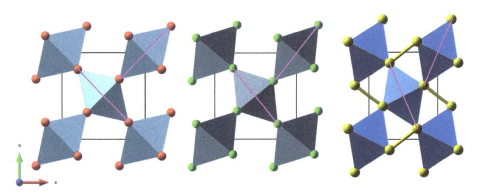

Figure 5.46: Comparison of the structures of TiO_2 (rutile) (left), $CaCl_2$ (middle), and FeS_2 (marcasite) (right). The mutual orientation of the neighbouring octahedra strands is emphasized by pink lines.

closest-packed layers are present in only one direction or orientation. In the tetragonal distorted variant, the undulated dense-packed layers are present in two perpendicular directions (see Figure 5.47). Or to put it another way: the more strongly undulated layers in the **hcp** packing (oriented perpendicular to the densest layers, dashed lines in Figure 5.47, right) become partially straightened in the tetragonal packed arrangement, while the straight close-packed layers (solid lines in Figure 5.47) become slightly undulated. The final extent of the corrugation is identical for both perpendicular layers.

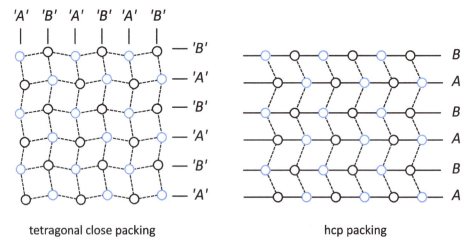

Figure 5.47: Arrangement of spheres in a tetragonal close (left) and a hexagonal-closest packing (right). Black and blue circles represent spheres at different heights along the c-direction, solid lines represent densest layers, dashed lines undulated/corrugated layers.

In that notion, the CaCl$_2$ structure can also be considered as a distorted variant of the rutile aristotype. The high tetragonal symmetry of rutile (space group $P4_2/mnm$, no. 136) is broken in CaCl$_2$ (space group $Pnnm$, no. 58, orthorhombic) due to the rotation of the edge-sharing octahedra chains; the mirror plane perpendicular to the [110] direction is removed. The space group $Pnnm$ is the space group of highest symmetry that is both a subgroup of $P6_3/mmc$ (the space group of the **hcp** packing) and $P4_2/mnm$.

The packing of the O atoms in rutile corresponds to one of the so called tetragonal (dense) sphere packing types [47, 48], with symmetry-equivalent spheres in contact with *eleven* others, meaning that the coordination number is reduced by one with respect to the **hcp** packing. This tetragonal packing was introduced by W.H. Baur in 1981 [49] as a "densest" such packing, but Michael O'Keeffe showed that this is not the case [50]. The space filling degree of Baur's tetragonal packing (71.87%) lies between that of a densest-packing (74.05%) and a body-centred packing (68.01%).

A useful description of a tetragonal sphere packing is that spheres are at both ends of rungs of an infinite ladder. Neighbouring ladders are rotated by 90° and are shifted along the ladder in such a way that spheres of one ladder are at the gaps between the rungs of the neighbouring ladders. Each sphere has now 4 + 2 + 2 = 8 nearest neighbours from the neighbouring ladders and three within its own ladder (two to the neighbouring rungs, one to the other end of the rung). Referring to the rutile structure, the equatorial oxygen atoms of the edge-shared TiO$_6$ octahedra form the rungs of the ladder and the titanium atoms are located exactly at the centre between two rungs (see Figure 5.48).

Numerous compounds belong to the rutile structure type:
- highly polar metal oxides, *e.g.*, TiO$_2$, VO$_2$, NbO$_2$, CrO$_2$, MoO$_2$, WO$_2$, MnO$_2$, RuO$_2$, OsO$_2$, IrO$_2$, GeO$_2$, SnO$_2$, PbO$_2$, and TeO$_2$,
- some metal (oxy)fluorides, *e.g.*, MgF$_2$, VF$_2$, MnF$_2$, FeOF, FeF$_2$, CoF$_2$, NiF$_2$, PdF$_2$, and ZnF$_2$,
- MgH$_2$,
- the high-pressure modification of SiO$_2$ known as stishovite, and
- the high-temperature modifications of CaCl$_2$ and CaBr$_2$.

Ternary compounds ABO$_4$ belonging to the rutile structure type with a statistical distribution of the cations among the octahedral interstitial sites are: AlTaO$_4$, AlSbO$_4$, GaSbO$_4$, TiNbO$_4$, TiTaO$_4$, TiVO$_4$, VTaO$_4$, VSbO$_4$, CrNbO$_4$, CrTaO$_4$, CrSbO$_4$, FeTaO$_4$, FeSbO$_4$, RhVO$_4$, RhNbO$_4$, RhTaO$_4$, and RhSbO$_4$.

The most common ordering of cations in rutile type compounds is exhibited by the so called trirutiles of type AB$_2$X$_6$ (X = O or F). They crystallize in the same space group type as the rutile itself ($P4_2/mnm$) but have a tripled unit cell along the c-direction. An example of this type, LiMo$_2$F$_6$, is shown in Figure 5.49.

An exhaustive account on the rutile structure type and all of its derivatives, including a scheme of the relationships of symmetry among them (Bärnighausen tree) can be found in two reviews by W.H. Baur [51] and H.P. Beck [52].

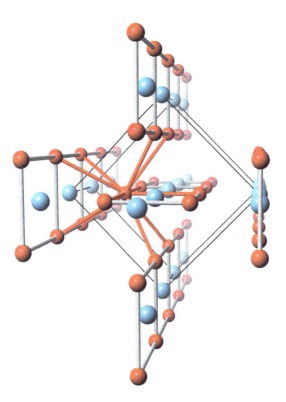

Figure 5.48: Alternative description of the structure of rutile as ladders of O atoms (red), in which the Ti atoms (light-blue) are located at the centre of the rungs. Each oxygen atom has 4 + 2 + 2 + 3 = 11 nearest O atoms in its surrounding; for one oxygen in the centre of the picture, the respective O-O connections are shown as red sticks.

5.4.5 BiI$_3$ (*Strukturbericht* type D0$_5$)

In the BiI$_3$ structure type, the iodine anions build an **hcp** packing (stacking sequence AB...) and two-third of every next octahedral layer is occupied by Bi atoms, so that the overall stacking sequence is A-$\gamma_{2/3}$-B-$\square$-... Every Bi ion is surrounded by six iodide anions and every iodide anion is linked to two Bi ions in an angular fashion (CN = 6:2). BiI$_3$ is the hexagonal counterpart of the AlCl$_3$ structure type. As in AlCl$_3$, also in BiI$_3$ sheets of edge-sharing octahedra are present in which holes of octahedral shape are present (compare Figure 5.18).

Compounds that crystallize in the BiI$_3$ structure type are:
- ScCl$_3$, FeCl$_3$, α-TiCl$_3$, FeCl$_3$, TiBr$_3$, FeBr$_3$, YI$_3$, VI$_3$, AsI$_3$, SbI$_3$, and BiI$_3$,
- low-temperature modification of CrCl$_3$.

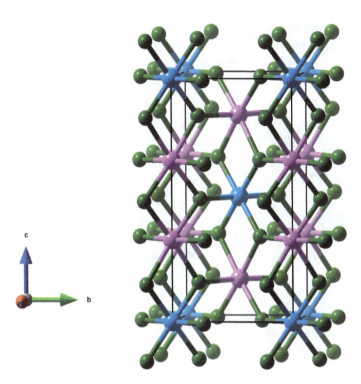

Figure 5.49: Crystal structure of the trirutile compound LiMo$_2$F$_6$. Li, pink; Mo, blue; F, green.

5.4.6 β-TiCl$_3$ and MoBr$_3$/ZrI$_3$/RuBr$_3$

For several other metal trihalides a structure type different to that of BiI$_3$ is realized: Again, the anions form an **hcp** packing and ⅓ of all available octahedral interstitial sites are occupied, but different to the type of BiI$_3$ in which every *second* octahedral layer has an occupancy of ⅔, here *every* layer has an occupancy of ⅓, leading again to a stoichiometry of MX$_3$; thus the stacking sequence is A-γ$_{1/3}$-B-γ$_{1/3}$... This structural type is reported under different names in literature, for instance β-TiCl$_3$ type or ZrI$_3$ type, although these compounds are not all isotypical. The principal feature is, however, identical: the structure contains an array of isolated, infinite chains of confacial octahedra (MX$_{6/2}$) running along the c-axis that are held together only by dispersion forces, in accordance with the very fibre-like habitus usually found for these crystals. Differences exist with respect (i) to the exact packing of the anions, i.e., if they actually form an ideal or distorted **hcp** packing and (ii) if there are metal-metal bonds between pairs of each two neighbouring octahedra (note that they are face-shared).

In β-TiCl$_3$ (space group $P6_3/mcm$, no. 193), there are no such metal-metal bonds detectable, as the interatomic Ti-Ti distances along the c-axis are identical (2.910 Å).

However, in MoBr$_3$, ZrI$_3$, and RuBr$_3$ (space group *Pmmn*, no. 59), alternately short (3.1725 Å) and long (3.5605 Å) metal-metal distances are present, indicative of metal-metal bonds between pairs of two neighbouring octahedra (see Figure 5.50, right).

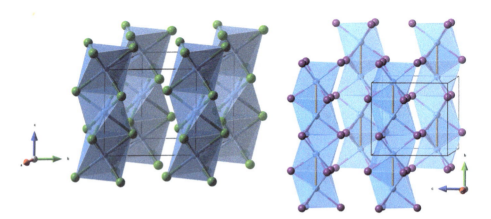

Figure 5.50: Crystal structure of β-TiCl$_3$ (left) and ZrI$_3$ (right); Ti and Zr, dark and light blue; Cl, green; I, purple. Zr-Zr bonds are indicated by orange sticks.

5.4.7 α-Al$_2$O$_3$ (corundum, *Strukturbericht* type D5$_1$) and FeTiO$_3$ (ilmenite, type E2$_2$)

In the prototypical structure of α-Al$_2$O$_3$ the oxygen atoms are arranged according to a (slightly distorted) **hcp** packing (stacking sequence *AB*...) and the Al atoms occupy ⅔ of all octahedral voids in every such octahedral layer, leading to a stoichiometry of M$_2$X$_3$. The edge-connected octahedra form layers identical to the ones in BiI$_3$ or AlCl$_3$, i.e., the edge-shared octahedra form hexagons with holes in the centre of six such octahedra (compare Figure 5.18). Along the stacking direction, there are pairs of face-shared AlO$_6$ octahedra that are separated by an unoccupied octahedron. Because of the confacial arrangement, the Al-Al distance (2.655 Å) is relatively short. The Al atoms (and therefore the unoccupied octahedral voids) can be at three different sites, which are realized sequentially along the stacking direction. This leads to a stacking sequence of *A*-γ$_{2/3}$-*B*-γ'$_{2/3}$-*A*-γ"$_{2/3}$... The Al ions are surrounded by six oxygen atoms in form of a slightly distorted octahedron, and the oxygen atoms are surrounded by four Al atoms in the form of a distorted tetrahedron (CN = 6:4). The unoccupied octahedral sites lead to a considerable distortion of the surrounding oxygen. Ideally, i.e., without those distortions, they would be surrounded by four Al atoms and two vacancies in the form of a regular trigonal prism (compare the structure of NiAs, Section 5.4.1).

Compounds that crystallize in the corundum structure are:
- α-Al$_2$O$_3$, γ-Al$_2$S$_3$, Ti$_2$O$_3$, V$_2$O$_3$, Cr$_2$O$_3$, α-Fe$_2$O$_3$, Co$_2$As$_3$, Rh$_2$O$_3$, and α-Ga$_2$O$_3$.

Corundum crystallizes in the trigonal space group $R\bar{3}c$ (no. 167) with six formula units per unit cell. Different views of the structure are depicted in Figure 5.51, in which also the centres of the unoccupied octahedral voids are visualized as black spheres.

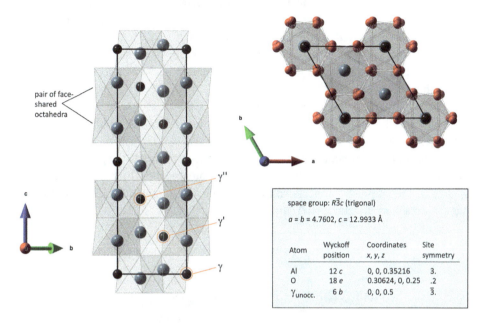

Figure 5.51: Crystal structure and crystallographic parameters of α-Al_2O_3; Al, grey-blue; O, red; the centres of the unoccupied octahedral sites are shown as black spheres.

Ilmenite ($FeTiO_3$) is a ternary and ordered derivative of corundum with two metal ion types. It is derived from corundum in a way that the Al layers of the corundum are alternately replaced by Ti and Fe layers. It crystallizes in the trigonal space group $R\bar{3}$ (no. 148; $a = b = 5.0884$ Å, $c = 14.0855$ Å). It is one of the important structure types of the stoichiometry ABO_3 (or MM'O_3), and the ilmenite type is favoured over the competing perovskite type (see Section 5.3.6), if the following two prerequisites are fulfilled: (i) the ratio of the ionic radii $r(A^{2+})/r(O^{2-})$ is smaller than 0.7 and (ii) the A and B ion types should be of similar size.

5.4.8 ZnS (wurtzite) (*Strukturbericht* type B4)

After having discussed compounds that are based on a hexagonal-closest packing in which octahedral interstitial sites are occupied, we now move on to compounds in which tetrahedral sites are occupied. An important thing to remember is that it is virtually impossible to arrive at compounds in which all tetrahedral sites of the **hcp** packing are occupied, contrary to the **ccp** packing, where this is a very common structure type.

The reason for this is that the tetrahedral voids in the **hcp** packing are present as pairs with a common face (see Section 4.3). According to electrostatic considerations, the occupancy of such neighbouring face-shared tetrahedra in highly polar/ionic compounds is avoided because those sites come very close to each other in such a configuration.

In the ZnS (wurtzite) structure type, the sulphur atoms form the **hcp** packing (stacking sequence ABAB...) and the Zn ions are in half of the tetrahedral interstitial sites, or to emphasize the foregoing statement, they occupy all T sites of only one type (T+ or T−). The sulphur atoms are likewise surrounded by four Zn ions in a tetrahedral fashion (CN = 4:4).

Wurtzite (β-ZnS) is the hexagonal counterpart of the cubic sphalerite (α-ZnS) modification. Wurtzite is the high-temperature modification that can be obtained by heating the α-ZnS modification to about 1,200 °C. The difference of the stacking sequence has consequences for the orientations of the tetrahedra with respect to the plane perpendicular to the stacking sequence. While the orientation is identical for all tetrahedra in sphalerite, the tetrahedra are rotated by 180° when going from one to the next T layer along the stacking direction (see Figure 5.52).

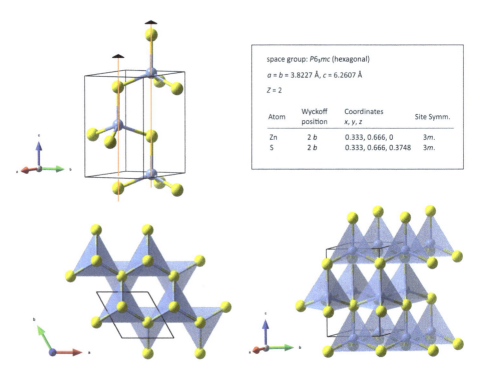

Figure 5.52: Crystallographic parameters (top-right) and crystal structure of ZnS (wurtzite) in different views; slightly extended unit cell (top-left), threefold axes of rotation are shown as thin orange lines, running through the tips and centres of the ZnS$_4$ tetrahedra; polyhedral representation (bottom left and right); all tetrahedra are oriented in the same way with respect to the c-axis, but from one tetrahedral layer to the next tetrahedral layer, they are rotated by 180° in the (a,b) plane.

Compounds that belong to the ZnS (wurtzite) structure type are:
- β-AgI, BeO, ZnO, CdSe (red), γ-BN, AlN, GaN, and α-SiC, and the
- high-temperature modification of CuI.

Two ternary compounds in which two metal cations are statistically distributed over the Zn sites are AgInS$_2$ and CuInS$_2$. An ordered variant with two metal cations is realized in the crystal structure of Cu$_3$AsS$_4$ (the mineral is known as enargite). Here, two types of rows of tetrahedra are present, running along the a-axis of the orthorhombic superstructure (space group $Pmn2_1$, no. 31), in which every second row is exclusively occupied by Cu atoms, while in every other row, the two metals occupy the positions in an alternating fashion, see Figure 5.53.

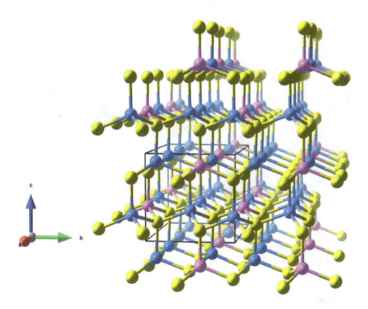

Figure 5.53: Crystal structure of Cu$_3$AsS$_4$ (enargite), a wurtzite-type structure with cation ordering.

5.4.9 Olivine (Mg,Mn,Fe)$_2$SiO$_4$ (*Strukturbericht* type S1$_2$)

Olivine is a short form for the olivine *group*, consisting of the end members forsterite (Mg$_2$SiO$_4$), tephroite (Mn$_2$SiO$_4$), and fayalite (Fe$_2$SiO$_4$) of the complete (i.e., without a miscibility gap) solid solution series (Mg,Mn,Fe)$_2$SiO$_4$. In this type of structure, the oxygen atoms build a slightly distorted **hcp** arrangement (stacking sequence AB ...), ⅛ of the tetrahedral sites are occupied by silicon atoms and half of the octahedral sites are occupied by (one or several of) the metals ions. In Figure 5.54 the structure is shown perpendicular to the stacking sequence of the densest layers. Two successive layers along the *a*-direction of the orthorhombic unit cell (space group *Pbnm*, no. 62, $a = 4.756$ Å, $b = 10.207$ Å, $c = 5.980$ Å) are shown in the bottom part of Figure 5.54.

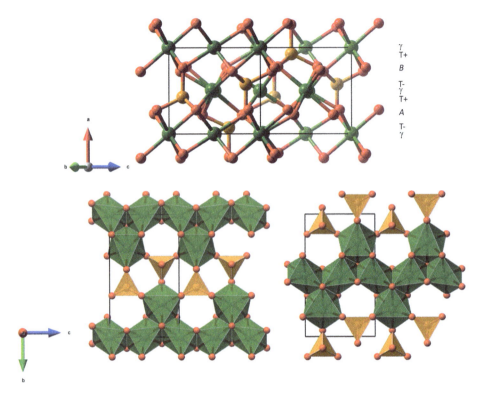

Figure 5.54: Crystal structure of olivine viewed perpendicular to the stacking sequence (top), and two successive layers perpendicular to the (*b,c*) plane (bottom); metal ion, green; Si, orange; O, red.

As only isolated SiO_4 tetrahedral units occur, olivine belongs to the group of *nesosilicates*. The olivine structure can be regarded as hexagonal counterpart of the spinel structure, with the difference that the divalent metal ion (Mg, Mn, Fe) is at the octahedral site. It is therefore also not surprising that the olivine structure transforms into the spinel structure at high pressures.

5.5 Other important structure types not based on densest packings

5.5.1 CsCl (*Strukturbericht* type B2)

The structure of CsCl can be regarded as derived from the tungsten (W) structure type in which either the central atom or the atoms at the corners of the cube of the **bcc** packing are substituted with an atom of different type, leading to the stoichiometry MX. Note, however, that the structure is of course no longer body-centred but cubic primitive. Every Cs atom is surrounded by eight Cl atoms in the form of a cube, and vice versa (CN = 8:8), see Figure 5.55.

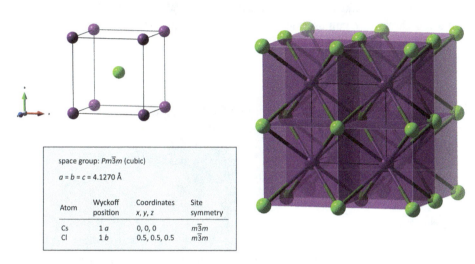

Figure 5.55: Unit cell of CsCl (top left), visualization of the coordination polyhedra (right) and the most important crystallographic descriptors (bottom left); Cs, purple; Cl, green.

Ionic compounds that crystallize in the CsCl structure type with the space group $Pm\bar{3}m$ (no. 221) are:
- CsCl, CsBr, CsI, and CsSH
- TlCl and TlBr,
- NH_4Cl, NH_4Br, and NH_4I
- high-temperature modifications of CsCN and TlCN.

Furthermore, more than 200 intermetallic compounds (see Chapter 11) crystallize in the CsCl structure type.

5.5.2 MoS$_2$ (*Strukturbericht* type C7)

The structure of MoS_2 can be regarded as derived from the NiAs structure (see Section 5.4.1). In NiAs the As atoms build an **hcp** arrangement with the stacking sequence AB... and the Ni atoms occupy all octahedral voids. The Ni atoms are primitively stacked and form trigonal prisms that share common edges within neighbouring trigonal prisms within the same layer. The As atoms are in the centre of these trigonal prisms. Now, if we substitute the Ni atoms with sulphur and the As atoms with molybdenum and remove one-half of the molybdenum atoms by deleting all Mo atoms in every second layer completely, we arrive at the structure of MoS_2. In other words: the S atoms form hexagonal layers with the stacking sequence *AABBAABB*... or *AABBCC*... or other stacking variants. In each pair of congruent layers, for example *AA*, there are edge-linked trigonal prisms in which the Mo atoms are located (see Figure 5.56). The most common stacking variants/polytypes are MoS_2-3*R* (space group *R*3*m*, no. 160) and MoS_2-4*H* (space group *P*6$_3$/*mmc*, no. 194).

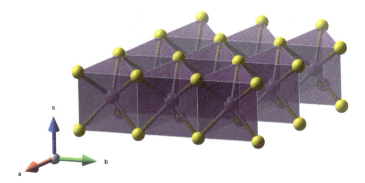

Figure 5.56: One layer of edge-linked trigonal prisms in the structure of MoS_2; Mo, pale-purple; S, yellow.

The structure has a distinctly layered character since there are only very weak van der Waals interactions between the double layers of sulphur atoms, so that the layers can be shifted against each other very easily. For this reason, MoS_2 (the mineral is known as molybdenite) is also used as a lubricant.

Further reading

U. Müller, *Inorganic Structural Chemistry*, 2nd ed. Wiley, Chichester, **2006**.
M. O'Keeffe, B. G. Hyde, *Crystal Structures. Patterns and Symmetry.* Dover, Mineola, **2020**.

6 Defects in solids

Defects play a major role in a large number of physicochemical processes and can drastically change the properties of solids. For example, very small traces of the impurity atom chromium can turn colourless aluminium oxide (Al_2O_3) into a beautiful red ruby. The entire semiconductor industry is based on doping silicon with impurity atoms. The ductility of metals is strongly influenced by the presence of defects called dislocations. Segregation processes in solids take place preferentially at grain boundaries, which are defects themselves. And there are many more examples that make defects an interesting topic to study.

A perfect crystal consists of a single piece – it is a single crystal – in which all atoms are at their correct sites. There are no missing atoms and never is an atom substituted by another type of atom that should not be in the compound, i.e., there are no impurities. Such a perfect crystal does not exist in reality; it can be obtained only theoretically and only at a temperature of absolute zero. Above absolute zero, defects are an *entropic necessity*. This results from the fact that, for example, so-called point defects (these might be vacancies), at least in a small number, lower the free energy, because defects increase the entropy. Or to put in another way: as long as the enthalpy that is needed to create defects is not larger than the corresponding entropy term in the Gibbs-Helmholtz equation ($\Delta G = \Delta H - T\Delta S$), the free energy will decrease. This also means that with increasing temperature, the number of defects will increase. At a certain temperature, an equilibrium situation will arise in which the number of defects can be derived from the minimum free energy (see Figure 6.1). The number of defects, n_D, is proportional to $e^{(-\frac{\Delta H}{RT})}$, where ΔH is the enthalpy required to create a defect. A typical value for the enthalpy that is required to create those defects, which often occur in ionic solids, the Schottky and Frenkel defects (see Section 6.1.2), is approx. 10^{-19} J.

The different types of defects that occur in solids can be classified into different classes, and there are several different classification schemes. For example, one distinguishes stoichiometric defects from non-stoichiometric ones. Stoichiometric defects are those in which the composition of the compound does not change. Such defects are also called intrinsic defects. Extrinsic defects are those that are accompanied with a change in composition.

Another classification scheme is based on the spatial dimension of the defect. A distinction is made between:
- 0D defects, i.e., point defects
- 1D (line) defects, i.e., dislocations
- 2D (plane) defects, e.g., layer defects, stacking faults, grain boundaries, and
- 3D (volume) defects, e.g., precipitates.

Since the distinction between the dimensions is not clear for some types of defects, some authors also prefer to distinguish only between point and extended defects.

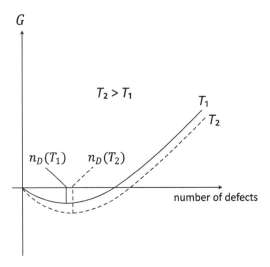

Figure 6.1: The Gibbs energy of a crystal as a function of the number of defects. At a given temperature at equilibrium, the number $n_D(T)$ of defects will be present. At higher temperatures the number of defects is also higher.

Furthermore, it should be clear that the differentiation based on the dimension of the defect can only be of a formal nature, since even point defects influence the direct surroundings of the defect. This means that defects always create a 3D disturbance of the structure.

6.1 Point defects

6.1.1 Point defects in crystals of elements

The simplest kind of localized defects in a crystal can be considered a "mistake", a deviation from the regular lattice-like arrangement of the motif at a *single* site in a pure crystal that is composed of only one kind of atom, i.e., an element, such as silicon or iron. Such a defect is called a point defect. Two different types of simple point defect can occur in a pure crystal of an element, E. The first type is called in the language of defect chemistry a vacancy, V_E, a position that is usually occupied where the atom is absent (see Figure 6.2). A second kind of point defects that can occur is that an extra atom is forced to be incorporated at a site that is usually not a regular atom site of the crystal; this extra atom is then called an interstitial atom and the defect site an interstitial site, abbreviated as E_i, with E being the type of atom at that interstitial site. If this extra atom is of the same kind as the other atoms of a monoatomic crystal, then it is called a self-interstitial atom. Such vacancies and interstitial sites can arise in a number of ways. One possibility is that they are formed during crystal growth. This happens especially when the crystallization process is fast and far from thermodynamic equilibrium. If these defects have developed without external influence, they are called *native* defects. But there are also other processes that can lead to these point defects, for instance, if the crystal is subjected to irradiation of high-energy electromagnetic waves or particles. In this case, the defects that are generated are called *induced* defects. The number of vacancies and interstitial sites – regardless of the way they were introduced – can be gradually decreased if the crystal is heated at a moderate temperature for a long period of time, a process that is called annealing. However, no matter how long the sample is annealed, a certain population of point defects will always remain, see above.

Even if you try very hard to create an ultrapure compound, you will never succeed in creating a 100% pure compound. And even in nature, there is no piece of matter that does not contain certain impurities. If foreign atoms are present by accident or undesirably, they are termed impurities, but if they have been added deliberately, to change the properties of the material on purpose, they are called dopant atoms. Foreign atoms can either form interstitials or they can occupy a position that is usually occupied by the host or parent material. The latter is called a substitutional point defect. The four kind of point defects considered so far are depicted schematically in Figure 6.2.

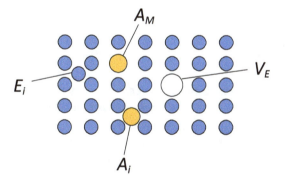

Figure 6.2: Four kind of point defects in a pure monatomic crystal of an element E; a vacancy, V_E, and a self-interstitial site, E_i, a substitutional defect A_M, and a non-self-interstitial, A_i, the latter two of which can be either an impurity or a dopant.

6.1.2 Point defects in ionic compounds

If we turn to ionic compounds, the situation is slightly more complex than in pure elements because the ionic charges must remain balanced when point defects are introduced into the crystal, due to the electroneutrality principle. The two most frequently observed point defects that occur in ionic crystals are Schottky (after Walter Hans Schottky, 1886–1976, a German physicist) and Frenkel (after Yakov Il'ich Frenkel, 1894–1952, a Russian physicist) defects.

Schottky defects

A Schottky defect, which is a stoichiometric defect, consists of a *pair* of vacant sites, a cation vacancy and an anion vacancy (see Figure 6.3, left). This defect pair may be caused by crystal growth or by diffusion of this ion pair from the bulk phase to the surface. Schottky defects are typical point defects that are predominant in alkali halides and alkaline earth oxides. Although there is some structural reorganization near the vacancies, these are not sufficient to fully compensate for the vacancy volume. Therefore, Schottky defects lead to density reduction.

The Schottky type of vacancies may be distributed randomly but there is a tendency to associate into larger clusters because the vacancies carry an effective charge, and oppositely charged vacancies attract each other. An anion vacancy in NaCl has a net positive charge of +1 because the vacancy is surrounded by six Na^+ ions, each with partially unsatisfied positive charge. Analogously, a cation vacancy has a charge of −1.

The number of Schottky defects present in a structure depends – like all defects – on the temperature. To give an example: at room temperature, in sodium chloride approx. one out of 10^{15} possible sites are unoccupied, while near the melting point, one out of 10^5 cation and anion sites remains unoccupied. The larger number of vacancies at higher temperatures favours the mobility of atoms, facilitating diffusion (reactions) and ionic charge transport in the solid.

Frenkel defects

Frenkel defects are formed when atoms leave their normal lattice sites and occupy interstitial sites, leaving a vacancy behind. Since the size of the ions plays an important role the sublattice of the smaller ion is energetically preferred; usually this is the cation. Silver bromide, which also crystallises in the NaCl type, shows Frenkel defects with vacancies and occupation of interstitial sites by silver cations (see Figure 6.3, right). The disordered silver ions occupy cubic voids with four nearest Ag^+ and Br^- neighbours, both tetrahedrally surrounding the interstitial silver ion. In the case of Frenkel defects of the anion lattice, the anions occupy interstitial sites and leave anion vacancies. Defects of this type occur in compounds with a fluorite structure (e.g., alkaline earth fluorides, CeO_2). In the structure of CaF_2, the anions occupy all tetrahedral voids in the **ccp** packing of the cations. In the case of an anion Frenkel defect, some anions occupy octahedrally coordinated interstitial sites.

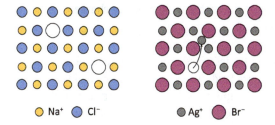

Figure 6.3: A Schottky defect in NaCl, consisting of a pair of vacancies (a cation and an anion vacancy) (left), and a cation Frenkel defect in AgBr, where a silver ion moves from its normal position to an interstitial site (right).

6.1.3 Kröger-Vink notation for point defects

A well-established set of notations for point defects and reactions that involve the generation or annihilation of point defects was established by Kröger and Vink [53]. The chemical or electronic (electron or electron hole) species is notated as

$$M_{Site}^{Charge}$$

where M corresponds to one of the following species:
- an atom, e.g., Na, Cl, Ag, I, Si, etc.
- a vacancy, which is given the symbol V or v (a lower case v is preferable, because V is also the element symbol for vanadium)
- an interstitial, which is given the symbol i
- or an electron, e, or an electron hole, h.

The superscript *Charge* is indicated by three different symbols:
- A fat dot, •, represents a single positive charge, multiple dots multiple positive charges.
- A tick mark (apostrophe), ′, represents a single negative charge, multiple tick marks multiple negative charges.
- A lowercase x is used to indicate neutral species; that should not be misunderstood in the way that this is a neutral ion, which would be a contradiction in terms, but it rather means that the charge is balanced by the surrounding ions.

Finally, the subscript *Site* indicates the lattice site that the species occupies. It might also be an interstitial site, i. Sometimes an "s" is specified to express that it is a site at the surface of the material. For electrons and electron holes, usually no site is specified.

To give a few examples:
- Zn_{Cu}^{x} – a zinc ion at a copper site with neutral charge
- $v_{Cl}^{\bullet}$ – a chloride vacancy with a single positive charge

- O_i'' – an oxygen anion on an interstitial site with double negative charge
- e' – an electron (without a specified site)

In principle, the generation of defects can be described in a completely analogous manner to chemical reactions. The mass, charge, and number of sites must be identical on both sides of the equation. Sometimes the number of the vacancy sites is not identical on both sides, which is then slightly incorrect. The reason for this is laziness, which makes it a bit simpler for all of us. For instance, the formation of a cation Frenkel defect in AgBr is often formulated as

$$Ag_{Ag}^x \rightleftharpoons v'_{Ag} + Ag_i^\bullet$$

The number of interstitial sites is obviously unbalanced: zero on the left and one on the right side of the "equation". The correct notation would therefore be

$$v_i^x + Ag_{Ag}^x \rightleftharpoons v'_{Ag} + Ag_i^\bullet$$

There is a special feature in the formation of point defects from a previously perfect, defect-free material, namely that this defect-free material is symbolized by a so-called null reactant (∅). The other possibility is to specify the cation and anion sites explicitly. Some further examples should provide clarity on how such reactions involving defects are presented:

An equation for a Schottky defect formation in TiO$_2$ looks like:

$$\emptyset \rightleftharpoons v''''_{Ti} + 2\,v_O^{\bullet\bullet}$$

A Schottky defect formation in BaTiO$_3$ can be written as

$$\emptyset \rightleftharpoons v''_{Ba} + v''''_{Ti} + 3\,v_O^{\bullet\bullet}$$

The generation of a Schottky defect in MgO taking into account that the ions that leave vacancies behind move to the surface will be notated as

$$Mg_{Mg}^x + O_O^x \rightleftharpoons v''_{Mg} + v_O^{\bullet\bullet} + Mg_s^x + O_s^x$$

6.1.4 Colour centres – a special kind of point defects

Special varieties of point defects can turn ionic crystals that are usually colourless into compounds with a certain colour. The first experiments in that regard were carried out in the late 1920s and early 1930s by R.W. Pohl, who studied the colouration of synthetic alkali halide crystals [54]. One way to give those crystals a colour is by heating the crystals in the presence of vapour of an alkali metal. Interestingly, however, the colouring that can subsequently be observed is only dependent on the composition of the crystal, but not on the type of alkali metal present as vapour. Crystals of NaCl heated in the presence of Na vapour have the same colour as those heated in the presence of K vapour, namely yellowish. KCl heated in the presence of alkali metal vapour, on the other hand, is always violet. Pohl postulated simply that after the vapour treatment *Farbzentren* (German for Farbe = colour and Zentrum = centre) must be present, without having a definite conclusion about the concrete nature of this F-centre, as it was later abbreviated.

Schottky finally had the correct idea in 1934 and interpreted the colour centres *as electrons that reside in anion defects*, trapped there. This interpretation was quickly followed by quantum mechanical considerations and calculations that were in line with this hypothesis. However, the final experimental proof could only be provided much later with the help of electron paramagnetic resonance (EPR) measurements.

The generation of such F-centres in NaCl involves first the adsorption of some Na atoms at the surface of the crystal, as a result, a slightly non-stoichiometric compound is formed, $Na_{1+\delta}Cl$, with $\delta \ll 1$. These adsorbed atoms are then ionized, and the electrons diffuse into the crystal and occupy anion vacancies, which were either present before or were generated by the heat that is applied during the process. To counterbalance the excess charges of the extra Na^+ ions at the surface, an equivalent number of Cl^- anions move from the bulk to the surface, leaving new anion vacancies behind (see Figure 6.4, left).

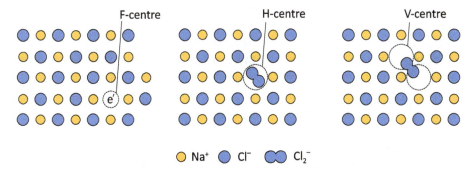

Figure 6.4: F-centre in an NaCl crystal, consisting of a trapped electron residing in an anion vacancy (left), a H-centre (middle), and a V-centre (right).

The trapped electron provides a classic example of a "particle in a box". Such an entrapped electron can absorb electromagnetic radiation in the wavelength range of visible light, which leads to a colouring of the crystal. The absorption spectrum of such a crystal has a relatively narrow band (F-band), whose position depends on the size and shape of the anion vacancy available to the electron, but not on where the electron comes from. In NaCl, the maximum of the F-band is at about 465 nm, in KCl at about 560 nm.

After the discovery of the F-centres, research of other types of colour centres remained a very active area in solid-state physics for a long period of time. Some other colour centres that were identified in alkali metal halides are:

- F′-centre – two electrons in an anion vacancy.
- F_A-centre – this is an F-centre in which one of the six surrounding cations is substituted by a foreign monovalent ion, for instance K^+ in NaCl.
- H-centre – this describes a molecular X_2^- anion, which is located at a single anion X^- site with two neighbouring X-ions, all oriented parallel to the <110> direction (see Figure 6.4, middle).
- V-centre – is similar to the H-centre, but here the X_2^- anion occupies two normal X^- sites (see Figure 6.4, right).
- M-centre – is a pair of nearest neighbouring F-centres, i.e., two electrons in two nearest-neighbouring anion vacancies.
- R-centre – is a cluster of three nearest-neighbouring F-centres located in the (111) plane.

6.1.5 Swapping places – order-disorder phenomena and the relation between superstructures and sublattices

In certain crystalline compounds, pairs of atoms of different kinds that are located at different crystallographic sites may swap their places. In particular, such atom sets are prone to swapping places that have a similar size and are chemically similar. Therefore, such phenomena are particularly common in alloys, intermetallic compounds, or in ionic compounds in which more than one kind of cation or anion is present.

If the number of swapped atoms is large and, especially, if it increases significantly with increasing temperature, there is a transition from an ordered to a disordered phase. Complete disorder is reached if there is no longer any preference for a certain crystallographic site for the atoms involved in this exchange of positions. In alloys, atoms can occupy crystallographically different sites in an ordered manner or be disordered over all available positions. Usually, at low temperatures, the phase is ordered, while at higher temperatures the phase becomes more and more disordered. The conversion back to the ordered phase on cooling is often possible, but usually proceeds fairly slowly.

Let us look at the example of the 1:1 alloy between Cu and Zn, also known as β-brass.[4] Below 468 °C, CuZn crystallizes in an ordered structure of the CsCl structure type (see Section 5.5.1), in which one of the atom types occupies the centre of

---

[4] Although quite a few brass instruments are actually made of brass, the definition of brass instruments does not refer to the material they are made of, but to the way the sound is produced. Thus, one finds brass instruments made of wood, like the alphorn, the serpent and the didgeridoo, while some woodwind instruments are made of brass, like the saxophone or the Western concert flute. The peculiarity of brass instruments is that the vibration is generated by the lips of the player and thus a

the cube and the other type of atoms the corners of the primitive cubic unit cell; this is the β'-phase (space group: $Pm\bar{3}m$). Above 468 °C, however, any preference for one of the two crystallographic sites is lost and the two atomic species are statistically distributed between the two sites. A completely disordered **bcc** phase (β-phase) is formed, analogous to the structure of α-Fe (space group: $Im\bar{3}m$), see Figure 6.5.

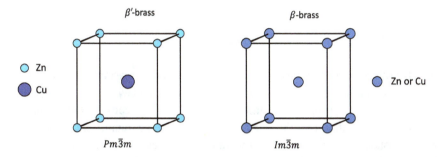

Figure 6.5: Ordered (low-symmetry) β'-brass phase of the alloy CuZn (left) and the corresponding disordered (high-symmetry) β-brass phase (right) that is formed above 468 °C.

During the transition from the disordered β- to the ordered β'-phase, a so-called *superstructure* is formed. The high-symmetry phase is called *basic structure*. A superstructure is characterized by the fact that it has *reduced* translation symmetry compared to the basic structure, i.e., the space group of the superstructure is a *subgroup* of the space group of the basic structure. The point group, on the other hand, can be the same or also a subpoint group. Furthermore, at least one of the Wyckoff positions of the basic structure is split into two (or more) independent Wyckoff positions of the superstructure, and the corresponding positions are occupied by chemically different atoms. In the present case, the Wyckoff position 2a (0,0,0) of the space group $Im\bar{3}m$ (jointly and statistically occupied by Zn and Cu) splits into the two Wyckoff positions 1a (0,0,0) and 1b (½, ½, ½) of the space group $Pm\bar{3}m$. To complete the terminology around this topic of superstructure and basic structure, which often gives rise to confusion [55]: the unit cell of the ordered structure is a *supercell* (it is *larger* compared to the *primitive* one of the disordered high-symmetry phase), while the lattice of the ordered phase is a *sublattice* of the disordered phase. A sublattice is a lattice that is obtained from a superlattice by removing lattice points (here the lattice point that makes up the body-centring, i.e., the one at (½, ½, ½) is removed). One has to be a bit careful when changing from real to reciprocal space, because here, it is the other way round: The reflections that additionally show up in X-ray diffraction images at the transition to the ordered phase are called superstructure reflections, but they belong

---

human organ becomes part of the instrument. Brass instruments are also called labrosones or labrophones, from Latin and Greek words meaning "lip" and "sound".

to the corresponding superlattice in reciprocal space, because the periodicity in reciprocal space has increased accordingly, or the periodicity length has decreased. This means that a *sublattice* in real space becomes a *superlattice* in reciprocal space. Of course, one can also express it the other way round: The centring that is additionally present when the disordered phase has been formed leads to the corresponding extinction of reflections in the diffraction pattern of the formerly ordered phase.

6.1.6 Non-stoichiometric compounds and defect clusters

Stoichiometry refers to the relationship between the amounts of reactants and products before, during, and after chemical reactions. It is based on the law of definite proportions, the law of multiple proportions, and finally, on the law of conservation of mass. This leads to the realization that the relationships between the quantities of reactants and products usually form a ratio of positive integers. However, this is often not valid for solids. Therefore, compounds exist with apparently non-rational compositions. Examples of non-stoichiometric compounds are the sodium tungsten bronzes, Na_xWO_3 with $0 < x < 0.9$, which were presented in Section 5.3.6. Compounds that have a variable composition over a certain range x and do not change their overall structure belong to the non-stoichiometric compounds or phases. In non-stoichiometric compounds, the number of atoms in the unit cell is not identical to the number of equivalent lattice sites. This results in a deficiency or excess of one type of atom. The deviation from the ideal composition, i.e., the respective stoichiometric compound, is formally compensated by a change of the oxidation state of one of the species that leads to mixed-valence compounds. For Na_xWO_3 the correct formula is $Na_xW^V{}_xW^{VI}{}_{1-x}O_3$. However, since the altered charges are usually not localized, cation excess means the presence of additional electrons (accompanied by n-type conduction) and cation deficiency the occurrence of defect electrons/holes (accompanied by p-type conduction).

One of the most common and widely studied non-stoichiometric compound is wüstite, usually represented by the formula $Fe_{1-x}O$, with $0.04 < x < 0.12$. It turned out that wüstite is an iron-deficient rather than an oxygen-excess phase as density measurements suggest. Wüstite plays an important role in geological contexts, as it forms a solid solution with periclase (MgO) and magnesiowüstite – (Fe,Mg)O – which is also called ferropericlase and a major component of the Earth's lower mantle. Furthermore, it is also of technical importance because it is present in the iron catalysts used for the Haber-Bosch process and is produced in the blast furnace process by the reduction of magnetite (Fe_3O_4).

The "ideal" stoichiometric compound FeO crystallizes in the cubic NaCl structure type (space group $Fm\bar{3}m$). However, rock salt-like FeO seems to be unstable below 10 GPa (100,000 bar). All wüstite samples synthesized below 10 GPa are iron-deficient. However, the relatively simple formula $Fe_{1-x}O$ does not reflect the structural diversity in which wüstite can occur. The first thing that has to be taken into account is that –

due to requirement of charge equalization – for every Fe^{2+} vacancy, two Fe^{3+} ions must be present. A more representative formula would be, therefore, $Fe^{2+}_{1-3x}Fe^{3+}_{2x}v_xO$. The second question is which sites the Fe^{3+} ions occupy. Initially one might think that they are statistically distributed over the octahedral sites in the **ccp** packing of oxide anions (i.e., substitute Fe^{2+} ions on their sites). However, through a series of investigations using X-ray diffraction experiments and Mössbauer spectroscopy, among others by Koch and Cohen, it has been found that the Fe^{3+} ions not only occupy octahedral sites, but also occupy the normally unoccupied tetrahedral voids in the densest packing of the oxide ions (i.e., occupy *interstitial sites*) and do so preferentially in a locally clustered fashion, forming defect *aggregates* or defect *clusters* of various types. For the description it is useful to divide the face-centred unit cell of the rock salt structure into eight octants, similar to the way it was done to describe the spinels. The basic defect unit then consists of exactly one octant, in the centre of which there is an Fe^{3+}, which is surrounded tetrahedrally by four vacancies (and by four oxide ions), meaning that the vacancies are also clustered. In the nomenclature of such defect clusters this is a 4:1 unit, where the first figure indicates the number of vacancies and the second the number of Fe^{3+} ions of the defect or defect cluster. This basic unit can now be linked via corners (type 1 cluster), edges (type 2), or faces (type 3) to such neighbouring basic units. Note, that face-shared linked octants are equivalent to edge-shared $Fe^{3+}v_4$ tetrahedra etc. In wüstite, mainly type 2 and 3 clusters are present, one of which is known as Koch-Cohen cluster (13:4) (see Figure 6.6). Other defect clusters that are present in wüstite and magnetite comprise 7:2 and 16:5 corner-shared, 7:2 and 16:6 edge-shared, and 9:4 and 13:8 face-shared clusters (see Figure 6.7).

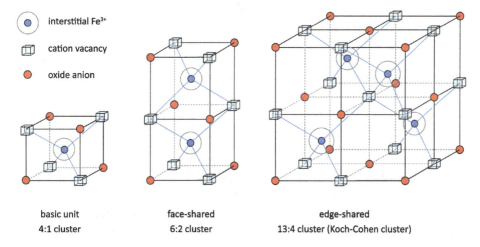

Figure 6.6: Basic defect/vacancy unit (left) and defect cluster association of iron-deficient wüstite $Fe_{1-x}O$ and magnetite; two basic units combined to form a face-shared 6:2 cluster (middle) and four basic units combined to form a corner-shared 13:4 cluster complex known as Koch-Cohen cluster (right).

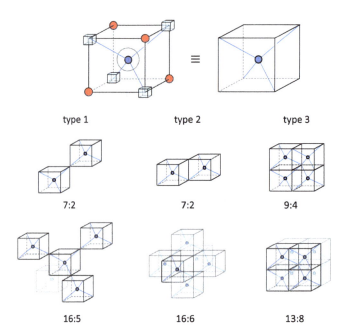

Figure 6.7: Various corner-, edge-, and face-shared defect clusters in wüstite and magnetite. Note that the central cube/octant in the 16:6 defect cluster of type 2 contains no Fe^{3+} ion.

Such defect clusters are highly negatively charged due to the aggregation of Fe^{2+} vacancies. To compensate this charge, extra Fe^{3+} ions are distributed over octahedral sites around the clusters to preserve electroneutrality. The concentration of these defect clusters increases with increasing x, i.e., with increasing iron deficiency, and hence the average separation between the clusters decreases. Experimental evidence from X-ray and neutron diffraction is given that the clusters order into a regularly repeating pattern resulting in a superstructure for wüstite.

6.1.7 Substitutional and interstitial solid solutions

In the section on point defects, impurities and dopants have already been discussed, the difference being that the latter are added intentionally. Both terms are used rather for systems in which the fraction of foreign atoms is very small (usually < 1%). In this subchapter, we will deal with systems in which the fraction of a second, third, etc. type of atom is typically considerably larger: for those materials, it is common practice to refer to them as *solid solutions*. For newcomers to solid-state chemistry, the term may be somewhat confusing at first, as the term solution usually implies a liquid state of aggregation. Apart from this decisive difference regarding the state of aggregation, however, there are many analogies to liquid solutions, including the famous aphorism "like dissolves like", originally expressed in the Latin language as "similia similibus solventur". A solid solution is a uniform and homogeneous mixture of two (or more) crystalline materials that share a common crystal structure. The word "solution" is used to describe the intimate mixing of components at the atomic level and distinguishes these homogeneous materials from those mixtures in which only a physical blend of the components is present. The general rule for the occurrence of substitutional solid solutions is that the pure compounds should have the same crystal structure, similar radii of the constituting atoms, and chemical behaviour. In the case of interstitial solid solutions, on the other hand, the most important aspect is, whether the solving component provides interstitial sites that fit to the size of the species that is supposed to be dissolved.

In some systems, there is complete miscibility between the pure components A and B over the entire range of possible compositions. This is the case, for example, with the Ag-Au system. The corresponding phase diagram then resembles the schematically drawn one in Figure 6.8 (left). Much more frequent, however, is the case of only limited miscibility, i.e., when the composition exceeds or falls below a certain level, phase separation occurs. The area of compositions in which the components are not miscible with each other is called the miscibility gap. One example is the Ag-Cu system: only about 15% Cu dissolves in Ag and only about 5% Ag dissolves in Cu. The phase diagram for systems with a miscibility gap looks similar to that shown in Figure 6.8 (right). Phase diagrams are discussed in detail in Chapter 7.

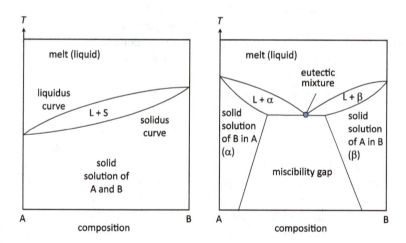

Figure 6.8: Schematic phase diagrams of a simple binary system composed of A and B without (left) and with a miscibility gap (right); L, liquid; S, solid.

Structurally relatively simple types of solid solutions are either substitutional solid solutions or interstitial solid solutions. In substitutional solid solutions, the atoms or ions of a second component that is introduced into a parent compound replace atoms or ions of that parent compound at their sites. In interstitial solid solutions, the introduced species occupies a site that is unoccupied in the parent compound.

Substitutional solid solutions

Many alloys are substitutional solid solutions. The previous mentioned example of Ag-Au is one of them, Cu-Au and Cu-Ni are two others. As already indicated, especially those systems preferentially form solid solutions in which the components are very similar. For alloys, W. Hume-Rothery (1899–1968, a British metallurgist and material scientists) was able, based on numerous investigations, to establish a set of empirical rules indicating when the probability that two metals will form a substitutional solid solution with a wide composition range is high, known as Hume-Rothery solubility rules:
1. The metals/elements have the same crystal structure.
2. The atomic sizes of the two metals/elements differ not more than by 15%.
3. The metals/elements have roughly the same electronegativity.
4. The metals/elements have the same valence state.

If another rule is added to this set, this extended set also applies surprisingly well to ionic compounds or minerals, respectively:
5. The ions that replace each other should have the same charge.

MgO (periclase) and NiO (bunsenite) form a substitutional solid solution over the entire compositional range; both compounds crystallize in the NaCl structure type, and the size of the cations is very similar (Mg^{2+}: 72 pm, Ni^{2+}: 69 pm). By contrast, MgO and CaO form only a partial substitutional solid solution. Both compounds also crystallize in the NaCl structure type, but the size of the cations is significantly different (Mg^{2+}: 72 pm, Ca^{2+}: 100 pm).

A prominent example for substitutional solid solutions of the corundum structure type is the system Al_2O_3/Cr_2O_3. These compounds are isotypical. The oxide ions build a slightly distorted **hcp** packing and two-thirds of the octahedral voids are occupied by the M^{3+} cations. Due to their similar size (Al^{3+}: 54 pm, Cr^{3+}: 62 pm) they form a complete substitutional solid solution. The sum formula can be formulated as $(Al_{2-x}Cr_x)O_3$, with $0 \leq x \leq 2$. While pure Al_2O_3 is a colourless compound, a small amount (< 1%) of Cr^{3+} ($3d^3$ system) replacing Al^{3+} results in the typical red colour of ruby. If the substitution degree gets larger and larger, the colour of this solid solution turns slightly into green (above 8% Cr^{3+}); pure Cr_2O_3 (also known as the rare mineral eskolaite) has an olive-green colour and is used as a synthetic pigment called chromium oxide green, not to be confused with chrome green, which is a mixture of Prussian blue and chrome yellow ($PbCrO_4$).

Due to the slightly larger ionic radius of Cr^{3+} compared to Al^{3+}, the lattice constants of $(Al_{2-x}Cr_x)O_3$ increase linearly with increasing x. This behaviour corresponds to *Vegard's rule*, according to which the lattice constants within a solid solution series change linearly with the composition (see Figure 6.9).

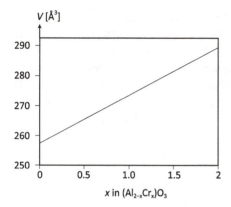

Figure 6.9: The linear increase of the unit cell volume in the Al$_2$O$_3$/Cr$_2$O$_3$ system with increasing chromium content.

Aliovalent substitutional solid solutions

If, in a substitutional solid solution, the species that replaces another from the parent structure has the same charge or oxidation state, it is called a *homovalent* (sometimes also called *isovalent*) substitutional solid solution. The simple system Al$_2$O$_3$/Cr$_2$O$_3$ is an example of that type. If, on the other hand, the exchanged species have a different charge, we speak of *heterovalent* or *aliovalent* (alio from Latin "belonging to something different") substitutional solid solutions. In those types of solid solutions, additional changes involving vacancies or interstitials (ionic compensation) or electrons or holes (electronic compensation) are needed to maintain electroneutrality.

Anion vacancy compensation

A technically important example for ionic compensation involving anionic vacancies is the doping of zirconium dioxide (ZrO$_2$, zirconia) with CaO, MgO, or Y$_2$O$_3$; these materials are used in various applications as high-performance ceramics (see also Chapter 12). The additives are used to stabilize the high-temperature cubic phase of zirconia, which in the absence of additives forms only at temperatures above 2,370 °C. Stabilization means that the cubic phase is retained also at room temperature. The cubic phase that crystallizes in the fluorite (CaF$_2$) structure type (space group $Fm\bar{3}m$, no. 225) has advantageous properties in terms of mechanical strength and translucency.

For each Ca^{2+} or Mg^{2+} ion replacing a Zr^{4+} ion, one oxide vacancy is formed, if CaO or MgO is used as an additive. If Y$_2$O$_3$ is used as a stabilizer, then one oxide vacancy is created for each two Y^{3+} ions replacing Zr^{4+} ions, which can be formulated in Kröger-Vink notation as follows:

$$Y_2O_3 \rightarrow 2\, Y'_{Zr} + 3\, O^x_O + v^{\bullet\bullet}_O$$

Usual doping levels for Y$_2$O$_3$ are between 9% and 10%. This results in a relatively high number of oxide vacancies, and above temperatures of 600 °C the oxide anions in yttrium-stabilized zirconia (YSZ) can relatively easily diffuse through the crystal. YSZ is, therefore, a very good ionic conductor, while it simultaneously shows a high electric resistance. The optimal transport properties of this material are used, for example, in high-temperature solid oxide fuel cells (SOFC). Furthermore, YSZ is used as a solid electrolyte for measuring oxygen partial pressures, e.g., in the form of the lambda (O_2) sensor, where lambda refers to air-fuel equivalence ratio, usually notated as λ. It is used to control and to optimize catalytic exhaust gas purification. The aim here is to minimize the emission of pollutants such as nitrogen oxides, hydrocarbons, and black carbon. The sensor compares the residual oxygen content in the exhaust gas with the oxygen content of a reference, usually the current atmospheric air. From this, the air-fuel ratio λ can be determined and thus adjusted.

The difference of the oxygen partial pressures on both sides of a YSZ oxygen sensor is compensated by diffusion of oxygen ions towards the lower partial pressure. Oxygen molecules accept electrons when entering a porous metal electrode and can migrate through the solid electrolyte as O^{2-} to donate electrons at the metal electrode of the opposite side. The electrical voltage measured during the reaction is proportional to the difference in partial O_2 pressures. In the temperature range of approx. 500–1,000 °C, partial pressures as small as 10^{-16} bar can be measured.

Cation interstitial compensation

Unlike the example of zirconium dioxide just discussed in which no suitable cation interstitial sites are available to ensure electroneutrality, the situation is different with the so-called "stuffed silica" phases. These are aluminosilicate phases in which Si^{4+} is partially replaced by Al^{3+}. In order to maintain electroneutrality in this case, the substitution is accompanied by the incorporation of (mostly) monovalent cations at interstitials sites. The size of these sites is dependent on the density of the parent silica phase, i.e., quartz, tridymite, or cristobalite.

Quartz has a comparably high density and stuffed quartz phases can only accommodate the smallest alkaline metal cation, Li^+. $Li_{1-x}Al_{1-x}Si_{1+x}O_4$ solid solutions ($0 \leq x \leq 1$) crystallize either in the α- or β-quartz structure and are of considerable interest for their unique physical properties and for the insight they provide into general crystal-chemical systematics. The endmember with $x = 0$, β-eucryptite ($LiAlSiO_4$), is a one-dimensional solid electrolyte. Intermediate compositions are common components of high-temperature glass-ceramic products due to their near-zero thermal expansion and appropriate viscosity for high-speed glass-forming. Cristobalite and tridymite are less dense than quartz and their interstitial sites are considerably larger. They can accommodate not only larger alkali metal cations, such as Na^+ and K^+, but even larger alkaline earth metal cations like Ba^{2+}.

Of enormous economic importance are, of course, the framework-like zeolites (see Section 14.1), which are also aluminosilicates; they occur and are synthesized in different cation-/proton-exchanged variants.

Interstitial solid solutions

A very interesting example of a solid solution of the interstitial type is the system Pd/H. As early as 1866, Thomas Graham (1805–1869, a British chemist) found out that palladium is capable of absorbing large amounts of hydrogen. In fact, the volume is so large that palladium is considered a metallic hydrogen sponge, which readily absorbs surrounding hydrogen. Palladium hydride, PdH_x, $0 < x < 1$, is – as opposed to what the name suggests – not an ionic compound. Instead, the hydrogen molecules reaching the palladium surface dissociate and the hydrogen atoms then diffuse into the bulk phase, where they occupy the octahedral voids, i.e., octahedral interstitial sites within the unaltered **ccp** structure of the Pd atoms. The amount of hydrogen that Pd absorbs depends on the pressure of the surrounding hydrogen and the temperature. At room temperature and 1 bar H_2 pressure, x reaches a value of approx. 0.7, i.e., approx. 70% of the octahedral voids are occupied by hydrogen atoms.

PdH_x forms two phases: The α-phase is a solid solution in which the hydrogen atoms are distributed completely statistically over the available octahedral voids. But already at proportions of $x > 0.017$, the β-phase forms, in which areas that correspond to the NaCl structure type are built, i.e., areas in which *all* octahedral voids in the Pd lattice are occupied. If x exceeds 0.58, only the pure β-phase is present. Thus, the two-phase region lies in the composition range from 0.017 to 0.58.

Since the hydrogen uptake of PdH_x is completely reversible, it was one of the first model systems for future hydrogen storage systems, even though it is not suitable for economic use in a hydrogen economy, at least in the mobile sector, due to the low gravimetric storage densities. For example, it has been shown that the kinetics, i.e., the speed of the H_2 uptake and release, and also the thermodynamics can be influenced by the particle size and particle morphology.

The interstitial solid solution with the greatest technological importance is the Fe/C system, as it is the starting point for steel production. The production volume of steel exceeds that of all other metallic materials by more than 10 times. By definition, steel is iron with a carbon content of less than 2 wt%; iron with a higher carbon content is called cast iron, which has a low melting point, and the melt has a low viscosity. These are advantageous properties of cast iron; however, in contrast to steel, cast iron cannot be forged; it is hard and brittle.

The element iron exists in three different modifications, and their ability to form solid solutions with carbon is significantly different: α-Fe, stable up to 910 °C, has a **bcc** structure and can take up only a very low amount of approx. 0.02 wt% carbon. γ-Fe, stable between 910 and 1,400 °C, has a **ccp** structure and can dissolve up to 2.06 wt% carbon. Finally, above 1,400 °C up to the melting point of 1,534 °C, the δ modification is present, which, again, has a **bcc** structure. δ-Fe can dissolve only approx. 0.1 wt% carbon.

The reason why the two **bcc** modifications of Fe can take up so little carbon is due to the size of the interstitial sites. In γ-Fe with its **ccp** structure, the carbon atoms can be accommodated reasonably well in the regular octahedral voids of the iron packing;

the minimum distance between C and Fe is half the lattice constant, i.e., 1.824 Å. The interstitial sites in **bcc** iron, on the other hand, are highly distorted/compressed octahedral sites, in which the shortest distance between C and the two nearest Fe atoms (those in the centre of the unit cell) is, with 1.433 Å, very small (see Figure 6.10).

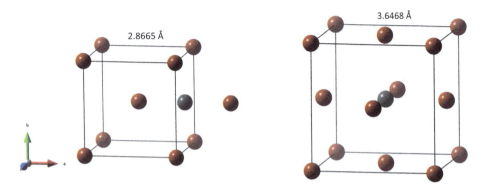

Figure 6.10: Strongly compressed octahedral interstitial site in **bcc** Fe (left) and regular octahedral interstitial site in **ccp** Fe (Fe, dark red; C, grey).

6.2 Line defects – dislocations

Dislocations are an important class of defects that have a great influence on a variety of properties of solids. As with point defects, the existence of line defects, to which dislocations belong, was predicted much before experimental evidence for them could finally be provided. Without dislocations, one could not explain the comparatively easy deformability of many metals, which is several orders of magnitude easier than theoretically for a perfect metal without any line defects expected. Another sign of (twisting) line-like defects are the spiral structures on the surface of some metals, sometimes even visible to the naked eye. Dislocation lines are not only important for mechanical properties, but they are also fast diffusion paths in the crystal. Furthermore, they are preferential nucleation sites for new phases, and reactions of solids often occur at active surface sites where dislocation lines emerge from the crystal.

Usually, dislocation lines are categorized into two main categories, namely edge and screw locations, but they can also have any degree of intermediate character.

6.2.1 Edge dislocations

In an edge dislocation, a lattice plane terminates *inside* a crystal. It can be thought of as the insertion of an extra half-plane into a crystal. The row of atoms at the tip of that extra half-plane that are perpendicular to the extra plane form this dislocation line (see Figure 6.11); this is where the extra plane terminates, which is often symbolized by a terminus sign (⊥, it can have, of course, any orientation, depending on the orientation of the extra plane). In the closer vicinity of the dislocation, there is a local perturbation of the lattice-like order of the motifs/atoms. Several stress fields can be identified: in the areas highlighted in green, the atoms are subject to shear stress, in the area highlighted in dark red, there is compression stress (the atoms are squeezed together) and in the purple-coloured area, there is tensile stress (the atoms are pulled apart). The magnitude of the distortion decreases with increasing distance from the dislocation until it vanishes completely far away from the dislocation where the crystal is again almost perfect.

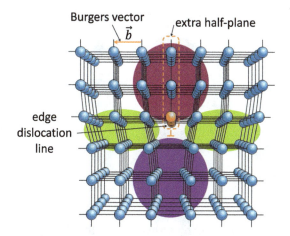

Figure 6.11: Schematic representation of an edge dislocation line in a primitive cubic crystal that runs perpendicular to the drawing plane at the location of the terminus symbol (⊥) of the extra half-plane; green areas symbolize regions with shear stress; in the dark red area compression stress is present, and in the purple area tensile stress prevails (redrawn and adapted from [56]).

The edge dislocation is further characterized by its Burgers vector $\vec{b}$ (named after Johannes Martinus Burgers, a Dutch physicist, 1895–1981). In order to derive this vector, we can draw a continuous line from atom to atom that encloses the dislocation core in a clockwise fashion (right-hand rule, RH), and in each of the lattice direction we should go forwards an equal number of steps. There are two scenarios, depending on the location of the atom we choose as the starting atom (S). The final or finishing atom (F) is not connected to the starting atom (see Figure 6.12, left), the Burgers circuit is not closed, or the F atom is identical with the second atom of the Burgers circuit (see Figure 6.12, right). In both cases we finally draw a line from the S to the F atom (this is called the SF convention; together with the RH rule it is the RH/SF convention). The direction from S to F is the direction of the Burgers vector! The Burgers vector and the dislocation line in edge dislocations are perpendicular to each other. The magnitude of the Burgers vector is the interatomic distance, i.e., the unit cell length in that direction. You are free to choose any other convention, for instance, also the RH/FS convention is sometimes used. But it is very important to consistently apply the same convention if you consider different dislocations in the same system.

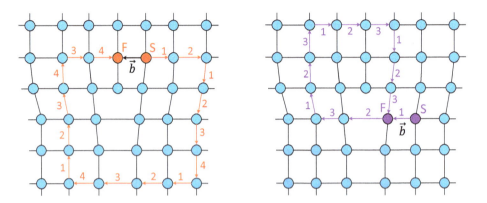

Figure 6.12: Procedure to determine the Burgers vector of an edge dislocation. Here the RH/SF convention was used (see text).

6.2.2 Screw dislocations

Another type of dislocations that is often present in metals and alloys is called screw dislocation. It may be thought of as being formed by a shear stress that is applied from opposite directions (say upper front and lower back part of the crystal sample) to produce a distortion in which one part of the crystal is shifted one atomic distance relative to the other part of the crystal (see Figure 6.13, upper part). The screw dislocation is again (as the edge dislocation) a *line* running along the row of atoms through the crystal where the atoms are displaced to each other (see Figure 6.13, lower left). When the *screw* dislocation is a *line*, why it is then called a screw dislocation? The answer is that if we follow the atoms *around* that screw dislocation line, they lie on a helical path like in a spiral staircase. This can also be seen when constructing the Burgers circuit in order to derive the Burgers vector (see Figure 6.13, lower right): again, using the RH/SF convention we start at the atom S, going two steps down, four to the left, four up, four to the right, and, finally, two further steps down (so that the number of steps in each direction is identical), reaching the atom F. This derived Burgers circuit is not closed and in order to close it we have to draw a vector from S to F – the Burgers vector. Interestingly, in contrast to edge locations, where the dislocation line and the Burgers vector are oriented perpendicular to each other, here the dislocation line and the Burgers vector are parallel.

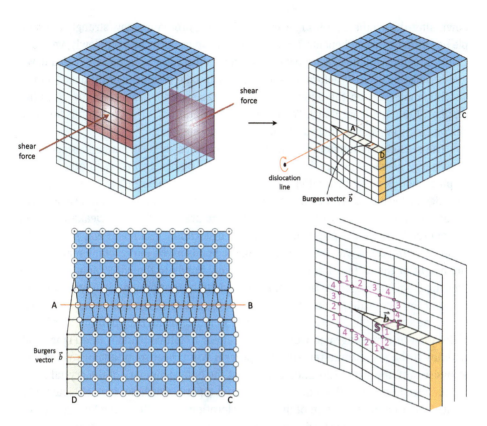

Figure 6.13: A screw dislocation may be thought of as being formed by a shear stress that is applied from opposite directions, here, at the upper front and lower back part of the crystal sample (top left) to produce a distortion in which one part of the crystal is shifted one atomic distance relative to the other part of the crystal (top right). The graphic in the lower left panel shows the projection of the crystal as viewed from above; the dislocation line is running from A to B; atoms above the plane that contain the dislocation line are shown as open white circles whereas those that are below that plane are shown as small black dots. In the graphic shown in the lower right, the derivation of the Burgers vector using the RH/SF convention is illustrated (see text) (graphics are redrawn and adapted from [56]).

6.2.3 Movement of dislocations – plastic deformation

Dislocations are primarily responsible for the fact that metals and alloys are much more readily deformable than theoretically thought, assuming a perfect crystal. The ease with which materials can be plastically deformed is directly dependent on the density of dislocations. This density is alternatively expressed either as total dislocation length per unit volume or as a number per cross-sectional area. The density in turn depends on the nature of the material and the conditions under which it was produced, e.g., what thermal stresses it has been subjected to, how quickly it cooled down after the synthesis, if it underwent a sintering procedure, how strongly the sample has been deformed/bent etc. For metals that have been cooled very slowly, values of 1,000 per square millimetre are typical. For metals that have been rapidly cooled down and have been strongly deformed, the density can take on values of up to 10^{10} mm^{-2}. Ceramic materials, on the other hand, are characterized by relatively low densities (approx. 100–10,000 per mm^2). The lowest density, approx. 0.1–1 per mm^2, is found in silicon single crystals produced by the Czochralski method.[5]

The reason why many metals and alloys are so easily deformable is because far less energy needs to be applied to displace lattice planes against each other, if this displacement is along dislocations. This is illustrated in Figure 6.14. We first assume the edge dislocation is at the end of the lattice plane labelled A. Now, if a shear force is applied according to the direction shown in the drawing, then the weakest point in the material will be the plane containing the dislocation. The metal will react to the shear force in such a way that it will break the bonds in the lattice plane marked B at the height of the dislocation line and form new bonds to the corresponding neighbouring atoms in the lattice plane A, i.e., to the atoms that previously formed the dislocation line; typically, a dislocation line contains a few hundred to approx. 1,000 atoms. As a result, the dislocation line will be in lattice plane B, or in other words, the dislocation line has migrated from lattice plane A to lattice plane B (see Figure 6.14, middle). If the shear force remains, the process will repeat in an analogous manner stepwise until the edge dislocation emerges at the outer boundary of the crystal forming an edge of a unit cell width, leaving behind a perfect ordered lattice (Figure 6.14, right). Note that the migration of the edge dislocation is parallel to the Burgers vector. The process by which plastic deformation is produced by the movement of a dislocation line is called slip. The crystallographic plane along which the dislocation line moves is the corresponding slip plane, which is indicated in green in Figure 6.14.

Why it takes so much less effort to deform a metal crystal along such a slip line can be explained by the following analogy: in order to slide a flat carpet along the floor in its entirety, all points of adhesion to the floor must be loosened at the same time. On the other hand, it is much easier to lift the carpet slightly on one side perpendicular to the direction of displacement, so that a carpet fold is created, and then to spread out this fold until you reach the other end. This process is visualized in the top part of Figure 6.14.

---

[5] The Czochralski method, also Czochralski technique or Czochralski process, is a method of crystal growth used to obtain single crystals of semiconductors like silicon, germanium and gallium arsenide, noble metals like palladium, platinum, silver, and gold, some salts and certain synthetic gemstones. The method is named after the Polish scientist Jan Czochralski (1885–1953), who invented the method in 1915 while investigating the crystallization rates of metals. According to one of his nephews, he made this discovery by accident: while taking some notes, instead of dipping his pen into his inkwell, he dipped it in molten tin, and drew a tin filament, which later proved to be a single crystal [57]. The most important application may be the growth of large cylindrical ingots, or boules, of single crystal silicon used in the electronics industry to make semiconductor devices like integrated circuits.

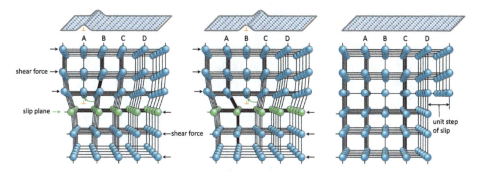

Figure 6.14: Schematic representation of the atomic rearrangements that are involved when an edge dislocation travels through a crystal by applying a shear force. The comparison with spreading of a carpet fold is shown at the top part of the graphic (graphics are redrawn and adapted from [56]).

Unlike in edge dislocations, where the direction of motion and the associated Burgers vector are oriented parallel, screw dislocations move perpendicular to their Burgers vectors. Nevertheless, the result of a deformation by moving screw dislocations is the same as in edge dislocations; a step on the surface of the crystal is formed (see Figure 6.15). In mixed dislocations, the direction of movement and the Burgers vector build an angle, depending on the curvature of the dislocation line.

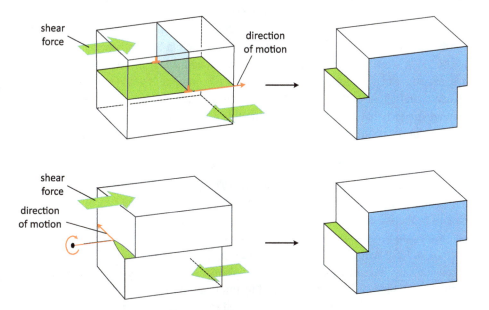

Figure 6.15: The result of applying a shear force is identical for edge (top) and screw dislocations (bottom), although the direction of movement is parallel for edge dislocations but perpendicular to the shear force for screw dislocations (redrawn and adapted from [56]).

6.2.4 Crystallographic shear planes

Defects, regardless of their nature or dimension, have structural influences on their local environment. In the iron-deficient $Fe_{1-x}O$ phase, for example, the interstitial sites occupied by the Fe^{3+} ions are being aggregated to form clusters (cf. Section 6.1.6). However, there are also many possible and known vacancy-induced local distortion patterns. The simplest are local relaxations of atoms towards a vacancy (for metals), or vice versa, away from the vacancy (for ionic compounds). In certain structures that have a particularly large number of anion vacancies, a kind of cooperative effect occurs that is based on the formation of so-called crystallographic shear (CS) planes. The anion vacancies are eliminated during the formation of such shear planes and the local coordination environment at the periphery of the CS planes change. In this way, areas of different compositions and structures are being formed in the crystal. The interesting thing about these phases is that these planar defects usually do not occur in an isolated manner (if this is the case, they are called *Wadsley defects*) or that they are statistically distributed throughout the crystal but are present rather in an ordered or periodically structured way.

Such structures with CS planes occur, for example, in the oxides of some early transition metals, such as TiO_2, V_2O_5, MoO_3, and WO_3, when they are *partially* reduced, either by heating in the presence of the corresponding metal in absence of oxygen or under a stream of gaseous hydrogen.

For a long time, it was thought that the resulting coloured, oxygen-deficient phases (TiO_{2-x}, VO_{2-x}, MoO_{3-x}, WO_{3-x}) were non-stoichiometric compounds with a large phase width. However, starting with the work of Arne Magnéli (a Swedish chemist, 1914–1996) in the 1940s, electron microscopy and X-ray diffraction studies have revealed that they are instead a range of stoichiometrically composed compounds with closely related compositions; in honour of the work of Magnéli they are also known as Magnéli phases. The common pattern that Magnéli discerned in some of these structures led him to formulate the concept of "recurrent dislocations", which were later termed crystallographic shear planes. These are a way of changing the stoichiometry without relinquishing the cation's primary coordination requirements. In oxygen-deficient rutile (TiO_2), a homologous series of physically separate phases with formula Ti_nO_{2n-1}, $4 \leq n \leq 9$, occurs.

Within the crystal structures of compounds with CS planes, the coordination of the cations remains essentially unchanged, but the anion coordination number increases to accommodate the lower metal to oxygen ratio. For the oxygen-deficient rutile-related structures (cf. Section 5.4.4) of TiO_{2-x} and VO_{2-x} the oxygen deficit is compensated through an increased number of edge- and also face-shared MO_6 (M = Ti, V) octahedra within the CS plane, which are separated by unmodified rutile-analogous blocks.

The simplest case, which is also best suited to visualize CS planes and their formation, is given by the oxygen-deficient phases of tungsten and molybdenum oxide, which have an ReO_3-analogue parent structure in which all MO_6 (M = Mo, W) octahedra are

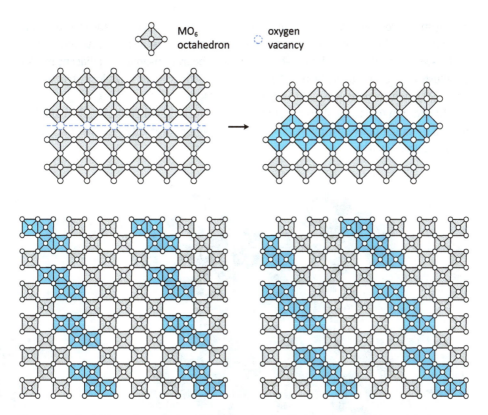

Figure 6.16: (Top) Schematic mechanism of the formation of a CS plane (highlighted in cyan) in the oxygen-deficient phases MoO_{3-x}/WO_{3-x} (right) out of the parent structure in which oxygen vacancies are concentrated along a line/plane (left). (Bottom) Two further examples of CS planes in MoO_{3-x}/WO_{3-x}, lying along the {102} (left) and {103} crystallographic plane (right).

exclusively corner-linked. If one now imagines that along a line/plane oxygen vacancies arise, then this structure will be unstable and as a result the blocks separated by the vacancies will collapse, so that within the CS plane the octahedra are now systematically edge-linked (see Figure 6.16, top). CS planes can occur in different orientations, and this also entails structural changes and affects the concrete linkage scheme of the octahedra within and in the vicinity of the CS plane, see Figure 6.16, bottom. With increased degree of reduction of the parent phases, the number of CS planes increases and the average spacing between adjacent CS planes decreases.

In partially reduced Nb_2O_5 and partially reduced mixed oxides of Nb and W, CS planes occur in two orthogonal sets and the regions of the parent (unreduced) structure are reduced in size from infinite sheets to infinite blocks or pillars. These block (sometimes also called double shear) structures are characterized by the length and width (expressed as $m \times n =$ number of octahedra, which are corner-connected in the cross-section) and manner of connectivity of the blocks of the unreduced ReO_3-analogeous parent structure. The metal cations in the resulting structure retain octahedral coordination within the blocks but they are at two heights in the third direction, usually c ($z = 0$ and $z = \frac{1}{2}$), being displaced by one-half of the octahedron diagonal in adjacent blocks. An illustrative example is given in Figure 6.17, depicting the (4 × 4) block structure of $Nb_{14}W_3O_{44}$, which crystallizes in the space group $I4/m$.

In addition to having phases that are built of blocks of only one size, the complexity can be much increased by having blocks of two or three different sizes arranged in an ordered fashion.

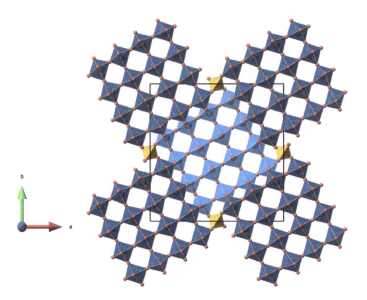

Figure 6.17: (4 × 4) block CS structure of $Nb_{14}W_3O_{44}$ (space group $I4/m$). Light and dark blue coloured blocks of MO_6 octahedra are offset by half the lattice parameter perpendicular to the plane of drawing. Note that the (4 × 4) blocks are connected at the corners of the blocks via tetrahedral MO_4 units, highlighted in orange.

6.3 Planar defects

6.3.1 Stacking faults, turbostratification, and interstratification

All crystals that have a layered structure are susceptible to stacking faults. A stacking fault is a type of defect in which the periodicity of a layer sequence is disturbed. These may include: One or more "wrong" layers inserted into the sequence, a change of the layer sequence, or layers inserted whose distance to neighbouring layers is larger or smaller than usual. Compounds that occur in different polytypes are particularly prone to stacking faults, see also Section 4.5. But note that stacking *variants* and stacking *disorder* are two different things.

Stacking disorder is frequently found in phyllosilicates, graphite, or molybdenum sulphide, compounds in which the chemical bonding within the layers is much stronger than between the layers. However, there are also well-known examples of stacking disorder that occur in compounds in which the strength of chemical bonding is similar (if not identical) within and between the layers, such as in ice I, diamond, or II–VI semiconductors. Stacking disorder (sd) and the associated phenomenon of polytypism can have profound consequences for the physical and chemical properties of the material in question. For example, in the case of ice I, whose equilibrium structure is hexagonal, the extent of stacking disorder (i.e., the ratio of hexagonal to cubic layers) affects its vapour pressure and crystal shape. It was also found out that stacking-disordered ice crystallites up to a size of about 100,000 molecules are more stable than hexagonal ice due to entropic reasons.

A special type of stacking faults that can occur in layered compounds is characterized by either a *random displacement* of the layers from their ideal positions by a shift within the layer planes (x,y plane) and/or by a *random rotation* of the whole layer around their normal vector (z-direction), while the stacking distance along the normal of the layers does not have to be affected (see Figure 6.18). This type of stacking disorder is called turbostratic-like disorder. It is often found in graphitic carbons (carbon black) resulting from the carbonization of suitable carbon precursors, where the carbonization temperature was relatively low (below 2,000 °C) or the time duration relatively short. Upon heating above 2,000 °C, the turbostratic disorder is relieved in a more or less continuous way, decreasing to zero at approx. 3,000 °C.

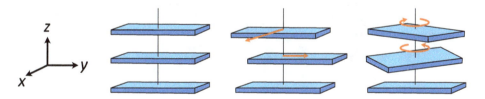

Figure 6.18: Perfectly aligned stack of three layers (left), stack of three layers in which the layers are randomly displaced in the (x,y) plane (middle), and stack of three layers in which the layers are mutually rotated around the z-axis (right).

Yet another type of stacking disorder is observed in layered compounds that are able to intercalate ions or molecules between their layers. The intercalation process is usually accompanied by an increase of the interlayer distance. If the intercalation is inhomogeneous, i.e., some layers are intercalated and others are not or not to the same degree, the stacking vector along the z-direction is not uniform anymore. It is also possible that the layers are slightly bent or undulated. Such stacking faults that affect the stacking order along the stacking direction are called *interstratification*. Interstratification and turbostratic-like disorder often occur simultaneously.

6.3.2 Internal boundaries in single crystals – mosaicity

So far, we have pretended that single crystals are perfect entities that consist of a single coherent block throughout. And although real single crystals look even under the light microscope as they do, instead they consist of so-called individual mosaic blocks (also called sub-grains) that are slightly tilted or twisted against each other (see Figure 6.19). Within the mosaic blocks, which are approx. 1,000 to 10,000 Å in size, the crystalline order is perfect, but the boundaries of these blocks represent 2D defects. Evidence that "single crystals" are by no means made up of a single block emerged from theoretical considerations in the early days of the development of X-ray diffraction. Nowadays, the extent and anisotropy of mosaicity can also be determined experimentally. The extent of mosaicity is often expressed as the *mosaic spread* that reflects the degree of orientation divergence of the mosaic blocks. Crystals of good quality have mosaic spreads of 0.1 degrees or less; metals and compact inorganic compounds often show values of several orders of magnitude lower. Organic and, in particular, protein crystals, on the other hand, can have mosaic spreads of more than 1 degree.

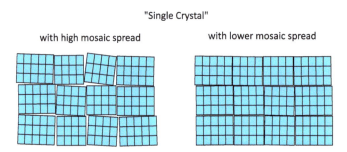

Figure 6.19: Schematic illustration of the mosaicity of single crystals.

Low-angle (sub-)grain boundaries

As long as the misalignment between mosaic blocks in a single crystal or grains in a polycrystalline material is not too large (up to approx. 15 degrees), the interfacial region between two (sub-)grains is called a small- or low-angle (sub-)grain boundary. These boundaries can be described as an array of dislocations. An appropriate array of edge dislocations – as depicted in Figure 6.20 – leads to a *tilt boundary*, while an array of screw dislocations leads to a misalignment parallel to the boundary, which is then called a *twist boundary*. These concepts of tilt and twist boundaries represent idealized cases. The majority of boundaries are of a mixed type, containing dislocations of both types, in order to create the best fit between the neighbouring sub-grains.

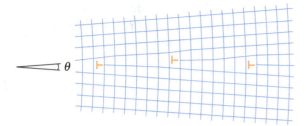

Figure 6.20: An array of edge dislocations that constitute a low-angle (sub-)grain boundary.

6.3.3 Grain boundaries

The grain boundaries between crystallites, or *grains*, of a polycrystalline material can only be partially described by the low-angle-boundaries (see preceding section), since the orientation of the assembled crystallites can be arbitrary and therefore the extent of the misalignment can also be very large. Usually, such high-angle grain boundaries (misalignment > 15°) include more extended areas in which there is an offset of the atoms between adjacent crystallites. Therefore, grain boundaries have a comparable "open" structure and can, after chemical etching of the surface, often be observed with the light microscope (see Figure 6.21 for an example) or even with the naked eye, provided the grains themselves are not too small.

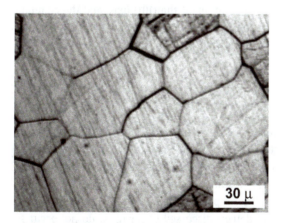

Figure 6.21: Micrograph of a specimen of a polycrystalline metal alloy called VT22 ($Ti_5Al_5Mo_5V_{1.5}Cr$); grain boundaries evidenced by acid etching (CC BY 3.0 Edward Pleshakov).

The interfacial energy between two adjacent highly misaligned crystallites is larger than for those having only a minor mismatch. This is the reason why grain boundaries are more chemically reactive than the grains themselves. Most grain boundaries are also preferred sites for the aggregation of impurities or precipitates of new or separated phases that form from the solid. As for mechanical properties, grain boundaries disrupt the motion of dislocations through a material. This means that reducing the crystallite size is a common way to improve the mechanical strength of a material.

6.3.4 Twin boundaries

Twin boundaries are a special variety of domain boundaries that occur in twinned crystal structures.[6] Twins are crystals that consist not of a single domain but of two domains that are intergrown; this means their respective lattices have different orientations, but there is a common interface, where the two domains meet. This interface is often a well-defined, exposed outer crystal face of the single crystals, but this must not be the case. Since the two domains are oriented differently, this causes considerable problems in deducing the crystal structure of twins from X-ray diffraction experiments, since the corresponding diffraction patterns overlap. Only when one has found out the so-called twin law or twin operation, which specifies how the two domains are oriented to each other, there is a real chance of solving the structure.

The two most common cases are that the two domains are mirrored along their common interface or that they are rotated around a twofold axis of rotation, i.e., rotated against each other by 180°. An example is given in Figure 6.22, in which an atomistic model of a twinned rutile (TiO_2) crystal is depicted based on a model of Takeuchi & Hashimoto [58]. The twin operation that describes the observed crystal lattice orientation of twin domains 1 and 2 is a 180° rotation about the [101] axis, and the common interface is the (101) plane. This operation produces mirror symmetry for the cation substructure, whereby all Ti atoms remain octahedrally coordinated by oxygen atoms.

6.4 Volume defects

As already mentioned in the introduction of this chapter, every kind of defect represents an already spatially extended disruption of the crystal structure, i.e., they are intrinsically already volume defects. Nevertheless, the distinction between 0, 1, 2, and 3D defects is useful, as the dimension is a suitable criterion to indicate the *origin* of the defect. However, it is difficult to draw a precise distinction line between the dimensions, if only because the types of defects discussed so far often do not occur in isolation but in aggregates or clusters, thus affecting a certain *volume* of the compound. However, there are "real" volume defects that cannot be ascribed to the type

---

[6] According to the author's experience as a crystallographer in charge of an X-ray service department where predominantly small-molecule crystallography is performed, every tenth to twentieth crystal is a twin.

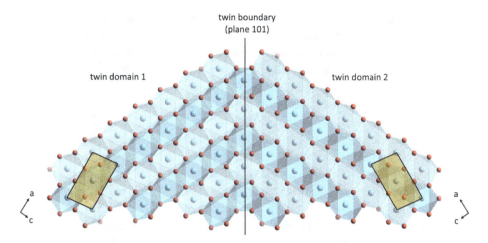

Figure 6.22: An atomistic model of a twinned crystal structure of rutile (TiO$_2$). The twin operation is a rotation around the [101] axis, the twin boundary is the (101) plane. The unit cells for each domain are highlighted in orange; Ti, light-blue; O, red (atomistic file kindly provided by Aleksander Recnik, see also Ref. [59]).

of defects already discussed. These comprise, for instance, cracks, pores, and, technically very important: precipitates.

Precipitates are formed in a variety of circumstances. For instance, phases that are stable at high temperatures may not be stable at low temperatures, and decreasing the temperature slowly will frequently lead to the formation of a new phase as a precipitate, as small, uniformly dispersed particles within the matrix of the original crystal structure. This is the basis for the very important technical process called precipitation hardening that is applied for a number of alloys, for instance, Al/Mg, Al/Cu, Cu/Be, and Cu/Zn alloys in order to improve their mechanical strength. Usually, technical alloys comprise more than two components, but the basics of precipitate hardening can be explained by assuming we have a simple binary system, A/B. The phase diagram of this binary system must meet at least the following three requirements, in order that precipitation hardening will work. (1) The solubility of the minor component, say B, in the other component, say A, should not be too small, at least of the order of several mol%. (2) The solubility limit of B in A should rapidly decrease with decreasing temperature. (3) The desired final composition of the alloy must contain a lower proportion of B than the proportion that corresponds to the maximum solubility of B in A. Such a model-like phase diagram, together with highlighted temperature points for the two-step process of precipitation hardening explained in the following paragraph is shown in Figure 6.23.

The precipitation hardening is accomplished in a two-step process. In the first step, an alloy consisting of two solid phases (α + β) with the composition $C_{initial}$ is heated to a temperature (T_2) in which the phase diagram shows only one single phase, say α, which is a solid solution of the minor component B in the major component matrix A. The temperature is then maintained until any β-phase (a phase with a minor component A in the major component B) has dissolved. Subsequently, the alloy is cooled down to a temperature at which the diffusion rates are so low that the β-phase is not formed again (T_0). Instead, a non-equilibrium structure forms, in which only the α-phase, a solid solution supersaturated with the component B is present. Subsequently, the alloy is heated again only to an intermediate temperature (T_1) that still lies in the two-phase region but allows for a considerable diffusion rate. Now the β-phase with composition $C_β$ starts to form as a precipitate, finely dispersed within the α-phase matrix with the composition $C_α$. This process is also called "ageing". After an appropriate ageing time the alloy is cooled to room temperature. The character (size, dispersity) of these β-particles, and subsequently the strength and hardness of the alloy, depend on both the precipitation temperature and the ageing time at this temperature. Usually, the formation of the β-phase particles within the α-matrix does not take place in a single step but proceeds via several transitional states until the final equilibrium β-particles are formed. Such a process is shown schematically in Figure 6.24.

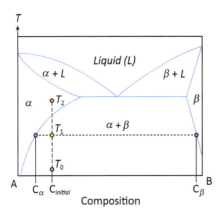

Figure 6.23: Hypothetical phase diagram of a binary alloy A/B that undergoes a precipitation hardening process (see text).

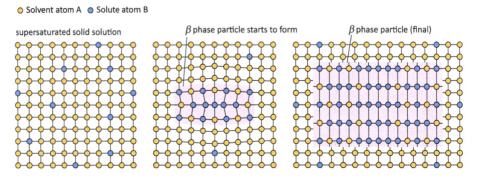

Figure 6.24: Schematic process depicting the two-step precipitation hardening (see text) (redrawn and adapted from [60]).

Conclusion and further reading

In this chapter we have discussed defects, i.e., anomalies or deviations from perfectly monocrystalline bodies. We have seen that defects can dramatically change diverse properties of materials. Since most materials are in solid form, defects are a very important subject of research in materials science and engineering. Defect engineering has now become a sub-discipline in its own right, concerned either with adding such defects to materials so that their useful properties are optimized or with minimizing defects. However, these things no longer pertain to the core area of solid-state chemistry, so they will not be discussed in more detail here.

The interested reader may refer to the following references for in-depth study of defects:

M. W. Roberts, J. M. Thomas (Eds.), *Surface and Defect Properties of Solids*, Vol. 1 to 6. Royal Society of Chemistry, Cambridge, **1972–1977**.
Point Defects Part I + II, Special Issues of *MRS Bulletin*, **1991**, *16*, 11 + 12.
D. A. Drabold, S. K. Estreicher, *Theory of Defects in Semiconductors*. Springer, Berlin, Heidelberg, **2007**.
D. B. Holt, B. G. Yacobi, *Extended Defects in Semiconductors – Electronic Properties, Device Effects and Structures*. Cambridge University Press, Cambridge, **2007**.
R. J. D. Tiley, *Defects in Solids*. John Wiley & Sons, Inc., Hoboken, NJ, **2008**.
W. Cai, W. Nix, *Imperfections in Crystalline Solids* (MRS-Cambridge Materials Fundamentals). Cambridge University Press, Cambridge, **2016**. doi:10.1017/CBO9781316389508

7 Phase diagrams

Phase diagrams are a graphical representation of ranges of external conditions (temperature and/or pressure) in which the various phases of a given system – that may consist of only one or more components – exist and are stable under thermodynamic equilibrium conditions. As we are usually limited to 2D illustrations, a phase diagram can only contain both state functions (p and T) if we are looking at only *one* component. For binary (or higher) systems, the phase regions are plotted as a function of either the temperature (more frequent) or pressure (less frequent), with the other state function being constant; in case the regions are plotted as a function of temperature, this is often the standard pressure (1.00 bar, as defined by the IUPAC) or normal pressure (1.01325 bar, used by the National Institute of Standards and Technology, NIST).

The ability to obtain and interpret phase diagrams is particularly important for researchers and developers in materials science, metallurgy, combustion and energy related science, corrosion engineering, environmental engineering, geology, and glass technology. Therefore, this short introduction to phase diagrams is supposed to serve as a sufficient foundation to understand more complicated scenarios (i.e., ternary and higher systems) and applications of phase diagrams.

The most fundamental law on which all phase diagrams are based upon is the phase rule of Gibbs. It reads:

$$P + F = C + 2$$

with P being the number of phases, F the degrees of freedom, i.e., the number of variables of state (temperature, pressure, or composition) that can be varied independently without changing the phase, and C the number of components (which can be elements or compounds). Let us have a look at a schematic phase diagram of a one-component system first.

7.1 One-component systems

Due to the fact that the composition in a one-composition system is fixed, the independent variables are reduced to temperature and pressure. The phase rule then reads $P + F = 3$. If we assume that our one-component system forms two distinct phases in the solid state, i.e., two polymorphs, and can exist as liquid and gas as well (i.e., we assume that it does not decompose at the melting or boiling point), the phase diagram might look like as schematically drawn in Figure 7.1.

If we are within an area where only one phase is present, then the value for $F = 2$; this means that we can independently vary the pressure *and* the temperature without changing the phase; the system is bivariate. If we are at the border between two phases, i.e., exactly on a line (AB, BC, BE, CD, or CF) where two phases meet, the system becomes

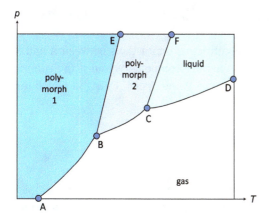

Figure 7.1: Schematic phase diagram of a one-component system, which exists in two different solid states.

univariate ($F = 1$): we can freely choose, for instance, the temperature, but in order to not leave the line, the two-phase coexistence condition, the pressure is then determined by the chosen temperature (and vice versa). Finally, if we are at points where regions of three phases meet (this is the reason why they are called triple points, here at points B and C), there is no degree of freedom left ($F = 0$), the system is invariant.

7.2 Two-component systems

In a two-component system, there are three independent variables: pressure, temperature, and composition. However, for many systems in the solid state, the pressure dependency of the phases is of minor interest and the phase diagram only depicts the phases dependent on the temperature. This leads to the so-called condensed phase rule, $P + F = C + 1$, and hence, since $C = 2$, $P + F = 3$. Furthermore, the vapour phase is also often neglected.

In principle, all phase diagrams of binary systems can be understood as derived from three basic types of phase diagrams resulting from whether the attractive interaction between the components, say A and B, is comparable to, stronger, or weaker than the interaction of the species within the pure components A and B:

– If the attractive interactions between A and B are comparable to the interactions within the pure components, then a solid solution over the whole range of compositions will form; this case is analogous to two liquids that are completely miscible. Examples in the solid state are the systems Cu-Ni, Sb-Bi, NiO-MgO, or Mg_2SiO_4 (forsterite)-Fe_2SiO_4 (fayalite).
– If the attractive interactions between A and B are significantly weaker than those within the pure components, then a singly eutectic system is present, in

which neither a compound is formed, nor solid solutions occur. Examples are the systems Al-Si, Sb-Pb, or KCl-AgCl.
- Finally, the third type is present, if the attractive interactions between A and B are significantly stronger than within the pure components and if the components will form compounds (AB). An example is the system Ca-Mg in which the compound $CaMg_2$ is formed.

7.2.1 Complete miscibility – solid solutions

A schematic phase diagram for a two-component system that forms a solid solution over the entire range of compositions is shown in Figure 7.2. This kind of phase diagram is realized for systems in which the two components form isostructural crystals (if the components are elements, then they should also have comparable atom sizes). This means that in components A and B, the respective other component can occupy the atomic position in any proportion.

In the imaginary case depicted in Figure 7.2, the melting point of the component A should be lower than that of B. The upper part of the diagram corresponds to the range of existence of the liquid phase in which both components are completely miscible. The lower part corresponds to the range of existence of solid solutions. In the region in between, delimited by the liquidus line on the upper side and the solidus line on the lower side, the two-phase region (shaded in blue) is present, in which the liquid and the solid solution coexist. The most important thing to remember when interpreting phase diagrams is that the two phases in the two-phase region have different compositions dependent on the chosen temperature. This is exemplified in Figure 7.2 by the orange dot. The overall composition (i.e., taking the liquid and solid solution *together*) is C(total), but the

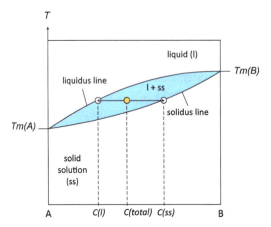

Figure 7.2: Schematic phase diagram of a two-component system, which forms a solid solution over the entire range of possible compositions. The two-phase region is shaded in blue.

liquid has a higher content of A and a composition C(l), while the solid solution has a higher content of B and a composition C(ss). In other words, the compositions of the two phases can be always inferred by drawing a horizontal line from the temperature point chosen until the cross-points with the liquidus and solidus line are reached.

Complete miscibility with thermal maxima/minima at intermediate compositions
If the attractive interaction between the components A and B are slightly stronger or slightly weaker than those within the pure components but still, they do form a solid solution in the solid state, the liquidus and solidus line may exhibit a minimum or maximum, respectively, at a certain point of intermediate composition as shown schematically in Figure 7.3, marked by the orange dot. These points are called indifferent points and they correspond to compositions at which the melting or freezing occurs without any change of composition. The analogous cases for the transition between a liquid and a vapour phase would be systems with a positive and negative azeotrope, respectively.

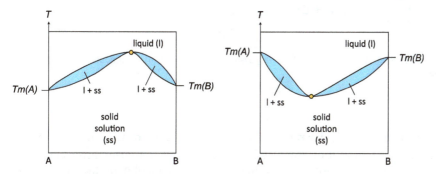

Figure 7.3: Schematic phase diagrams of a binary system that forms a solid solution but exhibits a thermal maximum (left), because the attractive interaction between the components A and B are stronger than those within the pure components, or a thermal minimum (right), because the interactions between A and B are weaker than those within the pure components.

7.2.2 Singly eutectic systems

If the components A and B form neither solid solutions (except for extreme low concentrations) nor a compound with one another but are miscible in the liquid state, typically a phase diagram like the one depicted in Figure 7.4 is obtained. Such phase diagrams contain an indifferent point called eutectic point (from Greek, eu = good, and teco = to melt). It marks the melting or solidification point, respectively, of the eutectic mixture. This mixture has a lower melting/solidification point than that of the pure components A and B as well as mixtures of any other conceivable compositions. Furthermore, the eutectic mixture is characterized by the fact that the phase transition at this point occurs without change in the overall composition. The eutectic line is the horizontal line that passes through the eutectic point and is identical with the solidus line. The area underneath is the two-phase region in which both components coexist as separate solids.

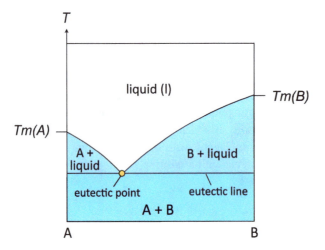

Figure 7.4: Schematic phase diagram of a binary system that neither forms a compound nor a solid solution.

Singly eutectic systems with solid solution formation at extreme compositions

A slightly modified phase diagram is obtained, if the two components are not over the entire range of composition completely immiscible but can form partly solid solutions, namely near the borders of the pure components A and B. In this scenario, the pure component A can form to a certain but limited extent, a solid solution with a little amount of B (often this phase is called α in order to distinguish it from the pure component A), and, vice versa, the component B can dissolve a certain but limited amount of A to form a solid solution β, see Figure 7.5. At intermediate compositions, both solid solutions are present as separate phases (α + β). Note that the limiting concentration of B in A and A in B increases with increasing temperature until the temperature reaches the temperature of the eutectic line.

Complete miscibility at higher temperatures and limited miscibility at lower temperatures

A further subtype of binary phase diagrams is often observed, if the components are moderately incompatible. Conceptually, it can be derived from a mixture of the two phase diagrams shown in Figures 7.3 (right) and 7.5. At low temperatures, the two components have a miscibility gap over a significant range of compositions, where

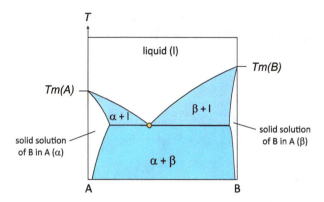

Figure 7.5: Schematic phase diagram of a binary system that shows the ability to form solid solutions only at extreme compositions near the border of the pure components A and B.

the solid solutions α and β are present as separate solids, but at higher temperatures again a solid solution will form over the entire range of compositions (see Figure 7.6). This solid solution range is separated by two two-phase regions with a thermal minimum. An example of such a system is Au-Ni.

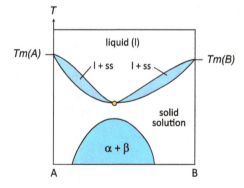

Figure 7.6: Schematic phase diagram of a moderately incompatible binary system (see text).

7.2.3 Systems with compound formation that melt congruently

The third basic type of phase diagrams of binary systems is obtained, if the interactions between the components A and B are stronger than those within the pure components and if they form a stoichiometric defined compound, a line compound. In the scenario depicted in Figure 7.7, the stoichiometric compound that forms has the composition A_2B. If the compound melts/solidifies without change of its composition into the liquid/solid phase it is said to have a *congruent melting/solidification* behaviour.

The liquidus curve at the composition of the compound has a maximum at that point, and the compound is represented as a vertical line (hence line compound).

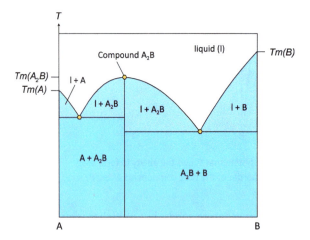

Figure 7.7: Schematic phase diagram of a binary system with compound formation and congruent melting behaviour of the compound.

Both parts of the phase diagram, left and right from vertical line of the stoichiometric compound A_2B, correspond to a singly eutectic system as shown in Figure 7.4. A real example that resembles the phase diagram of Figure 7.7 is that of Mg_2Ca (a Laves phase, cf. Section 13.5.2).

Note that a congruent melting behaviour is only present when both the solid and liquid phases are single-phase systems. This means that the melting/solidification of compositions at the eutectic point are not congruent processes as the liquid phase is a single-phase but the solid a two-phase mixture.

7.2.4 Systems with compound formation but incongruent melting

A more complicated type of phase systems is present if the components form a compound that melts/solidifies incongruently. This means that the solid compound and the melt have different compositions. For a better understanding, it is useful to look at the schematic phase diagram in Figure 7.8. Components A and B are supposed to form a stoichiometrically defined compound AB. If we move along the line to higher temperatures, then we reach the so-called peritectic point (from Greek, peri = to surround, teco = to melt). At this point (marked with a white dot in the figure) the compound AB does not simply form a liquid phase of that very same composition, but instead it *decomposes* into a solid A and a liquid of a different composition (richer in B), see the purple and red point, the end points of the peritectic line. As a new solid is formed, or, because the AB compound decomposes, this transition can be also called a peritectic *reaction*.

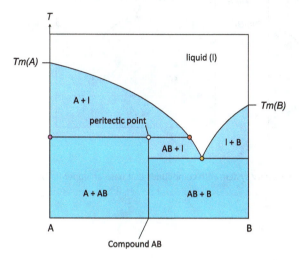

Figure 7.8: Schematic phase diagram with compound formation and incongruent melting (see text).

7.2.5 Systems with compound formation with an upper or lower limit of stability

Starting from a situation described in the previous subsection, one can imagine what happens when the stability of a compound formed from components A and B continues to decrease: It will not decompose into another solid and a liquid (incongruent melting). Instead, before melting occurs, it decomposes to form two solids. This scenario, a compound with an upper limit of stability, is depicted in the schematic phase diagram of Figure 7.9, left.

The opposite scenario also often occurs, namely that the binary system forms a compound, but it decomposes below a certain temperature into two solids. The compound has a lower limit of stability (see Figure 7.9, right). At the high-end temperature range of stability, the compound may behave as in one of the scenarios already described above.

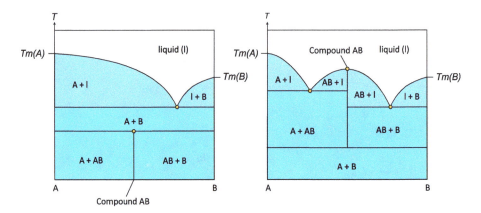

Figure 7.9: Schematic phase diagrams of a binary system with compounds that have an upper (left) and lower (right) limit of stability.

Further reading

M. Hillert, *Phase Equilibria, Phase Diagrams and Phase Transformations: Their Thermodynamic Basis*, 2nd ed. Cambridge University Press, New York, **2007**.

F. C. Campbell, *Phase Diagrams: Understanding the Basics*, ASM International, Ohio, **2012**.

A. D. Pelton, *Phase Diagrams and Thermodynamic Modeling of Solutions*, Elsevier, Amsterdam, **2018**.

H. Saka, *Introduction to Phase Diagrams in Materials Science and Engineering*, World Scientific, Singapore, **2019**.

8 Electronic structure of solid-state compounds

Most of the properties of solids depend directly on their electronic structure. When chemists think of an electronic structure, they usually consider spatially limited entities (molecules, clusters, etc.). They think essentially *locally* and apply the tools of their trade, be it the modern variant of valence bond (VB) theory or the molecular orbital (MO) theory, in order to explain structure-property relationships. This is no longer possible for solid-state compounds, which are quasi-infinitely extended. The orbital theory is replaced by the *band theory* for crystals. The application of the band theory requires quantum mechanical calculations (band structure calculations), in order to find – as in the molecular case – solutions for the Schrödinger equation, only that here not a limited number of atoms but quasi infinitely many atoms must be considered. The result of these quantum mechanical calculations is the *band structure*. Based on this band structure, a number of properties of solids can be inferred, for example, with respect to:
- mechanical properties (compressibility, elasticity),
- electrical properties (conductor, semiconductor, insulator),
- optical properties (colour, absorption and emission of light),
- magnetic properties (dia, para, ferro, antiferro, ferri, or helimagnetism),
- structural features (instabilities, structural distortions, phase transitions and accompanied symmetry reductions).

The difficulty for chemists to understand band theory and interpret band structures is that it is a domain of solid-state *physics*, and physicists often speak a very different language, often formal and very mathematical. This chapter will therefore be an attempt to introduce solid-state chemists to the basic features and concepts of band theory. The presentation is strongly based on two highly recommended books, namely the book of Roald Hoffmann *Solids and Surfaces: A Chemist's View of Bonding in Extended Structures* [61], and Richard Dronskowski's book *Computational Chemistry of Solid State Materials* [62]. For physically, mathematically educated readers, the presentation given here might seem far too detailed, but according to my teaching experience I know that already the presentation of the basic unit of imaginary numbers can cause headaches to some students. ☺

The endeavour to calculate solutions of the Schrödinger equation for quasi-infinitely extended atomic assemblies seems hopeless at first sight. However, it turns out that it is indeed laborious but not impossible, as long as we are dealing with *periodically* structured solids, i.e., *crystals*. Just as the structure of the overall crystal results from the repeated translation of the unit cell, it should be clear that the electronic structure also repeats periodically, and therefore, there are not an infinite number of combinations of orbitals or wavefunctions to consider. Conversely, however, this also means that finding a solution to the Schrödinger equation for *amorphous* solids is actually impossible. This is one reason why amorphous systems are not well understood and why their properties are particularly difficult to derive.

In order to approach band structures conceptually, it is not a bad idea to start from the MO-LCAO approach (molecular orbitals by linear combinations of atomic orbitals) and to extend it to a very simple, one-dimensional infinitely extended ensemble of atoms. To shortly recap the MO-LCAO approach: in this approach, the wavefunctions that are solutions of the Schrödinger equation ($\hat{H}\Psi = \hat{E}\Psi$) consist of a set of one-electron wavefunctions (ψ_i, molecular orbitals), which in turn consist of linear combinations of hydrogen-like atomic orbitals (ϕ_μ) of the atoms of the molecules:

$$\psi_i = \sum_{\text{Atoms}} \sum_\mu c_{i\mu} \phi_\mu$$

with $c_{i\mu}$ being the mixing coefficients of the linear combination. The term "hydrogen-*like*" means that we consider one-electron atomic orbitals but take into account the specific nuclear charge (Z_A) of the atoms. For the simple diatomic hydrogen molecule itself, there are two wavefunctions that satisfy the Schrödinger equation. They determine the shape of the molecular orbitals and their associated energy values. This results in a bonding MO in which the two 1s orbitals are in-phase ($1s_A + 1s_B$) and an antibonding MO in which the two atomic orbitals are in anti-phase relation ($1s_A - 1s_B$), see Figure 8.1. The MOs get labels that specify certain properties: the number of node planes in the direction of the conjunction line of the atoms (σ = 0 node planes, π = 1 node plane, δ = 2 node planes and so on); symmetry properties, for instance the behaviour upon inversion (g = even, from German

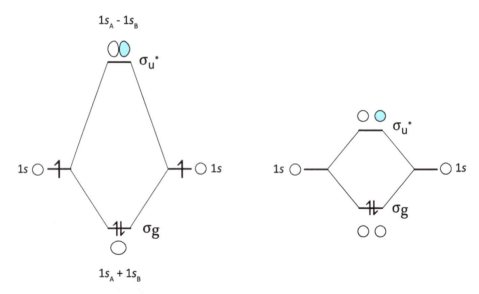

Figure 8.1: The construction of MOs for the diatomic hydrogen molecule according to the MO-LCAO approach (left) and according to the Hückel approximation (right).

"gerade", meaning that the sign of the phase does not change, or u = odd, from German "ungerade"), the degree of degeneracy of the energy levels (e = twofold, t = threefold, etc.), and it is specified if it is a bonding (no sign) or an antibonding state (*). The bonding MO of the hydrogen molecule is the σ_g orbital and the antibonding MO is the σ_u^* orbital. The associated energy values of these two MOs are not symmetric with respect to the initial value of the atomic orbitals, because the overlap integral must be taken into account. The lower energy level is less stabilized than the higher level is destabilized. In the framework of the Hückel theory, they are symmetrical, because in this theory the overlap integral is neglected. But even with this rough approximation, one recognizes that an He_2 molecule cannot exist.

The Hückel approximation can also be applied for semiquantitative estimation of the energy levels of the π MOs formed by the p_z orbitals in cyclic unsaturated hydrocarbons. To do this, the corresponding polygon is placed with the tip pointing downwards and then each vertex represents the location of one of the MOs, with increasing energy in the vertical direction upwards; finally, one fills in a corresponding number of electrons and can thus estimate, for example, whether the system is aromatic or antiaromatic.

After this brief recap of molecular orbital theory, we are now ready for the first step to consider crystals: we create a quasi-infinite chain of equidistant H atoms, a one-dimensional H-atom crystal with a lattice constant a. For this we proceed quite analogously to the case of cyclic hydrocarbons, except that we do not consider the p_z orbitals, but only the 1s orbitals (a completely analogous picture would emerge in the case of the cyclic hydrocarbons if the viewing direction was perpendicular to the plane of the polygon: then one would see only one lobe of the p_z orbitals, which would have the corresponding inverse phase below that polygonal plane). How does the distribution of the orbitals now look like? It can be estimated from the position of the orbitals of a polygon with very many atoms, as shown in Figure 8.2. Because for the asymptotic case of an n-gon with $n \rightarrow \infty$ the curvature approaches zero and therefore a segment of this n-gon is approximately a straight line and thus a good approximation for our 1D H-atom crystal.

What do we observe when we look at Figure 8.2? For an even number of atoms, there is only one orbital for the lowest and highest energy levels; for an odd number, the highest level is also twofold degenerated; and finally, pairs of twofold degenerate energy levels always lie between the extremes. The number of nodes increases from the bottom starting at zero and reaches the maximum value n at the top energy level. The allowed energy levels move closer together as the number of atoms increases, or in other words, the energy gap between two neighbouring energy levels becomes smaller. Moreover, one notices that the energy levels are not equidistant within the whole energy range. But: even if n becomes very large, the allowed energy levels remain discrete, of course, because we are still dealing with a quantized system. It is just a *quasi-continuous* energy range with very close energy levels. For chemists, this quasi-continuous range is an *energy band* or just a *band*.

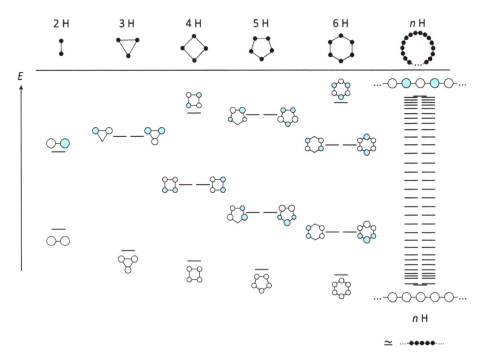

Figure 8.2: Development of the energy levels of a *n*-cyclic polymer of H atoms that can serve as an approximation for an infinitely extended, periodic, one-dimensional H-atom crystal; positive phases of the wavefunctions are symbolized by white, negative phases by cyan circles.

8.1 Bloch functions, Bloch's theorem, the quantum number *k*, and crystal orbitals

In a next step, we move to a real crystal and the crucial question is how the wavefunctions and thus the crystal orbitals look like. In other words, solutions to the Schrödinger equation are sought under a specific boundary condition, namely that the wavefunction, and hence the electronic structure, must reflect the periodicity of the crystal: its translational symmetry! The electronic structure and the value of the wavefunction at a specific location of the unit cell must be identical to the electronic structure, or the value of the wavefunction, respectively, at an equivalent location in all neighbouring unit cells. Bloch's theorem [63] makes use of this fact and formulates first of all how wavefunctions, which are solutions of the Schrödinger equation, must look like, so that they satisfy the demand of translational symmetry. The results are the Bloch functions named after Felix Bloch (a Swiss-American physicist, 1905–1983, who was awarded the Nobel Prize in Physics together with Edwards Mill Purcell in 1952 for "the development of new ways and methods for nuclear magnetic precision measurements"):

8.1 Bloch functions, Bloch's theorem, the quantum number k, and crystal orbitals

$$\psi_k(\vec{r}+\vec{T}) = e^{ik\vec{T}}\psi_k(\vec{r})$$

where $\vec{r}$ is the position vector specifying a particular point in the unit cell, $\vec{T}$ is the lattice vector with the translation period $|T|$, k is a quantum number rather unknown to most chemists in the form of a wave vector, which we will consider in more detail in a moment, and i is the fundamental unit of imaginary numbers ($\sqrt{-1}$). What does this equation basically mean? That a periodic wavefunction extending over the entire crystal – also called crystal orbital – which is a function of the lattice vector $\vec{T}$, can be represented by a wavefunction at a *particular* location [$\psi_k(\vec{r})$] of the crystal modified only by a phase factor [$e^{ik\vec{T}}$]. If the wavefunction at a particular location in the unit cell is known, it is also known for all other points that are equivalent with respect to translational symmetry throughout the whole crystal.

Bloch, who derived the theorem, had originally tried to solve the question of how to explain the very good electrical conductivity of metals with the aid of quantum mechanics. He assumed that there are two limiting cases: The first case pertains electrons that are very strongly bound to the atomic core and do not contribute anything to the conductivity. The second case includes electrons that are quasi-free and feel only a weak periodic potential of the atomic core. The different conductivities then result from a respective specific mixture of both limiting cases. The quasi-free electrons moving in the field of atomic nuclei can now be conceived as plane waves that are – in the language of Bloch – modulated in the rhythm of the lattice. In plane waves the wave fronts, i.e., surfaces of the same phase angle are planes that are extended perpendicularly to the direction of propagation of the wave.[7]

Mathematically, the Bloch equation is a (discrete) Fourier transform of wavefunctions located at a location $\vec{r}$ to a wavefunction that spans the entire crystal. This Fourier transformation transfers the local wavefunction from space into the reciprocal "k space" or into the frequency domain. Just as, for example, a simple sine wave with one particular wavelength can be represented by just one component in a *frequency spectrum* and all its higher harmonics are then just equidistant components in that spectrum, so do the different k values represent Bloch waves in reciprocal space. This also means that the allowed k values are equidistant in the k space.

The phase factor $e^{ik\vec{T}}$ is only dimensionless if k has the unit of a reciprocal length. Thus, in the first instance, k can be considered to be a wave*number*, which specifies how many whole wavelengths fit into the reciprocal lattice constant $1/\vec{a}$. However, the waves have also a certain direction in (reciprocal) space; therefore, k is not only a wavenumber but a wave *vector*. In the simplest case of a one-dimensional crystal with a lattice constant a along the x-direction we get: $\vec{k}=2\pi/\vec{a}$. But note that the phase factor is $e^{ik\vec{T}}$ with $\vec{T}=n\cdot\vec{a}$, which means that identical phases for equivalent points are also obtained if not only one whole wavelength fits in between two reciprocal

---

[7] One may also simply imagine laterally extended water waves on an (infinitely wide) beach.

lattice points, but also two, three, four and so on, i.e., any integer multiple of the wavelength.

Unique values of k are now obtained within the primitive reciprocal unit cell, also called the first Brillouin zone (BZ), after Léon Nicolas Brillouin, 1889–1969, a French physicist. The construction of Brillouin zones is very simple: In the first step, you draw vectors from a reciprocal lattice point to all nearest neighbouring lattice points. Then, in the second step, at the midpoints of these vectors you draw limiting planes (in 3D) or lines (in 2D) that are oriented perpendicular to the vectors. The area or volume, respectively, which is enclosed by these lines or planes, respectively, is the first Brillouin zone (see Figure 8.3). Brillouin zones that are obtained in this way are also parallelepipeds like the unit cells of real space and their volume is identical with the volume of the primitive reciprocal unit cells. The centre of the Brillouin zone is the origin of the reciprocal lattice, denoted by Γ; it corresponds to a k value of 0 or, in three dimensions, to a k value of $k_x = k_y = k_z = 0$. Other special symmetry points, points of high symmetry, of the reciprocal space are denoted with Latin capital letters, which depend on the symmetry, i.e., the space group; two examples are shown in Figure 8.4. Starting from this origin, all possible (or as many as possible) energy levels are now calculated along a reciprocal vector starting from Γ, and the plot of the energy values of these levels as a function of k is called *band structure*. It is easy to show that for the energy values, the relation $E(k) = E(-k)$ holds, and because it is irrelevant in a periodic structure whether one shows the range between k and $2\pi/a$ or between $-\pi/a$ to $+\pi/a$ the energy values are usually plotted as a function of $|k|$, often omitting that absolute values are meant by just writing $E(k)$.

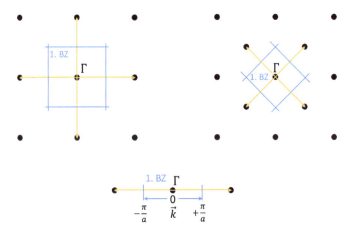

Figure 8.3: Construction of the first BZ for a reciprocal primitive square lattice (top left), a reciprocal square-centred lattice (top right), and a linear 1D lattice (bottom middle). The origin of the BZ is called Γ point.

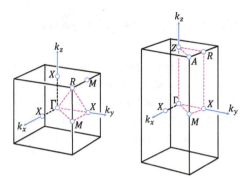

Figure 8.4: The Brillouin zones and high symmetry points, i.e., special k points, for the space group Pm$\bar{3}$m (left) and I4/mmm (right). The volume that is enclosed by the points Γ, M, X, R (left) and Γ, M, X, R, A, Z (right), respectively, are the *reduced* first BZ, the analogue of the asymmetric unit of the real unit cell (redrawn and adapted after [62]).

It is now possible to rewrite the Bloch functions representing the delocalized crystal orbitals in such a way that they can be understood as a weighted (according to the mixing coefficients) sum of atomic orbitals localized at atoms, quite analogously to the MO-LCAO approach:

$$\psi_k(\vec{r}) = \sum_n^N e^{i\vec{k}\vec{r}} \chi_n(\vec{r_n}) \tag{8.1}$$

This means that a crystal orbital can be represented as a sum of atomic orbitals located at a certain site within the unit cell [$\chi_n(\vec{r_n})$], which are differently weighted by the phase factor ($e^{i\vec{k}\vec{r}}$), namely as a function of the quantum number k. What the mixing coefficients are in the molecular MO-LCAO approach now become the Bloch phase factors in the solid state. This allows an orbital-like description of the 1D H atom crystal, to which we now return once again, that chemists too are able to understand. The state with the lowest energy level corresponds to the minimum value of k, namely $k = 0$. In that case, the rewritten Bloch function reads as

$$\psi_0(\vec{r}) = \sum_{n=1}^N e^0 \chi_n(\vec{r_n}) = \sum_{n=1}^N \chi_n(\vec{r_n}) = \chi_0(\vec{r_0}) + \chi_1(\vec{r_1}) + \chi_2(\vec{r_2}) + \chi_3(\vec{r_3})\ldots$$

with N being the total number of atoms of the crystal. In the case of the 1D H atom crystal, the atomic orbitals are the 1s functions, i.e., all 1s orbitals are combined with positive phase and there is no node plane perpendicular to the axis from atomic core to atomic core (compare Figure 8.2, far right, bottom).

The state of highest energy corresponds to the maximum value of k, namely $k = \pi/a$. One obtains:

$$\psi_{\vec{a}}(\vec{r}) = \sum_{n=1}^{N} e^{i\frac{\pi}{a}(na)} \chi_n(\vec{r_n}) = \sum_{n=1}^{N} (-1)^n \chi_n(\vec{r_n}) = \chi_0(\vec{r_0}) - \chi_1(\vec{r_1}) + \chi_2(\vec{r_2}) - \chi_3(\vec{r_3})\ldots$$

This means that a sum is built in which directly adjacent 1s functions have opposite phases (compare Figure 8.2, far right, top). What do the states between the extremes look like? The number of nodes increases by one with increasing k values. This means that k is also a kind of a node counter, which we can introduce as k'. The second lowest state has one node ($k' = 1$), the third lowest has two nodes ($k' = 2$), and so on. Thus, the allowed values for k are $k = n\pi/Na$, with n running from 0 to N. The limiting case in the first Brillouin zone is given for $n = N$, i.e., $k = \pi/a$.

The corresponding band structure for our 1D H atom crystal is schematically shown in Figure 8.5 together with the phase relations of the constituting 1s orbitals for the two lowest energy states, for the state at $k = +\pi/2a$ and for the highest energy level at $k = +\pi/a$.

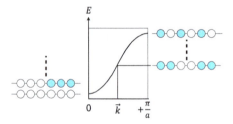

Figure 8.5: Band structure of a 1D H atom crystal together with the phase relations of the constituting 1s orbitals for the two lowest energy states, for the state at $k = +\pi/2a$, and for the highest energy level at $k = +\pi/a$.

Now we should also be able to mentally bridge the gap between real and reciprocal space. The Bloch functions are periodic functions in reciprocal space, which must satisfy the condition that the value of the wavefunction at a specific location of the unit cell must be identical to the value of the wavefunction at an equivalent location in all neighbouring unit cells. If we look at this in reciprocal space and assume that the wavefunction has its maximum amplitude at the gamma point (Γ), then it must of course also have the maximum amplitude at the neighbouring reciprocal lattice point, and $k = 0$, in a sense, indicates the minimum (inverse) "wavelength" that satisfies this condition, namely that exactly one wavelength fits between two reciprocal lattice points. This is visualized in Figure 8.6, bottom, with the wave marked in yellow. The next wavefunction that satisfies Bloch's condition is the one where there is space for two wavelengths; this is shown by the blue wave in Figure 8.6, and so on. If we look at the waves at the selected points of the boundary of non-redundant k values (+π/a or $-\pi/a$), we also see that the waves with even k values (including zero) have their minimum at this location, while those with odd k values just have their maximum there, which corresponds to nothing else but an anti-phase addition of the orbitals forming the highest energy state.

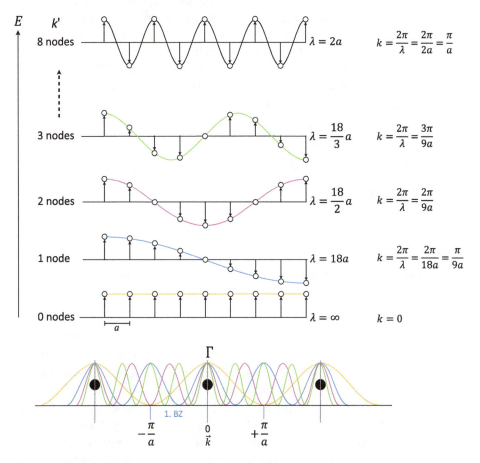

Figure 8.6: The first 4 Bloch waves with lowest energy in reciprocal space (bottom) and the corresponding standing electron waves in real space representation of a "crystal" composed of 9 H atoms with a lattice constant of a (top part); additionally, the wave with the highest energy is plotted.

How can we imagine this now in real space? That the electrons are standing waves, similar to the strings of a guitar, however with the difference that electron waves have open ends, so the situation is more like in acoustic waves in an open air tube. For an – admittedly quite small – linear H-atom crystal of 9 atoms, the current amplitudes at a given time for the four lowest energy vibrational states and additionally for the highest energy state are shown in the upper part of Figure 8.6. The state of lowest energy has no node, the wavelength is infinite, the amplitude is the same at all atoms (in-phase addition of the 1s orbitals!); for the state of highest energy the amplitudes are exactly inverse at neighbouring atoms (anti-phase addition of the 1s orbitals!).

Looking at Figure 8.5, we realize that the energy of the energy levels increases with increasing k values. But this does not have to be the case. It depends on the orbital we are looking at. If we, for instance, combine p_x orbitals instead of the 1s orbitals in the linear 1D H-atom crystal, we get the reverse case, the energy of the energy levels decreases with increasing k values, and the lowest energy state is the one where neighbouring orbitals have inverse phase relations, and the highest state is the one where all orbitals are aligned in the same way (see Figure 8.7). Furthermore, different bands can also overlap, i.e., the lower limit of one band can be at lower energy than the highest level of another band (see also next section).

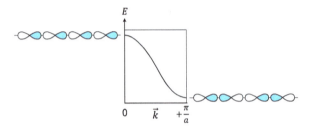

Figure 8.7: Band structure for a linear and equidistant arrangement of p_x orbitals.

8.2 Bandwidth, density of states, and the Fermi level

Another important feature of an energy band is its bandwidth, or dispersion. That is the energy difference between the highest and lowest energy level. The bandwidth is determined by the amount of overlap of orbitals from neighbouring atoms, completely analogous to the molecular case. The stronger the overlap, i.e., the stronger the interaction between neighbouring sites, the larger the splitting of the MOs of a molecule, and the larger the bandwidth in the solid state. This is illustrated in Figure 8.8, in which the bandwidth is plotted for our 1D H atom crystal for two cases, in which the H atoms are separated by 3 Å and 2 Å, respectively. With increasing bandwidth, the delocalization increases, too, while low dispersion is synonymous with weak interactions and increasing electron localization.

The reason why the dispersion is not exactly symmetrically centred around the level of −13.6 eV, i.e., the energy of an isolated H atom, is the same as in the molecular case: we have to take into account the overlap integral leading to a lower energy level that is less stabilized than the higher level is destabilized.

Looking again at the far right of Figure 8.2 we see that the energy levels are not evenly distributed among the band. There are more energy levels at the very low and top end of the band and lesser levels in between. And note that in this scheme, only a tiny number of levels are indicated compared to a real crystal where the number of levels is in the order of one mole. In order to get a more compact information of the

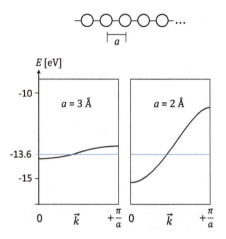

Figure 8.8: Band structure of a 1D H atom crystal in which the H atoms are separated by 3 Å (left) and 2 Å (right), respectively. The energy level of an isolated H atom is at −13.6 eV, marked with a blue line.

number of states that are present in a certain specific range of that band, frequently, the so-called density of states (DOS) is calculated and plotted (compare Figure 8.9):

$$DOS(E)dE = \text{number of levels between } E \text{ and } E + dE$$

Usually, the density of states of a particular band is inversely proportional to the slope of the band structure. The lower the gradient of a band the higher is the DOS. With a plot of the energy versus DOS(E)dE, the directional information of the band structure is eliminated, but a picture emerges that allows chemists to arrive at frontier orbital (i.e., HOMO-LUMO, highest occupied and lowest unoccupied molecular orbital) analogue considerations. The energy levels in the lower part of the band belong to bonding states, in the upper part to antibonding states. And the energy of the highest occupied level (at 0 K) is the so-called Fermi level, E_F. Whenever the Fermi level lies within one single band, the compound is a metallic electrical conductor. Only a minimal amount of energy is then required to move an electron from an occupied orbital below the Fermi level to an unoccupied orbital above it. At ambient conditions, even a fraction of the electrons is always above the Fermi level due to the thermal energy.

In many cases it is interesting to know the degree to which the different types of orbitals of the different elements in a compound contribute to the density of states. The procedure is analogue to the molecular case in which the mixing coefficients of the AOs to the MOs are calculated. After having done such a calculation, the so-called *partial* (or projected) density of states (PDOS) can be plotted. The principle is visualized for a polar diatomic molecule schematically in Figure 8.10.

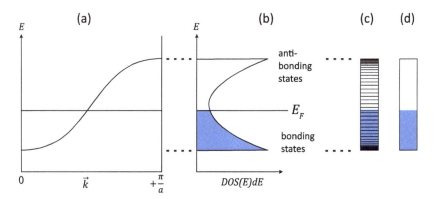

Figure 8.9: A sketch of an energy band structure (a), the respective plot of the density of states (b), and further reduced representations of bands that chemists are usually familiar with (c and d).

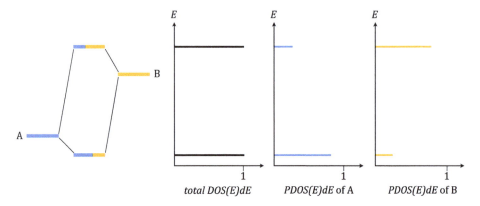

Figure 8.10: Schematic energy levels of the MOs of a polar diatomic molecule based on the individual contribution to these orbitals of atoms A and B and the respective total and partial DOS plots.

8.3 The Peierls distortion

According to our previous considerations of a linear chain of equidistant H atoms and the corresponding band structure, where the band is half-filled, one might think that hydrogen is a metal and shows a correspondingly good electrical conductivity. But this is obviously counterfactual. The atoms should come together to form separated pairs of atoms, i.e., H_2 molecules. And so they do, because the linear chain of equidistant H atoms is unstable and subject to the so-called Peierls distortion, after Rudolf Peierls (a British physicist, 1907–1995). This Peierls distortion can be regarded as the solid-state analogue to the Jahn-Teller distortion in coordination compounds or complexes. The Peierls distortion is accompanied by an energy gain. The starting point of this distortion is – as in the molecular case, too – an energetic degeneration of orbitals that is removed by lowering and elevating of the involved and formerly degenerate orbitals, which leads to an overall lowering of the total energy. Let us consider the linear chain of H atoms once again: this time, there should not be one electron between each H atom, but alternately two electrons in the form of an electron pair and no electron. However, the H atoms should still be arranged equidistantly. An equivalent description is that H_2 molecules interact with each other along a straight line such that the intramolecular distance is exactly the distance to the next molecule along the chain. As a consequence, the lattice constant a has doubled, $a' = 2a$. How does the band structure now look like? First, we have to consider that the H_2 molecule has two orbitals, which now interact with neighbouring orbitals along the chain, namely the σ_g and the σ_u^* orbital; thus, there will be two bands. The number of energy bands equals the number of orbitals in the unit cell. Since the two orbitals mentioned have the same orientation and the same symmetry properties as the 1s and a p_x orbital, respectively, the course of the energy band as a function of k should be clear. The energy of the band generated from the σ_g orbitals increases with increasing k values until there is a maximum number of nodes perpendicular to the molecular axis at the point $k = \pi/a'$, while the band generated from the σ_u^* orbitals starts with a maximum number of nodes at the Γ point and, thus, the highest antibonding state and then decreases with increasing k values until at the point $k = \pi/a'$ the number of nodes is identical to that of the increasing band generated from the σ_g orbitals! These two states are energetically degenerate (see Figure 8.11).

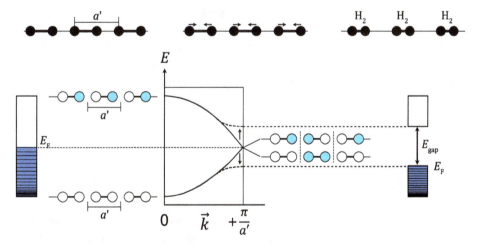

Figure 8.11: Band structure of a hypothetical chain of equidistant H atoms between which there is alternately a pair of electrons and no electron; this configuration is unstable and will undergo a Peierls distortion – the H atoms move towards each other to form pairs of H_2 molecules (shown in the top part of the figure), which leads to a removal of the energy degeneracy at the point $k = \pi/a'$.

Now, this degeneracy will be removed by a pairwise movement of the H atoms towards each other. This has no consequences for the gamma point ($k = 0$), because the energy gain at the lower end of the band for the atoms that move towards each other is compensated by the energy loss of the atoms that move away from each other. For the upper point of the band at the gamma point it is exactly the other way round. The situation is quite different for the right border of the band at $k = \pi/a'$. There are drastic consequences, because the branch coming from below is energetically lowered and the branch coming from above is raised accordingly. This means that the band splits into two sub-bands with bonding and antibonding states, respectively, separated by an energy gap (E_{gap}). This leads to an energy gain for the half-occupied band, the Fermi level is lowered, and the occupied states within the lower (sub)band move closer together. Depending on how large the energy gap is, there is a transition from a metallic behaviour to a semiconductor-like behaviour ($0.1 \leq E_{gap} \leq 4$ eV) or an insulator is formed ($E_{gap} > 4$ eV).

There is still one question open, namely, why, at first glance, the band structures of a 1D H atom crystal composed of N H atoms (compare Figure 8.5) and that of a 1D arrangement of $N/2$ H_2 molecules (compare Figure 8.11) look different. On closer inspection, however, it turns out that they do not differ at all. This is merely the result of the (artificial) doubling of the lattice constant from a to $a' = 2a$. That means that the band structure in Figure 8.11 is plotted only up to the point $k = \pi/2a$, corresponding to the centre of the plot in Figure 8.5. From that point on, the band is running back (from the right to the left) and increases again up to the gamma point at $k = 0$, a point that is quasi-identical with the point $k = 2\pi/2a = \pi/a$. It is also possible to obtain the original band structure quasi-optically geometrically by *unfolding* the band structure of the ensemble of H_2 molecules in the manner shown in Figure 8.12. This unfolding corresponds to a transition from a dimer to the monomers, and it is a useful mental exercise, especially for molecular chemists who are more used to MO diagrams than to band structures. The reverse process, the folding, which can also take place several times, is then the transition from monomer to oligomers and polymers. The process and the result of a threefold folding and the result of a fourfold folding process including the representation of the phase relations of the orbitals at selected $E(k)$ points are shown in Figure 8.13.

This linear chain of equidistant H atoms is, of course, only a model. It should give the reader an understanding how energy bands are formed. We know, of course, that in reality, the lowering and raising of the sub-bands continues to the point where isolated H_2 molecules are generated that are completely independent of each other. Therefore, the width of the bands decreases to zero and their dispersion disappears; what remains are the two localized bonding and antibonding MOs of the H_2 molecule as shown in Figure 8.1.

The Peierls distortion, or this type of metal-to-semiconductor or metal-to-insulator transition, is not the only way to stabilize a system. It is also not a question of the band structure alone. This distortion is mainly expected for steeply raising bands (i.e., those

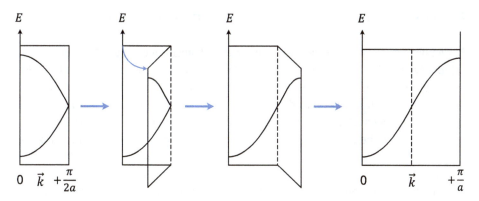

Figure 8.12: Unfolding of the band structure of a linear H_2 molecule polymer with a lattice constant of $a' = 2a$ into the band structure of the 1D H atom arrangement with the lattice constant a.

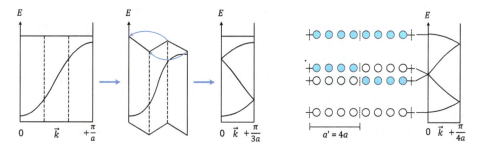

Figure 8.13: Threefold folding of a band structure of a linear 1D H atom arrangement into a band structure of a linear chain of H_3 trimers with a tripled lattice constant of $a' = 3a$ (left) and the fourfold folding of that band structure into a linear chain of H_4 tetramers together with the phase relationships of the constituting 1s functions at selected $E(k)$ points.

with a large dispersion), but its occurrence also depends on the filling degree of the band. As already shown for the chain of H atoms, the distortion has only slight consequences for the band at the gamma point. This means: if the band is filled only to a very low degree, it is irrelevant how steeply the band runs towards the edge of the reciprocal unit cell. Furthermore, the different spatial directions in a three-dimensional extended solid compound must be taken into account. While a distortion in one direction can have a stabilizing effect, it can have a destabilizing effect in another direction and can therefore be absent. Furthermore, it must be taken into account that the stabilizing effect of a Peierls distortion is small for heavy elements (from the fifth period onwards) and can easily be overcompensated by other influences. Therefore, undistorted chains and networks are observed especially in solid-state compounds of heavy elements.

8.4 Band structures in two- and three-dimensional systems

If we now extend our considerations and make a transition from our 1D H atom crystal to a system that is periodic in two or three directions, the changes are only subtle, because it can be shown that it is possible to separate the Schrödinger equation and hence the Bloch functions into their respective parts along the two or three spatial directions, respectively. All that changes is that we now need two or three k values, one for each of the reciprocal lattice vectors (or vectors of the first Brillouin zone, respectively). For the two-dimensional case we need a k number doublet, $k = (k_x, k_y)$, and a three-periodic system is characterized by a triplet of k numbers, $k = (k_x, k_y, k_z)$.

Let us consider now the simplest case for a two-periodic system, a square lattice, and let us place an s orbital on each lattice point. The resulting phase relations for selected points of the first Brillouin zone are shown in the upper row of Figure 8.14. At the gamma point ($k_x = k_y = 0$) a configuration with a maximal bonding combination is obtained. At the two equivalent points X ($k_x = \pi/a$, $k_y = 0$) and X' ($k_x = 0$, $k_y = \pi/a$), there is a bonding interaction in only one direction and an antibonding interaction in the direction perpendicular to it. Altogether, this should result in a non-bonding state. And, finally, at the point M ($k_x = \pi/a$, $k_y = \pi/a$) the maximum number of nodes is present; in both directions there are only antibonding interactions. However, there is a weak bonding interaction in the diagonal direction, i.e., the direction of the vector sum of k_x and k_y.

Instead of the s orbitals, we can now also place p orbitals at each of the reciprocal lattice points. For the p_z orbitals, the picture is qualitatively similar to the one of the s orbitals, except that the p orbitals are higher in energy terms, and that their interactions are not as strong, because only π-like interactions are prevalent. Therefore, the minima and maxima in the course of the band structure are not quite as pronounced as in the case of the s orbitals.

A different picture arises for the p_x and p_y orbitals (see Figure 8.14, middle and bottom row). Here we have to consider both σ- and π-like interactions. At the gamma point, the p_x orbitals have a σ-antibonding but π-bonding character. At the point X, the p_x and p_y orbitals differ particularly strongly, one has σ- as well as π-bonding character, the other σ- and π-antibonding character.

The course of the band for the s and p orbitals for the directions $\Gamma \rightarrow X$, $X \rightarrow M$, and $M \rightarrow \Gamma$ are shown schematically in Figure 8.15.

8.5 Examples of real band structures

With regard to the electrical properties of a solid, there are essentially three possibilities:
- The compound is an insulator if the valence band is completely filled and the lowest energy level of the (empty) conduction band is energetically clearly separated

8.5 Examples of real band structures — 179

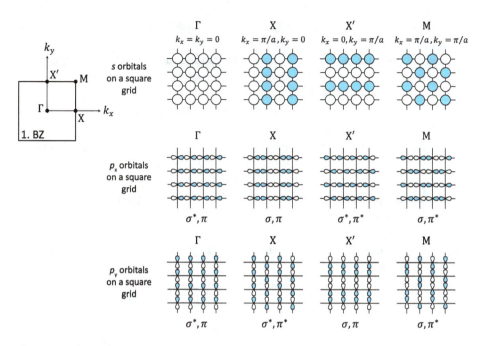

Figure 8.14: Phase relations at selected points of the first Brillouin zone of s (top row) and p_x as well p_y orbitals (middle and bottom row) placed on a square grid.

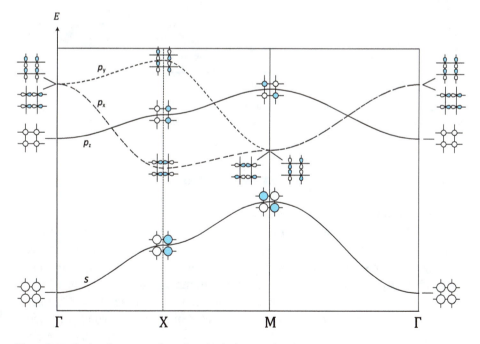

Figure 8.15: The band structure of s and p orbitals that are placed on a square lattice.

from it, i.e., if there is an energy gap between these bands that is approximately larger than 4 eV.
- A semiconductor is present if this energy gap lies in the range between approx. 0.1 and 4 eV. In the case of semiconductors, an additional distinction is made as to whether the energy gap can be overcome by the supply of thermal energy alone (up to approx. 1.5 eV) or whether this can be achieved only by the absorption of photons of corresponding energy (approx. 1.5–4 eV).
- Finally, a compound exhibits metallic properties and is a good electrical conductor if the valence band is either only partially filled, i.e., if there are accessible, empty energy levels within the valence band, or if the valence and conduction band overlap, i.e., if there is again a continuous, uninterrupted energy region that includes the Fermi level.

These three scenarios are shown again schematically in Figure 8.16. In the following, we will look at two concrete examples of how the band structures of real compounds look like and how their electrical properties can be derived.

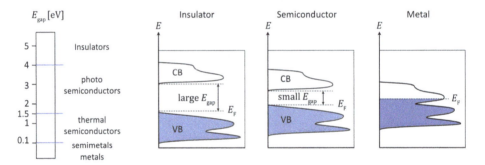

Figure 8.16: Classification of metals, semiconductors, and insulators based on the value of the energy gap between the valence and conduction bands (left) and the corresponding schematic band structures (right).

8.5.1 ReO$_3$ – a d^1 compound with metallic properties

Rhenium(VI)trioxide (ReO$_3$) crystallizes in the cubic space group $Pm\bar{3}m$ (compare Figure 5.37). The structure consists of corner-sharing ReO$_6$ octahedra arranged in a primitive cubic array. ReO$_3$ is a crimson-to-purple solid with a metallic lustre. It has also a very high specific electric conductivity (approx. 2×10^7 S/m at room temperature) and the conductivity increases with decreasing temperatures, also typical for metallic behaviour. In the oxidation state +VI Re has the electron configuration [Xe] $4f^{14}5d^1$. This one 5d electron is mainly responsible for its electric properties. Looking at the band structure and the DOS plot (Figure 8.17), they reveal that the valence band (mainly generated by the 2p levels of oxygen) and the conduction band (dominated by the Re

5d levels) severely overlap and that there is no energy gap between the Fermi level and higher energy states – characteristic of metals. The band width/dispersion of the Re 5d levels is not only a direct result of the Re(d)-Re(d) interactions but also partly caused by some Re(d)-O(p) mixing. The local octahedral symmetry (O_h) of the Re coordination environment leads to a splitting of its 5d levels into three (degenerated) t_{2g} orbitals and two (degenerated) e_g orbitals. This is also reflected in the band structure at the origin of the first BZ, at the Γ point, where the two sets of the analogous band states are degenerated but split into 5 bands at points of lower symmetry, best recognizable at a point midway between X and M. We learned in the previous sections that the number of bands is equal to the number orbitals per primitive unit cell. Neglecting the core orbitals of ReO$_3$, we have three oxygen atoms with three 2p orbitals each and one Re atom with five 5d orbitals, giving 14 bands altogether.

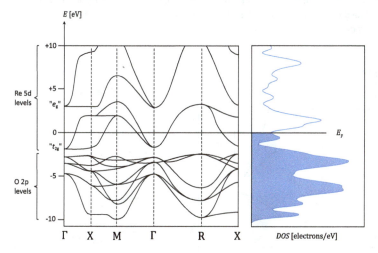

Figure 8.17: The relevant part of the band structure and DOS function of ReO$_3$ (the core levels are not shown). The notations "t_{2g}" and "e_g" are related to the analogous orbital levels of an isolated octahedral ReO$_6$ fragment. The DOS plot was smoothed by 0.3 eV.

8.5.2 MoS$_2$ – a d^2 semiconductor with an indirect band gap

In the series Zr(IV)S$_2$, Nb(IV)S$_2$, and Mo(IV)S$_2$, the first of these compounds, ZrS$_2$ with a [Kr] $4d^0$ electron configuration, is, in accordance with expectations, an isolator, and the second one, NbS$_2$, with a [Kr] $4d^1$ electron configuration, a metallic compound. At first glance, one might think, that MoS$_2$, with a [Kr] $4d^2$ electron configuration, must be a metallic compound, too. However, the band structure reveals that it is a semiconductor. In the crystal structure of MoS$_2$ (compare Section 5.5.2) the molybdenum cations are surrounded by six sulphide anions in a trigonal prismatic fashion. The respective ligand field splitting according to the crystal field theory of the five d orbitals of molybdenum(IV) in a trigonal prismatic coordination environment leads to an energetic low-lying d_{z^2} orbital (a'), filled with two electrons, and two sets of unoccupied double degenerated higher-lying states e' and e" ($d_{x^2-y^2}$, d_{xy} and d_{xz}, d_{yz}), see Figure 8.18, left. This splitting pattern is reflected in the band structure (Figure 8.19) at the Γ point of the hexagonal first BZ (Figure 8.18, right).

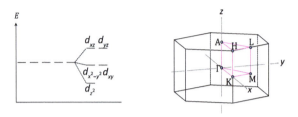

Figure 8.18: Ligand field splitting of the five d orbitals in a trigonal prismatic coordination environment according to the crystal field theory (left), and the first BZ of a hexagonal lattice and selected high symmetry points in the reduced BZ (right).

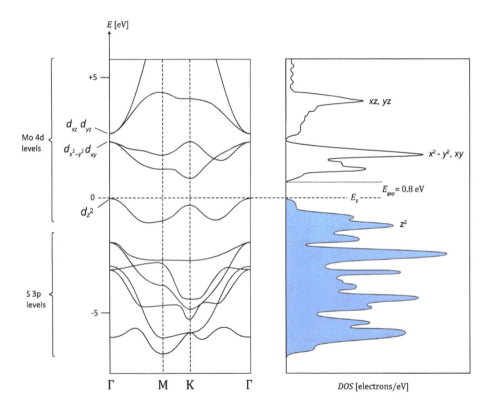

Figure 8.19: Band structure along a path of selected high-symmetry points in k space of MoS_2 (4H polytype) and the corresponding DOS plot.

Analysing the band structure and DOS plot of MoS$_2$, we recognize that the band that is mainly dominated by the z^2 contribution (in fact, some minor mixing with the six completely filled 3p bands of the sulphur is present) is well separated from higher-lying unoccupied bands (mainly dominated by the remaining d bands of Mo, which are partially mixed with some Mo s and p states). There is a small energy gap between the Fermi level and these higher states by about 0.8 eV, a typical value for a (thermal) (and indirect, see next section) semiconductor.

8.6 Direct and indirect band gaps

While the value of the energy gap as inferred from the DOS plot gives only an absolute value of the energy difference between the highest occupied state (the Fermi level) and the lowest unoccupied energy states, it ignores if the corresponding energy levels of the underlying bands are associated with the same or different k values. If the k values are (almost) the same, the energy gap is called a *direct* gap, while if the k values considerably differ, it is called an *indirect* band gap (see Figure 8.20 for a schematic sketch).

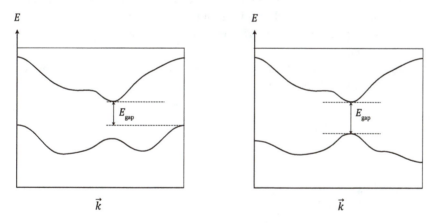

Figure 8.20: Schematic representation of the band structure of a semiconductor with indirect (left) and direct band gap (right).

The circumstance if the k values of the highest energy value of the valence band and the lowest energy value of the conduction band are different or identical is most important for processes that involve photo excitation. In order to understand this, we must look at another property of the wave vector $\vec{k}$. According to Einstein, each photon with a wavelength λ can be assigned an energy value ($E = h \cdot c / \lambda$, where h = Planck constant and c = speed of light) and a momentum ($p = h/\lambda$). De Broglie generalized this concept and did not consider only photons as waves but assigned each material particle with momentum p a wavelength according to the relation $\lambda = h/p$. In this way, the wave vector $\vec{k}$ is connected with a certain momentum, too, since $\vec{k} = 2\pi/\lambda$, where λ is the wavelength of an electronic plane wave in a crystal that is described by a Bloch function.

Now, a photon with an energy of approx. E_{gap} can force a transition of an electron from an occupied state at or near the Fermi level to the lowest unoccupied state very easily, if the associated k values of the two involved energy states match (the momentum transfer from the photon to the electron is negligible). However, in the case of an indirect band gap, the photon does not only have to have the energy E_{gap}, but the electron must also gain or lose momentum, depending on whether the associated k value of the excited state is connected with a higher or lower momentum. But how can the electron increase or decrease its momentum, if the transmitted momentum by the photon is negligible? By interacting with the lattice vibrations, the *phonons*. This electron-phonon coupling is possible, but since the momentum transfer must correspond to the different k values and involves the interaction between three entities (electron, photon, phonon), the process is much less likely than a transition in a semiconductor with a direct band gap. This is analogous to chemical reactions, where, in a particular reaction step, a reaction between two molecules will proceed at a much larger rate than a process involving three molecules. The same principle applies for the relaxation process, i.e., the transition back from the excited state to the ground state. This relaxation which is accompanied by the emission of a photon is much more efficient for a direct band gap semiconductor than for an indirect band gap semiconductor, where the process must be mediated by a phonon.

As a result of these considerations, devices that use photon-mediated electron transitions, for instance, light-emitting diodes (LEDs, see Section 10.3) are mostly based on semiconductors with a direct band gap.

Further reading

N. Ashcroft, N.D. Mermin, *Solid State Physics*, Saunders College Publishing, New York, **1976**.

R. Hoffmann, *Solids and Surfaces: A Chemist's View of Bonding in Extended Structures*, Wiley-VCH, Weinheim, **1989**.

R. Dronskowski, *Computational Chemistry of Solid State Materials: A Guide for Materials Scientists, Chemists, Physicists and Others*, Wiley-VCH, Weinheim, **2005**.

S. H. Simon, *The Oxford Solid State Basics*, Oxford University Press, New York, **2013**.

J.J. Quinn, K.-S. Yi, *Solid State Physics – Principles and Modern Applications*, 2nd ed. Chapter 3–7, Springer, Cham, **2018**.

9 Magnetic properties of solid-state compounds

When we talk about a magnet, usually we refer to a solid-state item that is a permanent magnet in which a particular type of magnetism – ferromagnetism – predominates. This can then be used, for example, as a compass needle or in a bicycle dynamo or as a component of the current generator of wind turbines. However, ferromagnetism is only one of many types of magnetism. All solids have diamagnetic properties to begin with. Only certain electronic properties add further types of magnetism, which then outweigh the diamagnetism and make up the overall magnetic properties of the material. No matter what kind of magnetism is present, it is always a many-particle property.

In this chapter, we will look at the cause of magnetism, the different types of magnetism, and what kinds of temperature dependencies occur.

9.1 Diamagnetism and paramagnetism

Of fundamental importance for the magnetic properties of a material is the electron configuration of the atoms of which it is composed of. Compounds containing only paired electrons, such as most organic compounds or typical inorganic salts like NaCl or CaF_2, exhibit *diamagnetic* behaviour. In contrast, the presence of unpaired electrons, which occur mainly in compounds of transition and rare-earth metals, causes *paramagnetic* behaviour (*Curie paramagnetism*). Metals that contain non-localized electrons and that show a high mobility possess a (nearly) temperature-independent *Pauli paramagnetism*. And finally, the temperature-independent *Van Vleck paramagnetism* can be caused by bonded atoms, whose charge distribution (in connection with excited states) deviates from the spherical symmetry. Thus, the basic distinction on the first level of magnetism type is to distinguish diamagnetism from paramagnetism. Independent of the type of paramagnetism, on the second level, one distinguishes certain collectively ordered states (like between ferro- or antiferromagnetism) of the spin states of their unpaired electrons, which are caused by interatomic magnetic interactions (see Section 9.2).

Whether a material is diamagnetic or paramagnetic can only be determined by examining how the materials behave in an externally applied magnetic field. An external magnetic field with field strength H induces a magnetisation M (size of the magnetic dipole per unit volume) in the material, which is proportional to the material-specific volume-related magnetic susceptibility χ_v:

$$\vec{M} = \chi_v \vec{H}$$

Note that $\vec{M}$ and $\vec{H}$ are usually defined as vectors, because for some non-isotropic solids, their direction does not have to be necessarily co-parallel or anti-parallel. The SI unit of $\vec{H}$ and $\vec{M}$ is amperes per meter $[A \cdot m^{-1}]$; χ_v is therefore a dimensionless quantity.

For diamagnetic materials, χ_v is small (in the order of ~10^{-6} to 10^{-4}) and negative. This means that the external field induces a magnetic field inside the compound, the direction of which is opposite to the external field, or in other words, the external field is weakened inside the body, which can also be expressed as a decrease in the density of the magnetic field lines – the magnetic flux density is decreasing (see Figure 9.1, left). The magnetic susceptibility of paramagnetic materials is also relatively small, but positive, i.e., the magnetisation inside the material is in the same direction with respect to the external field; so the magnetic flux density inside paramagnetic materials is increased (see Figure 9.1, right). A comparison of the order of magnitude and the temperature dependence of the magnetic susceptibility is given in Table 9.1.

Table 9.1: Typical magnitudes and temperature dependence of susceptibilities of magnetic materials.

Type of magnetism	Magnetic susceptibility χ_v	Temperature dependence
Diamagnetism	-10^{-6} to -10^{-4}	None
Pauli paramagnetism	10^{-5} to 10^{-3}	Almost none
Curie paramagnetism	10^{-6} to 10^{-1}	Decreasing with increasing temperature
Ferromagnetism	10^{-2} to 10^{6}	Strongly decreasing up to T_C, further but slower decreasing above T_C
Antiferromagnetism	10^{-2} to 0	Increasing up to T_N, decreasing again above T_N

T_C, Curie temperature; T_N, Néel temperature.

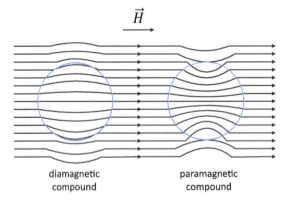

Figure 9.1: An external magnetic field $\vec{H}$ induces a magnetic field inside a material that is antiparallel with respect to the direction of the external field for diamagnetic materials and parallel for paramagnetic materials causing a decrease of the magnetic flux density inside the diamagnetic body (left) and an increase of the flux density inside a paramagnetic material (right).

The next thing we have to consider is the cause of the magnetisation inside a material. In principle, we need a kind of an electrical ring current, a flow of charge carriers inside the material that in turn generates a magnetic dipole (see Figure 9.2). This current can originate either from moving electrons along their orbits (μ_m) or from the eigen-rotation of the electrons, i.e., the electron spins (μ_{ms}). However, concerning the motion of the electrons along their orbits, the superposition of all magnetic dipole moments leads to a zero net magnetisation – what remains are the electron spins. In (inorganic) compounds consisting exclusively of atoms/ions with closed electron shells and in (organic) covalent compounds that possess only paired electrons, the spins equally do not lead to a net magnetic dipole moment.

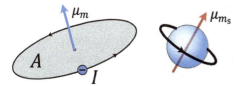

Figure 9.2: A ring current I generates a magnetic dipole moment μ_m normal to the orbit plane of the current that is proportional to the area A of that orbit plane (left). The eigen-rotation, i.e., spin of electrons also generates a magnetic moment (right).

When a diamagnetic substance is introduced into an external magnetic field, then this magnetic field disturbs the orbital motions of the electrons, which are now no longer perfectly spherically symmetrical. As a consequence, a weak magnetic field is generated by these slightly asymmetric orbits of the electrons, which is opposite to the direction of the external field. Interestingly, this diamagnetic effect is independent of the strength of the external field and almost independent of the temperature. A diamagnetic contribution to the total magnetic susceptibility occurs in all compounds, even in paramagnetic substances, due to the contributions of inner electron shells.

Paramagnetism is exhibited by substances whose total electronic spin is not equal to zero. The small magnetic dipole moments of the electron spins are initially completely randomly oriented in a solid if there are no interatomic exchange phenomena taking place, i.e., if these dipole moments behave independent of each other. The case is different if the dipoles interact with each other, giving rise to various collective magnetic ordering states. Therefore, no measurable net magnetisation results in the absence of an external magnetic field. By applying an external magnetic field, an energetically more favourable situation arises when the individual magnetic moments arrange themselves parallel to the external field. However, this ordering of the orientation of the spins is counteracted by the thermal energy. This means that the higher the temperature, the less complete is the ordering. This results in a strongly temperature-dependent course of the magnetic susceptibility of paramagnetic substances, which is described by Curie's law:

$$\chi_v = \frac{C}{T}$$

with C being the Curie constant, which has a specific value for each compound and contains the magnetic dipole moments of the individual particles the compound is composed of.

The Curie law is valid only for non-interacting magnetic moments. In the case of interacting magnetic moments – this is the case for ferro-, antiferro-, and ferrimagnetic materials – deviations of the Curie law occurs and the temperature-dependent behaviour of the magnetic susceptibility of such materials *in their paramagnetic state* must be described by the law of Curie and Weiss that introduces a Weiss constant θ:

$$\chi_v = \frac{C}{T - \theta}$$

According to the Curie-Weiss law, the tendency of a material to align the magnetic moments of the spins in a parallel or antiparallel fashion, below a certain temperature, as a result of their mutual interactions, is accounted for by a positive (for ferromagnetic materials in their paramagnetic state) or negative (for antiferromagnetic materials in their paramagnetic state) Weiss constant θ, respectively. Plotting the reciprocal susceptibility against temperature yields a straight line, which passes through the origin in the Curie case and intersects the temperature axis at θ in the Curie-Weiss case (see Figure 9.3). The Curie-Weiss law is strictly valid only at not too low temperatures; further details are considered in Section 9.2.

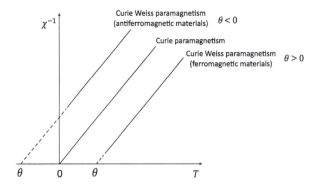

Figure 9.3: Plot of the inverse of the magnetic susceptibility as a function of the temperature for paramagnetic (Curie law), ferromagnetic ($\theta > 0$), and antiferromagnetic ($\theta < 0$) compounds in their paramagnetic state.

9.1.1 Quantifying the magnetic moments of paramagnetic substances

In the previous section, we looked at the basic phenomena of diamagnetism and paramagnetism and how such respective substances behave in a magnetic field, but we have not quantified any of the effects. This should be done now for the paramagnetic effect, which occurs only with substances that have unpaired electrons. We start first

9.1 Diamagnetism and paramagnetism

with isolated atoms or ions. The theoretical relationship between unpaired electrons and the resulting magnetic moments is provided by the equations of electrodynamics and quantum mechanics. According to these equations, magnetic moments are induced by charges moving in a circle. In the case of electrons, the orbit and the spin are responsible for this. Note that the contribution of the orbit is only important for those atoms/ions that have unpaired electrons because in this case, the spin and the orbit give rise to coupling effects, which in turn are dependent on the concrete spin state (see below). The angular momentum of an electron gives rise to a magnetic moment that is dependent on the azimuthal quantum number l (l = 0, 1, 2, ...), according to the equation:

$$\vec{\mu_l} = \mu_B \cdot [l(l+1)]^{1/2}$$

with μ_B being the fundamental unit of the magnetic moment of an electron called the Bohr magneton. This means that the electrons of the p, d, and f orbitals do generate a magnetic moment, but the electrons of the s shell do not. The Bohr magneton is derived from the following equation:

$$\mu_B = \frac{eh}{4\pi m_e} = 9.27 \cdot 10^{-4} \, \text{A m}^2$$

where h is the Planck constant, e the charge, and m_e the mass of an electron.

The magnetic moment of the spin contribution of a single electron is given by a similar equation that is dependent on the spin quantum number:

$$\vec{\mu_s} = g \cdot \mu_B \cdot [s(s+1)]^{1/2}$$

where g is the so-called g factor of the free electron. It is dimensionless and has a value of 2.0023193043617(15).

The individual spin and orbital moments now couple with each other to form a total magnetic moment. This coupling is described for 3d and 4f metals in a good approximation by the Russel-Saunders coupling (also called LS coupling). It assumes that the individual orbital moments $\vec{l}$ couple to a total orbital moment $\vec{L} = \sum \vec{l}$, and that the individual spin moments $\vec{s}$ couple to the total spin moment $\vec{S} = \sum \vec{s}$. In a secondary interaction, the total orbital moment and the total spin moment then couple to form the total magnetic moment $\vec{J}$, dependent on the total angular momentum quantum number J. The most stable electron configuration for an atom or ion in its ground state is $\vec{J} = \vec{L} - \vec{S}$ if the set of orbitals of one type (for instance the five 3d orbitals or the seven 4f orbitals) is less than or exactly half-occupied and $\vec{J} = \vec{L} + \vec{S}$, if the subshell is more than half-occupied. The most stable (ground) state of an atom or ion is then given by its respective term symbol in the form, $^{2S+1}L_J$, where for the total angular momentum L a special notation with capital letters is used; the first four letters for L = 0, 1, 2, and 3 are S, P, D, and F, respectively. For higher values of L the alphabetical sequence is followed from F onwards, except that J is omitted.

For a free atom or ion, we obtain a total magnetic moment that can be calculated by the following formula:

$$\mu_J = g \cdot [J(J+1)]^{1/2} \text{ with } g = 1 + \frac{J(J+1) + S(S+1) - L(L+1)}{2J(J+1)}$$

where g is the so-called Landé g factor. This g factor takes into account the spin-orbit coupling. For a spin-only magnetism, it would correspond to the g factor of the free electron, given above, with a value of approx. 2.

If we compare the values of the total magnetic dipole moment of the trivalent ions of the lanthanoid (4f) metals, calculated by the equation for μ_J with the experimental values (Table 9.2), we see a good agreement with the two exceptions of Sm^{3+} and Eu^{3+}. For these ions, it is necessary to consider not only the ground state, but the higher energy states of the LS multiplet as well, because the energy differences between the adjacent energy states at room temperature are not large compared to the thermal energy (kT).

Table 9.2: Electron configuration, term symbol, and calculated and experimentally observed magnetic moments of trivalent ions of lanthanoid metals.

Ion	Electron configuration	Ground term	μ_J (μ_B)	μ_{exp} (μ_B)
Ce^{3+}	$4f^1$	$^2F_{5/2}$	2.54	2.4
Pr^{3+}	$4f^2$	$^3H_4$	3.58	3.5
Nd^{3+}	$4f^3$	$^4F_{9/2}$	3.62	3.5
Pm^{3+}	$4f^4$	$^5I_4$	2.68	n.a.
Sm^{3+}	$4f^5$	$^6H_{5/2}$	0.84	1.5
Eu^{3+}	$4f^6$	$^7F_0$	0	3.4
Gd^{3+}	$4f^7$	$^8S_{7/2}$	7.94	8.0
Tb^{3+}	$4f^8$	$^7F_6$	9.72	9.5
Dy^{3+}	$4f^9$	$^6H_{15/2}$	10.63	10.6
Ho^{3+}	$4f^{10}$	$^5I_8$	10.60	10.4
Er^{3+}	$4f^{11}$	$^4I_{15/2}$	9.59	9.5
Tm^{3+}	$4f^{12}$	$^3H_6$	7.57	7.3
Yb^{3+}	$4f^{13}$	$^2F_{7/2}$	4.54	4.5

At first glance, it might be surprising that the calculated values are in such good agreement, because the ions are not "free", but rather influenced by the surrounding ligand field of the coordinated anions. However, the reason should be clear when you realize that the 4f shell is well shielded from any interaction with the surrounding atoms by the filled 5s and 5p orbitals.

In contrast, the ligand field of the d transition metal ions cannot be neglected. However, they usually have the effect that the orbital moments for these ions are almost completely (configurations: d^1-d^5) or at least partially (configurations: d^6-d^9)

suppressed. Therefore, the magnetic moments of the d transition metals, in particular for the $3d$ row, can be approximately calculated by the so-called *spin-only formula*:

$$\mu_S = g \cdot \mu_B \cdot [S(S+1)]^{1/2}$$

Note that for ions with an electron configuration of d^4 to d^7, there are two spin states possible (high-spin and low-spin), depending on the ligand field strength. In Table 9.3, the calculated and measured magnetic moments of some transition metal ions are compared.

Table 9.3: Electron configuration, term symbol, and the calculated and experimentally observed magnetic moments of some $3d$ transition metal ions in high-spin configuration.

Ion	Electron configuration	Ground term	μ_S (μ_B)	μ_{exp} (μ_B)
Sc^{3+}, Ti^{4+}	$3d^0$	$^1S_0$	0	0
Ti^{3+}, V^{4+}	$3d^1$	$^2D_{3/2}$	1.73	1.7–1.8
Ti^{2+}, V^{3+}	$3d^2$	$^3F_2$	2.83	2.8–2.9
Mn^{4+}, Cr^{3+}, V^{2+}	$3d^3$	$^4F_{3/2}$	3.87	3.7–4.0
Mn^{3+}, Cr^{2+}	$3d^4$	$^5D_0$	4.90	4.8–5.0
Fe^{3+}, Mn^{2+}	$3d^5$	$^6S_{5/2}$	5.92	5.7–6.1
Co^{3+}, Fe^{2+}	$3d^6$	$^5D_4$	4.90	5.1–5.7
Co^{2+}	$3d^7$	$^4F_{9/2}$	3.87	4.3–5.2
Ni^{2+}	$3d^8$	$^3F_4$	2.83	2.8–3.5
Cu^{2+}	$3d^9$	$^2D_{5/2}$	1.73	1.7–2.2
Zn^{2+}, Cu^+	$3d^{10}$	$^1S_0$	0	0

9.1.2 Pauli paramagnetism

In previous considerations, we were either dealing with systems of closed electron shells, i.e., systems without unpaired electrons (diamagnetism), or with rather isolated metal ions, which are present, for instance, in metal complexes, and lead to paramagnetism if the metal ions have unpaired electrons. Now, we consider solid state bodies where the pure local consideration of the electron and spin configuration of isolated particles is no longer sufficient, and we have to take the band structure into account in order to be able to understand the magnetic state of the metals.

Metals contain delocalized electrons that can move "freely" through the solid; these are also called itinerant electrons. This is where the term electron gas comes from. However, a comparison with an atomic or molecular gas is not quite accurate. One reason for this is the validity of Pauli's principle, according to which a single orbital can accommodate a maximum of two electrons with opposite spin. This means that most electrons occupy energy states far below the Fermi energy. And these electrons contribute neither to the electrical conductivity nor do they give rise to the occurrence of magnetic moments. Only a very small fraction of electrons that are thermally excitable (proportional to kT) are then above the Fermi level, E_F, and do so (see Figure 9.4, left). Because the energy of the Fermi level is extremely high compared to the provided thermal energy in usual conditions (the Fermi temperature, which is defined as $T_F = E_F/k_B$, amounts to approx. 6.5×10^4(!) K for the free-electron gas model of metals), the number of unpaired electrons that give rise to paramagnetic behaviour – if not special cooperative magnetic phenomena take place (see below) – is almost independent of the temperature, and the magnetic susceptibility, χ, is very small. This kind of paramagnetism is called Pauli paramagnetism.

However, the Pauli paramagnetism is field-dependent. The conduction band of metals can be subdivided into two sub-bands, one with electrons filled that have a spin-up state and another completely identical sub-band with electrons with spin-down configuration. The corresponding graph of the density of states is symmetrical for both spin configurations (see Figure 9.4, without magnet field). In the absence of cooperative magnetic interactions, there are equal numbers of electrons in the two sub-bands and the metal has, at least at $T = 0$ K, no overall magnetic moment. This situation changes if a magnetic field is applied. Let us assume that the direction of the field is along the spin-up direction of the electrons. Then, the sub-band with the electrons with spin-down configuration, antiparallel to the field direction, raises in energy. Subsequently, spin-down electrons close to E_F, which have slightly higher energy than the spin-up electrons close to E_F, spill over between the sub-bands to equalize E_F. The electrons that spill over change their spin orientation, giving rise to a net small spin-up magnetic moment (see Figure 9.4, with magnet field).

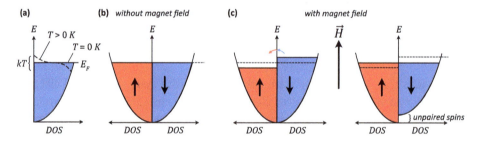

Figure 9.4: The conduction band of metals (a) can be subdivided into two sub-bands filled with electrons with either spin-up or spin-down configuration, which are energy-equivalent in the absence of a magnetic field (b); in the presence of a magnetic field parallel to the spin-up direction, the spin-down sub-band raises in energy, initiating a spill-over of spin-down electrons to the sub-band with spin-up configuration (c) leading to a net magnetic moment generated by the unpaired spins.

Actually, not only metals, but also some metal compounds show Pauli paramagnetism – the divalent transition metal oxides TiO and VO are examples for that. They are also metallic conductors. However, the compounds MnO, FeO, CoO, and NiO, all of which are semiconductors, are not Pauli paramagnetic but show an antiferromagnetic behaviour below the Néel temperature. An explanation for this is given in Section 9.2.2.

Now, one might think that Pauli paramagnetism should prevail in all metallic conducting compounds, especially in all metals. However, this is not the case. Several transition elements and most of the lanthanoids are either ferro- or antiferromagnetic. The 3d transition metals, Cr and Mn, are antiferromagnetic, Fe, Co, and Ni are ferromagnetic. Several of the lanthanoids are helimagnetic, namely Dy, Tb, Ho, and Er. The explanation for this can only be based on band structure considerations. Decisive parameters are the bandwidths, the energy of the Fermi levels, and the magnitude and sign of the exchange integral. Before an explanation is provided, however, let us first turn to the collective magnetic behaviours at the phenomenological level.

9.2 Cooperative magnetic phenomena

Among the paramagnetic substances, there are some that undergo a transition of their spin states below a certain temperature and realize a collective order of the orientation of their spins, and hence of their magnetic dipole moments. In this process, the magnetic moments form a one- to three-dimensionally ordered spin structure, which does not have to be consistent with the crystallographically determined unit cell and the symmetry of the crystal structure but can represent a superstructure of the crystal structure. This means that even for crystallographically equivalent atoms, the direction of their magnetic moments can be different. The most important states of order that can occur are:
- ferromagnetic order (strict parallel orientation of all spins),
- antiferromagnetic order (strict antiparallel orientation of neighbouring spins),
- ferrimagnetic order (antiparallel orientation of spins but with different size or number of the magnetic moments for the opposite directions),
- canted ferro- or antiferromagnetic order (the spins are not exactly co-parallel or exactly anti-parallel aligned but are tilted by a small angle), and
- helical ferro- or antiferromagnetic order (spins of neighbouring magnetic moments arrange themselves in a spiral or helical pattern).

This transition happens below a certain temperature, which is called the Curie temperature for substances that undergo a transition from the paramagnetic to the ferromagnetic or ferrimagnetic state and Néel temperature for those that become antiferromagnetic. The interesting thing about these transitions is that they cannot be understood on the basis of the direct interaction between the magnetic dipoles, i.e., on the basis of their magnetic forces alone – these forces would be far too small to generate (relatively large) areas of coherent order. And especially the ferromagnetic state of order would not be understandable at all, because poles of the same sign of a dipole repel each other and the question would be, why – in absence of an external magnetic field – should all dipoles align themselves in a parallel fashion? The deeper causes that give rise to these magnetic states of order are not fully understood yet. We will look at some aspects below, but for now, we will deal with some details of these different states of order on a more phenomenological level. We will focus on the three most important states of order, i.e., ferro-, ferri-, and antiferromagnetism. The former two show a similar behaviour, for instance, in an external magnetic field and with regard to the temperature-dependent susceptibility. However, with respect to the ordering of the magnetic moments and the coupling between them, the ferrimagnetic and antiferromagnetic ordering are similar, since in both cases, there are antiparallel oriented magnetic moments. For the ferrimagnetic case, the two orientations differ only in either the magnitude of the magnetic moments (larger/smaller) or the number of the magnetic moments is larger/smaller for one direction or the other.

Let us first look at the temperature-dependence of the different magnetic states. We already looked at the inverse of the susceptibility as a function of temperature for paramagnetic substances and ferro- as well antiferromagnetic substances in their paramagnetic state, and saw that they show a linear behaviour (well) *above* the Curie and Néel temperatures. Now, in Figure 9.5, the temperature dependence of the (direct) susceptibility is plotted against the temperature for para-, ferro-, and antiferromagnetic substances.

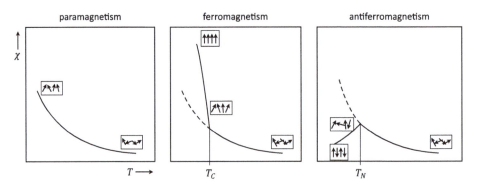

Figure 9.5: The temperature dependence of magnetic susceptibility for paramagnetism (left) and the occurrence of ferromagnetic (centre) and antiferromagnetic (right) ordered states in solids. Ferrimagnetic materials behave like ferromagnetic materials, but with an overall attenuated behaviour since partly antiferromagnetically coupled moments reduce the susceptibility.

At low temperatures and at high external field strengths, the ferromagnetic states of ferromagnetic materials are characterized by a perfect co-parallel alignment of their magnetic dipoles. The magnetic susceptibility is usually much higher than for paramagnetic materials. At higher temperatures, thermal motions increase and the parallel spin order is more and more disturbed. Above the ferromagnetic Curie temperature T_C, a transition to the magnetically disordered paramagnetic state occurs. For antiferromagnetic materials, the magnetic susceptibility at very low temperatures is close to zero, which is caused by a perfect antiparallel alignment of the magnetic dipoles. With increasing temperature, this order is broken up more and more and the external magnetic field partly causes a stronger co-parallel arrangement of the magnetic moments. As a result, a maximum of the magnetic susceptibility, which is still relatively small, is reached at the Néel temperature, T_N, before the transition to the magnetically disordered paramagnetic state takes place, and the usual decrease of the magnetic susceptibility with increasing temperature is observed. Néel and Curie temperatures for a few selected materials are gathered in Table 9.4.

Table 9.4: Selection of Curie and Néel temperatures of ferromagnetic and antiferromagnetic elements/compounds, respectively, after [64], if not otherwise specified.

Ferromagnetic compound	T_C (K)	Antiferromagnetic compound	T_N (K)
EuO	69	FeI_2	9
Dy	88	$CoCl_2$	25
Gd	292	$NiCl_2$	50
CrO_2	386	Mn	100
$Y_3Fe_5O_{12}$	560	MnO	116
$Nd_2Fe_{14}B$	573	MnS	160
Ni	627	FeO	198
Fe_3O_4	858	CoO	291
Fe_2O_3	948	Cr_2O_3	307
$SmCo_5$	973	Cr	308
Fe	1,043	NiO	525
Co	1,388	$LaFeO_3$	738 [65]

The reason for the magnetically ordered states: exchange interactions and band structure

Earlier, it was mentioned that the reasons for the cooperative magnetic phenomena are rather poorly understood. The ordinary magnetic dipole-dipole interactions are much too weak to explain them and, in particular, they can only be a part of the cause anyway, since they oppose the formation of ferromagnetic ordered states with their co-parallel oriented dipoles. We can prove this by a rough calculation of the interaction energy of two magnetic dipoles, which are separated by a distance of r:

$$E_{\mu_1\mu_2} = \frac{\mu_0}{4\pi r^3}\left(\vec{\mu}_1 \cdot \vec{\mu}_2 - \frac{3}{r^2}(\vec{\mu}_1 \cdot \vec{r})(\vec{\mu}_2 \cdot \vec{r})\right)$$

If we assume that the dipoles are in the order of 1 μ_{BM}, are parallel oriented to each other, and that they are separated by 2 Å, the interaction energy is $E = 2.1 \times 10^{-24}$ J. This corresponds to a temperature ($E = kT$) below 1 K. If we take into account the typical Curie temperatures (compare Table 9.4), it is clear that dipole-dipole interactions cannot be responsible for the long-range ordering of magnetic moments.

In fact, the interactions are of quantum mechanical origin and the decisive quantity is the exchange integral J of the Hamiltonian. For a simple two-electron system, the exchange integral determines if a singlet ($S_{total} = 0$) or a triplet state ($S_{total} = 1$) is energetically favoured. If the exchange integral J is positive, then the energy of the singlet state is higher and the triplet state is preferred, and vice versa. The situation is, of course, considerably more complicated if we turn to many-electron atoms and, in particular, to many-atom solids. The question arises, in which situations do the exchange integral become positive or negative. This also depends on the bonding situation. We know that in a free atom the ground state is determined by the first rule of Hund, which leads to a maximum of parallel spins: Systems with a symmetric spin wavefunction have lower energy if the spatial part of the wavefunction is antisymmetric. However, in chemical bonds, the potential energy of the electrons involved in that bond is lowered if the electrons are paired. Still, the exchange energy is present between electrons of the neighbouring atoms, which tends to make the spins parallel. This means that we have two opposing driving forces. And, whether a ferromagnetic or antiferromagnetic state has the lower energy exactly depends on the ratio of these forces. Usually, the bonding energy is much higher than the exchange energy, and this leads to paired electrons with antiparallel spins. However, if we have rather weak bonding forces, the exchange energy can become relevant.

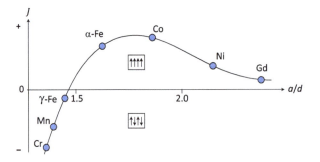

Figure 9.6: A Bethe–Slater curve, in which the magnitude of the exchange integral as a function of the ratio between the atomic distance a and the diameter d of the not completely filled electronic shell is plotted.

This is indeed the case for the 3d transition metals and also for some of the lanthanoids. The 3d orbitals are *inner* shells and the bonding between these orbitals is weak. And with increasing atomic number, these 3d orbitals are increasingly contracted so that the exchange interaction can become relevant. However, the atomic radius also changes. If we now plot the exchange integral as a function of the quotient of the atomic distance and the diameter of the not completely filled electron shell, we obtain the so-called Bethe–Slater curve (see Figure 9.6), showing that the conditions for a positive exchange integral are fulfilled only for a very few elements. For the 3d transition metals, the conditions are fulfilled only for Fe, Co, and Ni, which are ferromagnetic.

A stable ferromagnetic state can occur only if there is an excess of spins that are parallel and maintained in this state by the exchange energy of the system. For this, it is necessary that a certain amount of energy is provided to promote some electrons near the Fermi edge to states above the Fermi level, a process that is accompanied with an increase of kinetic energy of these promoted electrons. In a ferromagnetic metal, the exchange energy must be sufficient to counterbalance the increase in energy of the electrons at the Fermi level.

Whether a sufficiently high number of electrons can be promoted above the Fermi level depends decisively on the band structure, and in turn, on how the density of states curve looks like, in particular in the vicinity of the Fermi edge. And this depends heavily on how strongly the respective orbitals interact with each other. In the first-row transition elements, there are two energy bands to consider that result from the 3d, 4s, and 4p levels. The 4s and 4p orbitals, in particular, are diffuse and extended, and, therefore, overlap strongly with those on the neighbouring atoms, giving rise to a broad band with a high dispersion and a relatively small number of levels in the region of E_F. However, the electrons in the 3d orbitals are considerably more contracted, and with increasing atomic number, i.e., increasing nuclear charge, the degree of orbital overlap decreases, leading to relatively narrow d bands with a high density of states; the curve is very horizontally extended in the respective DOS plot (see Figure 9.7). If now the Fermi energy is located above or below the very narrow d band, there are an insufficient number of unpaired electrons available to realize a ferromagnetic state. For the early 3d transition elements, the Fermi energy is located below that narrow d band, and the unpaired d electron concentration is small. For the late 3d transition elements, the d band is effectively completely filled, and the Fermi energy is located above the narrow d band component in the DOS plot. Only for the three elements Fe, Co, and Ni, the Fermi energy lies within the narrow d band (with some contributions from the 4s and 4p levels), giving rise to a sufficient number of unpaired 3d electrons, which, together with the positive exchange energy, leads to ferromagnetism.

The number of unpaired electrons in these elements can be inferred from measuring the saturation magnetisation in an external magnet field. The results are listed in Table 9.5. It is interesting to note that these values do not match the expectations based on the electronic configuration of the isolated atoms (four unpaired electrons for Fe, three for Co, and two for Ni). The values are lower because a partial charge transfer from the 4s to the 3d state occurs. Furthermore, the transferred electrons are not necessarily equally distributed to the spin-up or spin-down population: for the 3d band of Fe, a population of approx. 4.8 electrons with spin-up and 2.6 with spin-down configuration is found.

The previously explained correlation of the degree of filling of the 3d band and the strength of the ferromagnetic effect is supported by the magnetic properties of many alloys. For instance, the ferromagnetic properties of an alloy of Ni (ferromagnetic) and Cu

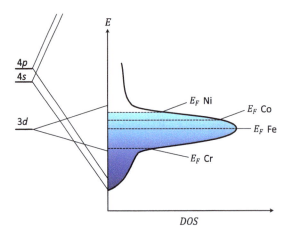

Figure 9.7: The energy levels of the contributing 3d, 4s, and 4p orbitals to the band structure and the density of states plot for the 3d transitions metals Cr, Fe, Co, and Ni. The coloured region represents the degree of filling for these elements.

Table 9.5: Electron configuration of the isolated atoms of the three ferromagnetic elements of the 3d transition metals and their number of unpaired spins in the ferromagnetic state.

Element	Electron configuration of isolated atoms	Electron configuration in the solid state	No. of unpaired spins
Fe	[Ar] $3d^6\ 4s^2$	[Ar] $3d^{7.4}\ 4s^{0.6}$	2.22
Co	[Ar] $3d^7\ 4s^2$	[Ar] $3d^{8.3}\ 4s^{0.7}$	1.72
Ni	[Ar] $3d^8\ 4s^2$	[Ar] $3d^{9.4}\ 4s^{0.6}$	0.60

(paramagnetic due to the completely filled 3d level) decreases with increasing content of Cu, until at a composition with approx. 31.5% Cu, the ferromagnetic property vanishes completely.

This low dispersion of the d bands is present only in the 3d transition metals. In the elements of the second and third row of transition metals, the 4d and 5d bands have a much larger bandwidth because the orbitals are much more diffuse. Therefore, only an insufficient number of electrons can be promoted above the Fermi level – so that no ferromagnetic state can be realized. These 4d and 5d transition metals are therefore paramagnetic, without exception.

This situation changes again for the lanthanoids because the 4f shell is sufficiently well buried within the outer 5s and 5p shells. These 4f levels are hardly involved in any chemical bonds, meaning that the atoms, even in solid state, have characteristics similar to the free atoms with respect to the electron configuration, with an integral number of electrons in the 4f shell. This leads to rather strong collective magnetic interactions if the orbitals of the 4f level are occupied by an uneven number of electrons. The resulting magnetic moment can be larger compared to the 3d transition metals because there are seven f orbitals, leading to a maximum of seven unpaired spins. The effects are similar for the actinoids.

Gadolinium is ferromagnetic while most other lanthanoids are antiferromagnetic below room temperature. Some lanthanoids also show interesting magnetic phase changes from ferromagnetic to antiferromagnetic states at different temperatures. The sequence of the magnetic states with increasing temperature is always: ferromagnetic → antiferromagnetic → paramagnetic, with the following Curie and Néel temperatures: Tb: T_C = 222 K, T_N = 229 K, Dy: T_C = 85 K, T_N = 179 K, Ho: T_C = 20 K, T_N = 131 K, Er: T_C = 20 K, T_N = 84 K, Tm: T_C = 25 K, T_N = 56 K.

9.2.1 Ferromagnetism

Although the magnetic susceptibility of ferromagnetic materials increases strongly below the Curie temperature, it is often the case that in the absence of an external magnetic field, there is initially no or only a very weak net magnetisation in a specific direction. The reason for this is that the spontaneous magnetisation, the co-parallel alignment of magnetic moments, that happens below the Curie temperature does not occur in a uniform direction over the entire volume of the material, but is extended to restricted regions – the so-called Weiss domains. While the direction within one domain is uniform, this direction of the magnetisation is different in individual domains so that there is only a small chance of a resulting net magnetisation (see Figure 9.8, left). These domains, which have a size of about 1–100 µm, are not identical with the grains of a microstructured material; they exist even in single crystals. The geometry of the Weiss domains also depends on the crystal structure. Most ferromagnetic solids are magnetically anisotropic. This means that the magnetic moments align more readily in some crystallographic directions than in others. These are known as 'easy' and 'hard' magnetisation axes, respectively. For instance, the <100> directions in α-Fe with a **bcc** structure are easy magnetisation axes, while the <111> directions represent hard magnetisation axes. In Ni, with its **ccp** (fcc) structure, it is the other way round; the <100> directions are hard and the <111> directions are easy axes of magnetisation. And finally, in Co, which crystallizes in the **hcp** structure, the magnetic moments preferably align parallel to the c-axis (see Figure 9.9). This means that the direction of magnetic moments within the Weiss domains will be in the direction of the easy axes in the absence of a magnetic field, and that an alignment along the hard axes can usually be achieved only by applying an external field in the appropriate direction.

The individual Weiss domains are delimited from each other by the so-called Bloch walls. Typically, the magnetic moments of two neighbouring domains point in exactly opposite directions. However, the direction does not change abruptly at the boundary, but extends over a certain region – the Bloch wall – over which a gradual change in the direction of the magnetic moment takes place (see Figure 9.8, top-right). The domain

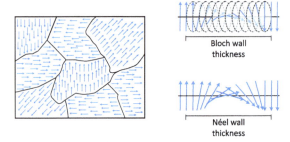

Figure 9.8: Schematic representation of Weiss domains of a ferromagnet material after spontaneous magnetisation below the Curie temperature (left); an example of the magnetic dipole structure within a Bloch wall where the orientation of the magnetic dipole changes gradually in a helix-like fashion from one domain (upwards pointing dipoles) to another with opposite direction (downwards pointing dipoles) (top-right); an example of the magnetic dipole structure in a Néel wall (bottom-right).

wall thickness depends on the anisotropy of the material, but on average, spans across 100–150 atoms; in α-Fe, the thickness is approximately 40 nm. The magnetisation always remains parallel to the wall plane, i.e., the magnetisation rotates helically. As a result, the magnetisation of the Bloch wall on the surface of a material points out of the plane and a magnetic stray field is created. The occurrence of so characterized Bloch walls is the dominant scenario in bulk ferromagnetic materials. However, in thin films, it might also be the case that instead of a Bloch wall, a Néel wall is formed, which is characterized by the fact that the gradual change of the magnetisation direction occurs in-plane (see Figure 9.8, bottom-right).

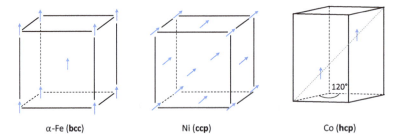

Figure 9.9: Preferred alignment of the magnetic moments in α-Fe (**bcc**), along the [100] direction (left), in Ni (**ccp**), along the [111] direction (middle), and in Co (**hcp**), along the [001] direction (right).

Ferromagnetic substances in an external magnetic field

The state of ferromagnetic materials immediately after spontaneous magnetisation is usually associated with no net magnetisability because the different orientations of the magnetic moments in the Weiss domains overall lead to mutual cancellation of the moments.

Of course, this changes when an external magnetic field is applied. In some domains there may already be an orientation of the dipoles, which more or less coincide with the direction of the external field. In other domains, however, as the strength of the external field increases, the moments whose direction do not coincide are forced in the direction of the external field, making the domains with unfavourable orientation smaller. At high field strengths, all moments will eventually be oriented co-parallel to the external field and the material will reach its saturation magnetisation.

Now, what distinguishes ferromagnetic from paramagnetic substances is that a net magnetisation is preserved even when the external magnetic field is turned off. The magnetisation that is retained is called remanence, which is specific to each material. If this remanence level is high, we speak of hard magnetic materials; if it is low, we speak of soft magnetic materials. Permanent magnets are made preferentially of hard magnetic materials.

In order to get a state inside a ferromagnetic material that is overall identical to the one after the spontaneous magnetisation, one has to apply an external magnetic field in the opposite direction. The field strength at which a zero net magnetisation – a complete demagnetisation – is reached, is called coercivity field strength H_C. If the field strength is further increased, saturation with the opposite direction is finally reached. Again, if the external field is turned off, a remanence level of magnetisation remains. This means that due to the remanence properties, when applying an alternating external magnetic field (with not too high frequency), in a plot of the magnetisation M as a function of the field strength of the external magnetic field H, a typical hysteresis (from Greek: lagging behind) loop will be obtained. Such a hysteresis loop, with a schematic sketch of the magnetic microstructure at characteristic points, together with the initial curve of magnetisation, is shown in Figure 9.10.

In terms of applications, hard magnetic materials that show a rather rectangular and very broad hysteresis with a very high level of remanence, are suitable for permanent magnets that can be used, for instance, in electrical generators of wind turbines. For other applications, more softer materials are needed, for instance, in magnetic-tape data storage: The magnetic material, often a thin film on a flexible support that receives the signal to be stored, must be magnetically soft enough to respond quickly to small signals but must be magnetically hard enough to retain the information for long periods. However, the shape of the hysteresis loop is also strongly influenced by the microstructure of the solid, as grain boundaries and impurities hinder domain wall movement. This in turn, has consequences for the coercivity and remanence of a material. In Table 9.6 values for the coercivity of some selected materials are gathered.

Superparamagnetism of ferromagnetic nanoparticles

A very interesting phenomenon can be observed if the particle size of a ferromagnetic (or ferrimagnetic) material gets smaller and smaller until the particle consists only of a

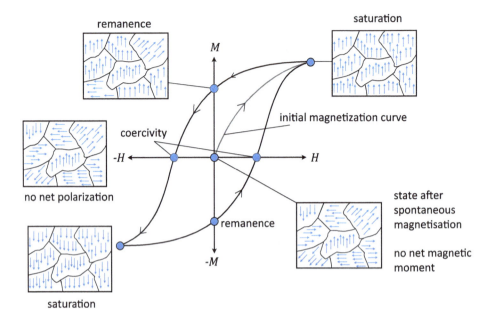

Figure 9.10: The initial magnetisation curve and a complete hysteresis loop of a ferromagnetic material, together with sketches of the microstructure at selected points. The sketches are simplified in the sense that also the Bloch walls migrate, or in other words, the Weiss domains of favourable orientation grow at the expense of those with unfavourable orientation.

Table 9.6: Overview of the coercivity of some selected materials.

Material	Coercivity (kA m^{-1})*
$Ni_{0.75}Fe_{0.2}Mo_{0.05}$ (supermalloy**)	0.0002
$FeNi_4$ (permalloy**)	0.0008–0.08
Fe	0.004–0.08
Ni	0.06
Co	0.8
$Al_xNi_yCo_z$	30–150
$Nd_2Fe_{14}B$	800–950
$Fe_{0.48}Pt_{0.52}$	980
$SmCo_5$	800–3,200

*Representative values
**Permalloys are alloys that have relative permeabilities much larger than of iron; superalloys have permeabilities that are again one to two orders of magnitude larger

single magnetic domain. Such particles are called Stoner-Wohlfarth particles. In such particles, the remanence is zero after turning off the external magnetic field. The reason is that in the absence of an external field, the thermal energy is sufficient to reverse the direction of the magnetisation in such a particle. This process is called Néel relaxation.

And, as the time for such a magnetisation flip (the Néel relaxation time) is comparable to the duration for which the magnetisation is measured, the magnetisation seems to have the value 0, on average. A collection of such particles, therefore, behaves macroscopically like a paramagnet, but still has the high magnetic saturation of a ferromagnet. Unlike in a paramagnet, it is not the individual atoms but these small single-magnetic domain particles that change their magnetisation direction independent of each other. The energy barrier that has to be overcome to change the direction of the magnetisation depends on the particle size and its magnetic anisotropy. Below a certain temperature, which is called the blocking temperature, the thermal energy will no longer be sufficient to reverse the orientation of the magnetisation.

The characteristics of superparamagnetic nanoparticles make them interesting, for example, for new magnetic data storage technologies, for future hard disk drives, and also for biomedical applications. Magnetic particle aggregates that are stabilized by hydrophilic (water-soluble) polymers are small enough to enter cells or pass borders like the blood-brain barrier. Their surface can be functionalized for selective interaction, and their magnetic properties make them controllable by an external magnetic field. This opens up new possibilities in the field of magnetic-controlled drug delivery [66] and for the treatment of cancer by magnetic hyperthermia therapy [67].

9.2.2 Antiferromagnetism and superexchange interactions

In the section on exchange interactions, it has already been discussed that for most elements the ratio of the atomic distance to the diameter of the electronic (sub)shell that is not completely filled is such that the exchange integral J is negative, and therefore show an antiferromagnetic behaviour below the Néel temperature.

However, besides the antiferromagnetic chemical elements, there are a number of compounds that also show an antiferromagnetic behaviour, although the distance between the magnetic species inside such compounds is far too large that a direct exchange interaction can cause this phenomenon. Examples of such compounds are the divalent oxides of some of the 3d transition metals, i.e., MnO ($3d^5$ system), FeO ($3d^6$), CoO ($3d^7$), and NiO ($3d^8$), all of which crystallize in an analogous NaCl structure.[8] In these compounds, a so-called superexchange of magnetic moments occurs, which is one of the *indirect* magnetic exchange mechanisms. Superexchange describes the antiferromagnetic coupling of magnetic moments of magnetic species via bridging diamagnetic species.

In the mentioned compounds, the metal cations are surrounded by an octahedral ligand field so that in all compounds the orbitals of the e_g set, $3d_{z^2}$ and $3d_{x^2-y^2}$, are each

---

[8] Note that in the monoxides of the early transition metals, i.e., TiO, VO, and CrO, the metal ions have more delocalized d orbitals, giving rise to partly filled t_{2g} bands, and in turn show metallic conductivity and are either diamagnetic or paramagnetic.

occupied by a single electron, which are aligned parallel. These orbitals are oriented parallel to the axes of the unit cell and, therefore, point directly at adjacent oxide ions; the metal-oxygen-metal angle along one axis is 180°. The unpaired electrons in these e_g orbitals couple with electrons in the p orbitals of the O^{2-} ions. This coupling may involve the formation of an excited state in which the electron moves from the e_g orbital of TM^{2+} (TM = transition metal) to the oxygen p orbital. Or, it may be regarded as partial covalent bond formation involving pairing of electrons between adjacent metal and oxygen atoms. The p orbitals of the O^{2-} ion contain two electrons each, which are antiparallel paired, according to Pauli's principle. Hence, provided that the TM^{2+} and O^{2-} ions are sufficiently close so that coupling of their electrons is possible, a chain coupling effect may occur, which propagates through the crystal structure, leading to a rather strongly antiferromagnetic coupled state (see Figure 9.11, left). To complete the picture, one must also take into account several other possible secondary exchange paths, for instance, the 90° angle configuration between two TM ions bridged by an oxide ion (this effect is rather weak and favours ferromagnetic order), the π-like interaction of the d orbitals that have their lobes in between the axes with the two lobes of an appropriately oriented p orbital (p-d-π superexchange, see Figure 9.11, right), and of course, the electron configuration of the TM ion is also of importance. While it should be clear that each two TM ions at different sides of a single oxide ion do have antiparallel orientation of their spins in their e_g set of orbitals, it remains to be clarified what happens to the electrons in the t_{2g} orbitals, which may be unpaired. The NiO case, which is discussed in almost all textbooks, is the least interesting one, because it is a d^8 system, in which all t_{2g} orbitals are doubly occupied. But what does the situation look like in the case of MnO, a d^5 system, in which all t_{2g} orbitals are singly occupied? Well, a state is realized, which – again according to the first Hund's rule – has a maximum spin multiplicity. And this means that the spins in all orbitals are pointing in the same direction. For the two indirectly coupling manganese centres on the left and on the right of the mediating oxide ion, this means that if, for example, all the electrons of the Mn on the left side do have spin-up configuration, all those of the Mn on the right side will have spin-down configuration.

Structural consequences

As a result of this antiferromagnetic coupling, the lattice constant of the *magnetic* unit cell (which can be determined with neutron diffraction techniques) of these transition metal monoxide compounds is *twice* as large, compared to the unit cell determined by conventional X-ray crystallography.[9] The crystallographically identical TM ions are no

---

[9] In most cases, the lattice constants of the magnetic unit cell are integer multiples of the lattice constants of the unit cell determined by X-ray crystallography. However, in some cases of hexagonal crystals with helimagnetism, for instance in dysprosium, in which the magnetic moments of the atoms form a helix with its axis pointing in the c-direction, the ratio of both lattice constants is not a multiple integer or, in other words: the pitch of the helix is not a multiple integer of the lattice constant c.

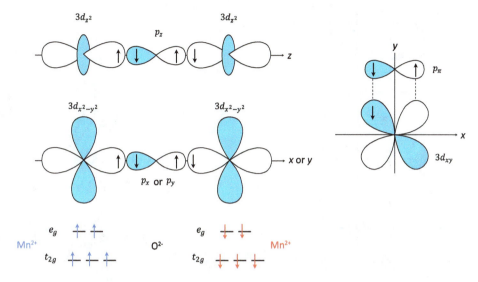

Figure 9.11: Illustration of the superexchange that occurs between neighbouring magnetic ions that are bridged by diamagnetic species, here, oxide ions. The exchange in linear TM-O-TM arrangements is strongly antiferromagnetic, where the main contributions stem from the σ-like on-axis interactions of the e_g orbitals with an appropriately oriented p orbital (left). The exchange in a π-like fashion between d and p orbitals favours a parallel alignment of the spins, but this effect is rather weak (right).

longer symmetry-equivalent if the orientation of their magnetic moments is taken into account (see Figure 9.12, left). The antiferromagnetic ordering has further structural consequences: While the TM monoxides mentioned are indeed perfectly cubic above the Néel temperature, i.e., in the paramagnetic state, rhombohedral compression or elongation along the [111] direction (the space diagonal) occurs (space group $R\bar{3}m$) below T_N because the magnetic moments in the layers (111) form ferromagnetic sheets, while the direction of the magnetisation in neighbouring planes is reversed (see Figure 9.12, right). For MnO, the rhombohedral angle is 90.62°, and the distance between two nearest Mn^{2+} ions with antiparallel oriented moments is 3.111 Å, while it is 3.114 Å between two nearest Mn^{2+} ions with parallel moments.

9.2.3 Ferrimagnetism and double exchange

In addition to the previously discussed exchange of magnetic moments, there is another mechanism that belongs to the class of indirect exchange mechanisms; it is called *double exchange*. There are two prerequisites for the double exchange: First, free-moving charge carriers must be present, which means that materials in which double exchange occurs are generally metallic conductors. And second, metal ions of different valencies must be present, e.g., Fe^{2+} and Fe^{3+}, as in the ferrimagnetic material, Fe_3O_4. This compound is

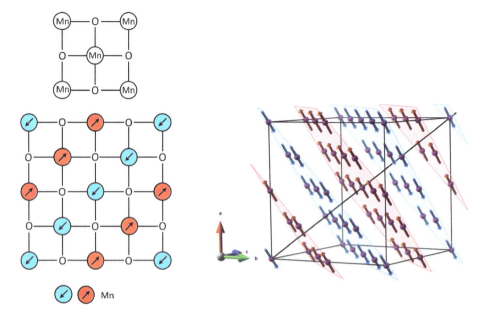

Figure 9.12: Projections of the X-ray (top-left) and magnetic (bottom-left) unit cells of MnO below the Néel temperature (116 K). In the magnetic unit cell, the antiparallel coupling of the magnetic moments of neighbouring Mn^{2+} ions are indicated by arrows. The magnetic moments are parallel aligned in (111) layers, while the direction of the magnetisation in neighbouring planes along the [111] direction (indicated by the black arrow) is reversed (right). The diamagnetic oxide ions are not shown.

naturally occurring as the mineral magnetite. Magnetised samples of magnetite are also known as 'lodestone'.[10]

Magnetite crystallizes in an inverse spinel structure that can be represented by the formula $(Fe^{3+})^{tet}[Fe^{2+}Fe^{3+}]^{oct}O_4$. The (Fe^{3+}) species occupy the tetrahedral sites (Wyckoff position 16d) and the species notated as $[Fe^{2+}Fe^{3+}]$ occupy the octahedral sites (Wyckoff position 8a), see Figure 9.13.

In the double exchange of magnetite, the exchange of electrons takes place mediated via the diamagnetic oxide ion between two neighbouring iron ions of different oxidation states. Because this exchange can be considered as a kind of two-stage process, it is called double exchange: if, for example, an electron jumps first from a Fe^{2+} centre to a bridging oxide ion, then it displaces an electron there. And this displaced electron

---

[10] Magnetite is one of only a very few minerals that is found in magnetised form. The phenomenon of magnetism was first discovered in antiquity by the fact that pieces of lodestone attract iron particles. The first magnetic compasses were also based on pieces of lodestone that were suspended so that they can freely turn. Their importance in the early history of navigation is also indicated by the literal meaning of the word: in Middle English lodestone means "course stone" or "leading stone", from the now-obsolete meaning of lode as "journey" or "way".

then subsequently jumps to a Fe^{3+} centre. Consequently, an electron transfer from Fe^{2+} to Fe^{3+} takes place, which reverses the oxidation states at the iron centres: Fe^{2+} becomes Fe^{3+}, and vice versa. These two states are energetically degenerate. And if this electron transfer takes place, the kinetic energy will decrease, thus lowering the total energy of the system.

But whether this electron transfer actually happens now depends on the spin states at the two metal centres involved. Referring back once again to Hund's first rule, it is only effectively possible if the spin multiplicity on both metal centres is maximal or, in other words, if a ferromagnetic order with parallel aligned magnetic moments is realized. In fact, the magnetic moments do not have to be 100% parallel aligned, but there is an angular dependence: The probability for this double exchange is maximum if they are perfectly parallel aligned, and it is zero, if the moments are antiparallel aligned.

In magnetite, the Fe centres that are octahedrally surrounded by oxygen anions tend to couple in a ferromagnetic fashion. However, for the total magnetic state, we also have to consider the Fe centres that are tetrahedrally surrounded by oxide anions. These also couple ferromagetically but their spin moments are aligned antiparallel to the Fe centres at the octahedral sites, i.e., couple antiferromagnetically to those. This means that such systems altogether can either be antiferromagnetic or ferrimagnetic if the magnetic moments do not completely compensate each other. This can be the case if the magnetic moments of the two species actually form an angle or if the number of the two species is not equal, but this, in turn, also depends on the degree of inversion of the spinel (see below). Magnetite, with its complete inverse spinel structure, is a ferrimagnet (below the Néel temperature of 850 K), because there are exactly twice as many tetrahedral sites as octahedral sites per unit cell.

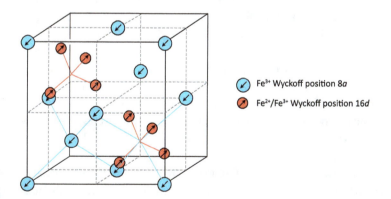

Figure 9.13: Magnetic structure of the ferrimagnetic compound magnetite. The iron cations are shown only for the four octants in the front; the oxide ions are not shown.

9.3 Some magnetic materials

9.3.1 Cubic spinel ferrites

The magnetite, Fe_3O_4, discussed above in Section 9.2.3 belongs to a larger class of materials, the cubic spinel ferrites, which have the general composition MFe_2O_4, where M is a divalent metal ion such as Fe^{2+}, Ni^{2+}, Cu^{2+}, Zn^{2+}, or Mg^{2+}. These compounds exhibit either a partial or complete inverse spinel structure. In the case of a complete inverse spinel structure, the formula is therefore better written as $(Fe^{3+})^{tet}[M^{2+}Fe^{3+}]^{oct}O_4$. The magnetic moments of all metal ions at the octahedral sites are coupled via double exchange ferromagnetically, i.e., are oriented parallel to each other. The metal ions at the tetrahedral sites are coupled antiferromagnetically via the superexchange mechanism. This means that spinel ferrites can be either antiferromagnetic or ferrimagnetic, depending on the degree of inversion and on the electron configuration of the divalent metal ion M^{2+}. For instance, $ZnFe_2O_4$, with a complete inverse spinel structure is antiferromagnetic below the Néel temperature of $T_N = 9.5$ K as Zn^{2+} has no magnetic moment. $MgFe_2O_4$ has an almost inverse spinel structure and is therefore slightly ferrimagnetic. $MnFe_2O_4$, with a degree of inversion of $\lambda = 0.1$, is ferrimagnetic with an overall magnetic moment of $\mu_B = 5$. Here, the magnetic moment is independent of the degree of inversion because both the Fe^{3+} and Mn^{2+} ions have d^5 configuration.

9.3.2 Garnets

Garnets are a huge group of nesosilicate minerals that have been used since the Bronze Age as gemstones and abrasives. The general formula of garnets is $A_3B_2(TO_4)_3$, with typical elements for A = Ca, Mg, Fe, Mn, and Na, and B = Al, Cr, Fe, Ti, Zn, and RE (RE = rare-earth), and T = Si, Ge, As, V, Fe, and Al. There are some synthetic garnets, in particular those in which rare-earth ions are incorporated that have interesting optical (see Chapter 10.2) or magnetic properties. For instance, yttrium aluminium garnet (YAG), $Y_3Al_2(AlO_4)_3$ was used for synthetic gemstones due to its fairly high refractive index. Nowadays, it is mainly replaced by cubic zirconia. Neodymium, erbium, or gadolinium-doped YAG phases find application in powerful lasers (see also Chapter 10) that are used for medicinal purposes in dermatology, ophthalmology, and dentistry.

Yttrium iron garnet (YIG), $Y_3Fe_2(FeO_4)_3$, and other rare-earth iron garnets, depending on the type of rare-earth ion, have tuneable ferrimagnetic properties with Curie temperatures around 550 K. YIG crystallizes in the space group $Ia\bar{3}d$ (no. 230), with a lattice constant $a = b = c = 12.375$ Å. The Y^{3+} ions are at the Wyckoff position 24c and are coordinated by eight oxide anions in a slightly distorted cubic fashion. The two types of iron ions are at the Wyckoff positions 16a and 24d, respectively, and are at the centres of FeO_4 tetrahedra and FeO_6 octahedra. Both types of Fe coordination polyhedra are exclusively corner-connected and have connections only to the other type of polyhedra.

Both the FeO$_4$ tetrahedra and FeO$_6$ octahedra have further common edges with the YO$_8$ cubes. The crystal structure of YIG is shown in Figure 9.14. The ferrimagnetic properties of YIG are the result of the antiparallel arrangement of the spins at the two Fe^{3+} sites, giving a net moment of 5 μ_B per formula unit.

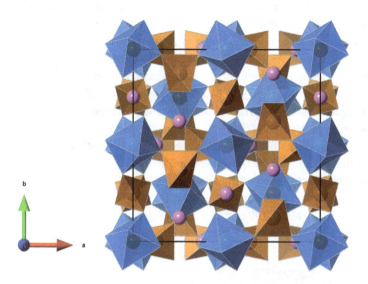

Figure 9.14: Crystal structure of the garnet phase YIG. Blue, FeO$_6$ octahedra; orange, FeO$_4$ tetrahedra; pink, Y ions.

9.3.3 Hexagonal ferrites – magnetoplumbites

A ferrimagnetic class of materials suitable for use as permanent magnets in transformer cores, is the hexagonal magnetoplumbite structural family. Materials for transformer cores should be magnetically soft with large power-handling capacity and low eddy current losses. Furthermore, they should have high permeabilities, should be magnetised easily at low applied fields, and should have a low coercive field. Soft magnetic materials are, mechanically, those in which the domain walls migrate easily.

Magnetoplumbites are a mineral group with the general composition AB$_{12}$O$_{19}$, where A is typically a large mono or divalent metal ion (K$^+$, Ca^{2+} Ba^{2+}, or Pb^{2+}) and B is another metal ion of medium size and various valences, but typically trivalent. The archetype is represented by the magnetoplumbite compound PbFe$_{12}$O$_{19}$, to which the commonly used hexagonal barioferrite, BaFe$_{12}$O$_{19}$, is isostructural. The structure is closely related to that of β-alumina (diaoyudaoite), ideally NaAl$_{11}$O$_{17}$. It is based on densest layers composed of oxide and A cations, in the voids of which the B cations are located. There are two fundamental building blocks that are stacked along the hexagonal c-axis. The first block, called S, consists of a two-layer sequence -cc- of cubic closed packed layers. A fraction of the voids is occupied by metal atoms in the same fashion as in the spinel structure, which gives an overall composition $\{B_6O_8\}^{2+}$ of the block. The second fundamental building block, denoted R, is built up of a three-layer sequence of hexagonal closest packed layers, -hhh-, where a quarter of the O atoms of the intermediate h layer is replaced by the cation A, resulting in a composition equal to $\{AB_6O_{11}\}^{2-}$. These two fundamental building blocks are then stacked along the c-axis in the fashion ...SRS*R*..., where an asterisk denotes a building block that is rotated by 180° around the a-axis. The magnetoplumbites crystallize in the space group $P6_3/mmc$ with two formula units per unit cell. The large A cations, ideally at the fractional coordinate (⅔, ⅓, ¼), in reality slightly disordered around that site, is surrounded by 12 oxide ions in the form of an anticuboctahedron. The B cations occupy five distinct crystallographic sites. The B1 cations are surrounded by six oxide ions in the form of a regular octahedron. The five-fold coordinated B2 cations, ideally located at the centre of a trigonal bipyramid (Wyckoff position $2b$) is, in reality, slightly displaced (split) into two statistically half-occupied, pseudotetrahedral Wyckoff $4e$ sites. The B3 cations are at the centre of a regular BO_4 tetrahedron. The B4 coordination polyhedra are trigonally distorted octahedra that occur in pairs, sharing a common face in a hematite-like arrangement, i.e., forming B_2O_9 dimers. The B5 cations are again at the centres of regular BO_6 octahedra. The total unit-cell contents (with $Z = 2$) can thus be expressed as $A_2^{[12anticuo]} B1_2^{[6o]} B2_2^{[5bpy]} B3_4^{[4t]} B4_4^{[6o]} B5_{12}^{[6o]} O_{38}$, see Figure 9.15 and Table 9.7.

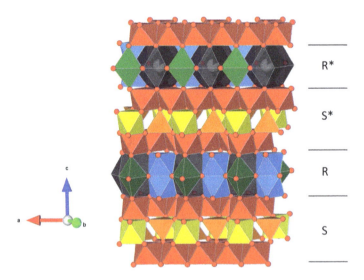

Figure 9.15: The structure of magnetoplumbites (for details, see text). Grey, AO_{12} anticuboctahedra; yellow, $B1O_6$ octahedra; green, $B2O_5$ trigonal-bipyramids; orange, $B3O_4$ tetrahedra; blue, $B4O_6$ octahedra; red, $B5O_6$ octahedra.

As stated above, the magnetic structure of magnetoplumbite is ferrimagnetic. The spin orientation of Fe^{3+} at each site is a result of the superexchange interaction through the O^{2-} ions. As the Fe^{3+} cation (d^5 system) has a spin-only magnetic moment of 5 μ_B, the

Table 9.7: Properties of crystallographic sites for A and B metal atoms in the magnetoplumbite-group minerals.

Site	Wyckoff position	CN	Point symmetry	Block	Spin direction for B = Fe^{3+}
A	2d	12	$\bar{6}m2$	R	
B1	2a	6	$\bar{3}m$	S	↑
B2	2b (4e)	5 (4 + 1)	$\bar{6}m2$ (3m)	R	↑
B3	4f	4	3m	S	↓
B4	4f	6	3m	R	↓
B5	12k	6	m	R-S	↑

total magnetisation per formula unit would be 20 μ_B, which is in good agreement with experimental results. Magnetoplumbite possesses a large magnetocrystalline anisotropy, which is related to a strong preference of the magnetic moments of the ions to align along the hexagonal c-axis.

Further reading

M. Getzlaff, *Fundamentals of Magnetism*. Springer, Berlin, Heidelberg, **2008**.
J. M. D. Coey, *Magnetism and Magnetic Materials*. Cambridge University Press, Cambridge, **2010**.
D. Jiles, *Introduction to Magnetism and Magnetic Materials*, 3rd ed. CRC Press, Boca Raton, **2015**.
J. M. D. Coey, S. S. P. Parkin (Eds.), *Handbook of Magnetism and Magnetic Materials*. Springer, Cham, **2021**.

10 Phosphors, lamps, lasers, and LEDs

10.1 Phosphors

A phosphor is a material that is capable of emitting light after absorbing energy, a phenomenon that is known as luminescence. The term is used for fluorescent or phosphorescent substances that emit light on exposure to electromagnetic radiation (X-rays, ultraviolet or visible light) – then it is termed photoluminescence –, for substances that emit light by excitation due to an electric current (electroluminescence) and for cathodoluminescent substances that glow when struck by an electron beam (cathode rays) in a cathode-ray tube (CRT). Phosphorus, the light-emitting chemical element, after which phosphors are named, emits light due to chemiluminescence; not phosphorescence.

The luminescence process comprises, first, the electronic excitation of the phosphor, and then second, the return to its electronic ground state, by which it emits the respective energy as light. Luminescence is formally divided into two categories, fluorescence and phosphorescence, depending on the nature of the excited state. Fluorescence is characterized by excited *singlet* states, the electron in the excited orbital is paired (by opposite spin) to the second electron in the ground-state orbital. Consequently, return to the ground state is spin-allowed and occurs rapidly by emission of a photon. A typical fluorescence lifetime is of the order of 10 ns. Phosphorescence is emission of light from *triplet* excited states, in which the electron in the excited orbital has the same spin orientation as the ground-state electron. Transitions to the ground state are spin-forbidden and the emission rates are slow, so phosphorescence lifetimes are typically milliseconds to seconds. Even longer lifetimes are possible, for instance in "glow-in-the-dark" toys. Following exposure to light, the phosphorescence substances glow for several minutes while the excited phosphors slowly return to the ground state.

The term fluorescence was coined by Sir George Stokes[11] and was derived from the mineral fluorite (CaF_2). Some samples contain traces of divalent europium, which serves as the fluorescent activator to emit blue light. But note that pure fluorite shows no fluorescence at all. A collection of minerals that fluoresces when exposed to UV light is shown in Figure 10.1.

A typical schematic photoluminescence spectrum is shown in Figure 10.2. Here, a phosphor is excited with higher energy, e.g., UV radiation, and spontaneously emits fluorescence radiation of lower energy, i.e., in the visible range of the e/m spectrum.

The processes that occur between the absorption and emission of light for *molecules* are usually illustrated by a Jablonski diagram (named after Alexander Jablonski (1898–1980), known as the father of fluorescence spectroscopy). A typical Jablonski

---

[11] The term "Stokes shift" is also attributable to Stokes. He established the law that the light re-emitted by fluorescent substances has a longer wavelength than the light previously absorbed.

Figure 10.1: Collection of various fluorescent minerals under ultraviolet light (photo: Hannes Grobe/Alfred Wegener Institute, CC BY SA 2.5).

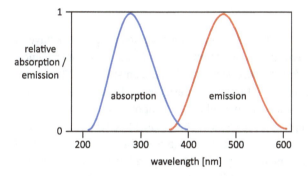

Figure 10.2: Schematic photoluminescence spectrum. A phosphor is electronically excited by UV radiation and emits fluorescence radiation in the visible range of the spectrum.

diagram is shown in Figure 10.3. The singlet ground, first, and second electronic states are depicted by S_0, S_1, and S_2, respectively. At each of these electronic energy levels, the fluorophores can exist in a number of vibrational energy levels, depicted by 0, 1, 2, etc. Following absorption, several processes usually occur. A fluorophore is usually excited to some higher vibrational level of either S_1 or S_2. With a few rare exceptions, excited phosphors in the condensed phase rapidly relax to the lowest vibrational level of S_1. This radiation-less de-excitation is called internal conversion and generally occurs within 10^{-12} s or less. De-excitation to S_1 involves collision with other molecules and vibrational excitation of other molecules. The return to the ground state typically occurs to a higher excited vibrational ground state level, which then quickly (10^{-12} s) reaches thermal equilibrium. Molecules in the S_1 state can also undergo a spin conversion to the first triplet state T_1; such a conversion is called intersystem crossing (ISC). Emission from T_1 is termed phosphorescence and is generally shifted to longer wavelengths (lower energy) relative to fluorescence. Transitions from T_1 to the singlet ground state are spin-forbidden, and as a result, the rate constants for triplet emission are several orders of magnitude smaller than those for fluorescence.

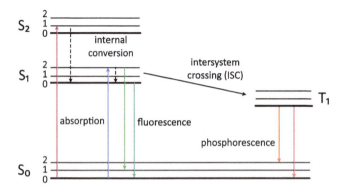

Figure 10.3: A typical Jablonski diagram (see text).

In the case of crystalline solids instead of molecules, the discrete energy levels have to be replaced by the continuous energy bands. One might think that, as a consequence, all solids show a quasi-continuous, at least very broad emission spectrum. This is indeed the case for some solids, but for intentionally designed and manufactured inorganic phosphors, the light emission and thus colour generation is not based on pure crystalline phases, but on the optical properties of the so-called activators, which are introduced as dopants into a suitable host structure. These activators are often ions of transition metals or rare-earth elements. Typically, the doping level is below 5 mol%. According to standard nomenclature, doping is indicated by a colon. For example, the doped compound La_2O_3:Eu is characterized by the composition, $La_{2-x}Eu_xO_3$ (typically with $x < 0.1$). The wavelength of emission and thus the colour of the luminescence can be influenced by the host structure. In this context, the environment in which an activator is located in a host structure determines the exact energy difference between the energy bands and thus the wavelength of emission. In general, it depends on how strong the interaction of the dopant ions with the surrounding atoms of the host structure is. For lighting applications, generally the transitions 5s-5p, 6s-6p, 3d-3d, 4f-5d, and 4f-4f are of importance. Broad emission bands result from strong interactions between the activators and the surrounding atoms of the host structure; this is often the case for s-p or 4f-5d transitions. In the case of f-f transitions, the arrangement of the surrounding matrix has a negligible influence on the values of the f-energy levels, which is why 4f ions always show almost unchanged spectra (emission colours). Activators with discrete 4f-4f transitions produce narrow emission lines with characteristic luminescent colours, such as in Sm^{3+} (red-violet), Eu^{3+} (red), Tb^{3+} (green), Er^{3+} (green), Dy^{3+} (yellow), and Eu^{2+} and Tm^{3+} (blue).

10.1.1 Fluorescent lamps

A fluorescent lamp is a low-pressure mercury-vapour gas-discharge lamp that uses fluorescence to produce visible light. An electric current passing through the gas inside a glass tube excites mercury vapour, which produces short-wave ultraviolet light that then causes a phosphor coating on the inside of the lamp to glow. A fluorescent lamp converts electrical energy into useful light much more efficiently than an incandescent lamp, i.e., a conventional lightbulb. The typical luminous efficacy of fluorescent lighting systems is 50–100 lumens per watt, while that of an ordinary light bulb is only about 16 lumens per watt.

The large-scale introduction of fluorescent lamps began in 1938. In the early years, the phosphor used was zinc orthosilicate with varying content of beryllium (Zn_2SiO_4:Be^{2+}). Zn_2SiO_4 is a nesosilicate that exhibits (in its undoped state) strong green fluorescence. It also occurs in nature as the mineral willemite. It crystallizes in the trigonal space group $R\bar{3}$ (no. 148) with lattice parameters $a = 13.948$ Å and $c = 9.315$ Å. Small additions of magnesium tungstate ($MgWO_4$) improved the blue part of the spectrum, yielding an acceptable broadband 'white' emission spectrum. After it was discovered that beryllium is toxic, it was replaced by the second generation halophosphate-based phosphors. They usually consist of trivalent antimony- and divalent manganese-doped calcium halophosphate ($Ca_5(PO_4)_3(Cl,F)$:Sb^{3+},Mn^{2+}). $Ca_5(PO_4)_3(Cl,F)$ occurs in nature as the minerals chloro- and fluorapatite, respectively.

The colour of the light can be adjusted by altering the ratio of the blue-emitting antimony dopant and orange-emitting manganese dopant. The colour rendering ability of these older-style lamps is quite poor. The colour rendering index (CRI)[12] of such lamps is only around 60. In particular, those lamps emit too little red light so that the human skin appears less pink, and hence pale and "unhealthy" compared with incandescent lighting.

For a long time, it was thought that to achieve a good CRI value, the emitted light should ideally cover the entire frequency range of visible light. But colour vision works differently. At a certain neurological level, the trichromatic theory (also known as Young-Helmholtz theory) can be considered valid, according to which there are three types of cones on the retina that are responsible for the perception of blue, green, and red light, respectively. The cone cells are usually designated S, M, and L, for short (peak maximum at 420 nm), medium (530 nm), and long wavelengths (570 nm), respectively. The perceived colour is then the result of additive colour mixing and the corresponding excitation level of the cones, whereby the human eye is particularly sensitive in the green range of light. Therefore, it should be possible to

---

[12] Colour rendering index (CRI) is a measure of how well colours can be perceived using light from a source, relative to light from a reference source, such as daylight or a blackbody of the same colour temperature. By definition, an incandescent lamp has a CRI of 100. Real-life fluorescent tubes achieve CRIs of anywhere from 50 to 98.

achieve a very high colour rendering index from a light source having a narrow triband spectrum if these bands correspond to the detection range of the red, green, and blue photoreceptors in the eye. New materials were synthesized with emissions close to target values and the modern family of 'Triphosphor' lamps was born, which conquered the market from the 1990s on. The CRI value of such lamps is typically 85. A typical triphosphor fluorescent lamp utilizes $BaMgAl_{10}O_{17}:Eu^{2+}$ (BAM) for blue (~ 450 nm), $LaPO_4:Tb^{3+},Ce^{3+}$ (LAP) or $Ce_{0.67}Tb_{0.33}MgAl_{11}O_{19}$ (CMAT) for green (~ 540 nm), and $Y_2O_3:Eu^{3+}$ (YOX) for orange-red (~ 650 nm) emission.

A comparison of the emission spectra of a typical halophosphate and a triphosphor fluorescence lamp is shown in Figure 10.4.

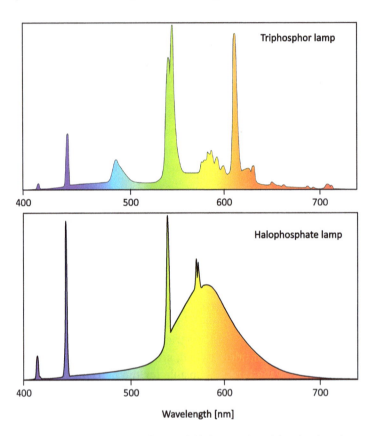

Figure 10.4: Emission spectra of a typical third-generation triphosphor-based (top) and second-generation halophosphate-based fluorescence lamp (bottom).

There are some other fluorescent lamps that have purposes other than illuminating rooms. In tanning fluorescent lamps, for instance, mixtures of $YPO_4:Ce^{3+}$ or $BaSi_2O_5:Pb^{2+}$ (UV-A) and $LaPO_4:Ce^{3+}$ (UV-B) are used to resemble the UV radiation component of the daylight. A further reduced spectrum containing either a relatively narrow emission line about 370 nm ($SrB_4O_7:Eu$) or 340 nm ($SrAl_{11}O_{18}:Ce$) or 310 nm ($MgSrAl_{10}O_{17}:Ce$) is produced by the so-called blacklight lamps. Such lamps are used for all kinds of fun purposes, such as indoor activities like blacklight miniature golf, or parties where guests wear fluorescent make-up, but also for medical (dermatology), forensic (detection of blood stains that are not visible to the naked eye), biological (insect attraction) and chemical (photo initialization of polymerization reactions) purposes.

10.1.2 Phosphors for CRTs of TVs and computer screens

In contrast to the author, most readers of the book probably know CRT-based television apparatuses or computer screens only from museums. However, the reader might be nevertheless interested about the kind of phosphors that were used for TVs in former times. Black-and-white television screens require an emission colour close to white, which is usually obtained by employing a mixture of two phosphors, one with a blue and the other one with a yellow emission spectrum. Common combinations are/were ZnS:Ag (blue) and (Zn,Cd)S:Ag (yellow), or ZnS:Ag and (Zn,Cd)S:Cu,Al (yellow).

Colour CRTs require three different phosphors, emitting in red, green and blue, patterned on the screen. For the excitation of the phosphors, the tube has one electron gun for each phosphor. The electron beams are deflected magnetically by means of deflection coils so that they sweep the entire screen line by line in 1/25 s (PAL, SECAM) or 1/30 s (NTSC). To ensure that they only excite the phosphors of the colour for which they are responsible, there is a mask about 20 mm in front of the screen that shades or catches the electrons that would hit the wrong colours. The phosphors that were used changed over time, with the exception of the blue-emitting phosphor, which was always ZnS:Ag. Initially, manganese-doped zinc orthosilicate ($ZnSiO_4:Mn$) was used as green phosphor. This was replaced step by step, first with (Zn,Cd)S:Ag,Cl then with (Zn,Cd)S:Cu,Al, then with (Zn,Cd)S:Cu,Al with a lower Cd/Zn ratio, due to environmental concerns, and finally with the entirely cadmium-free ZnS:Cu,Al. The red phosphor was originally manganese-activated zinc phosphate ($Zn_3(PO_4)_2:Mn$). It was replaced by europium(III)-activated hosts, for which first yttrium vanadate ($YVO_4:Eu$), then yttrium oxide ($Y_2O_3:Eu$), and finally yttrium oxysulphide ($Y_2O_2S:Eu$) was used.

Today, CRT monitors are practically irrelevant. They have almost been completely replaced by LCD or LED monitors.

10.2 (Solid-state) lasers

Laser is an acronym and stands for *light amplification* by *stimulated emission* of *radiation*. The light emitted by lasers is characterized by the fact that it is monochromatic, that it is possible to focus the beam very well (so that it has a very low divergence), that the coherence length is very large, and that a very high intensity can be achieved. It is also possible to generate very intense, extremely short pulses (in the femtosecond range) with accurate repetition frequency. The radiation emitted by a laser is by no means limited to the visible range of the electromagnetic spectrum. There are lasers for microwaves (then they are called MASERs), those that work in the infrared region (IRASER), in the UV range, and X-ray lasers are also available.

The three main components of a laser are: first, the laser or gain medium (*e.g.*, a gas, a diode, or a doped crystal); second, a component that accomplishes the pumping mechanism by supplying energy to the laser medium (for instance a flash lamp); and third, a resonator (also called optical cavity) that consists of a pair of mirrors at both ends of the laser device, one of which can be switched to a semi-transparent state, see also Figure 10.6.

Hitherto, we considered the generation of light by three principal processes: the *absorption* of energy that leads to an electronically excited state, the *spontaneous* emission of radiation from an excited state (involving a spin-allowed transition back to the ground state (fluorescence)), and the *decayed* emission from a much longer-lived triplet excited state (involving a spin-forbidden transition back to the ground state (phosphorescence)). In principle, the probabilities and, therefore, the transition rates for all these processes can be calculated by the respective Einstein coefficients. However, in a laser, the transition back to the ground state occurs by *stimulated* emission, where the emitted photon has the same phase, frequency, and direction (coherence) as the first, stimulating photon, which remains unchanged by this process. Now two coherent photons can stimulate the emission of two further photons if they hit along their pass other atoms in their excited states, and then four coherent photons are generated, and so on (see also Figure 10.5).

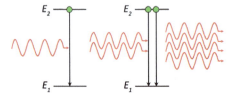

Figure 10.5: Stimulated emission leading to light amplification, where the light is coherent and has the same direction as the initial photon.

10.2.1 Operation conditions

To be able to operate such a laser system, there are two prerequisites. The first prerequisite is to achieve a so-called *population inversion* with respect to the normal population of the ground and excited electronic states in the gain medium, see also Figure 10.7. The population of the different available energy levels in thermal equilibrium is given by the corresponding Boltzmann distribution. If we assume two energy levels (say E_1 and E_2) whose energy difference is in the range of the energy of visible light, then the population ratio is given by the equation:

$$\frac{N_2}{N_1} = e^{-\frac{(E_2 - E_1)}{kT}}$$

with k being the Boltzmann constant. If we assume room temperature ($T \approx 300$ K), then $kT \approx 0.026$ eV, and with $E_2 - E_1 = 2.48$ eV (assuming light with a wavelength of 500 nm), the population ratio N_2/N_1 is vanishingly small; in other words, only the ground state is populated. With increasing temperature, the population of the high-energy state (N_2) increases, but N_2 never exceeds N_1 for a system at thermal equilibrium; at infinite temperature, the populations N_2 and N_1 become equal. This means that a population inversion ($N_2/N_1 > 1$) can never exist for a system at thermal equilibrium. In order to achieve a population inversion, the system has to be pushed into a non-equilibrated state, which is accomplished by the pumping device. However, with the help of the mentioned Einstein coefficients one can show that lasing operation is not possible if only two energy levels are involved. Graphically speaking, as soon as half of all particles in the laser medium are in the upper laser level, the probability that an atom in the lower laser level absorbs a photon is as high as the probability that an atom in the upper laser level emits a photon by stimulated emission. The additional spontaneous emission that still occurs further ensures that even this theoretical limit cannot be reached. Therefore, the second prerequisite for a laser operation is that in the laser gain medium at least a three-level system of electronic states is involved. With the help of the pumping device, the electrons are brought to the second excited state, which then pass to the first excited state, either by spontaneous emission or radiation-less (fast process) and then fall back from there to the ground level by stimulated emission (slow process). Even more efficient laser systems can be realized if a four-level energy system is involved. Here, the third excited state is the pumping level from which fast relaxation to the second excited state occurs. From this upper laser level, the actual laser emission takes place to the lower laser level, and finally the rapid decay from this first excited state to the ground state occurs. This ensures that the lower laser level depopulates quickly, which helps to maintain the population inversion.

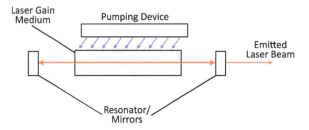

Figure 10.6: Schematic drawing of the main components of a laser.

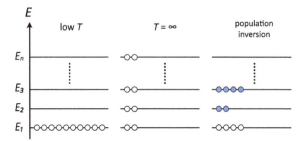

Figure 10.7: Population of electronic levels at low and at infinitely high temperature as well as the population of the levels that is needed for realizing a laser system (population inversion).

10.2.2 The ruby and the He-Ne laser

The first laser system that was ever realized in 1960 was the ruby laser made by Theodore H. Maiman and his assistant Irnee D'Haenens [68], i.e., a laser based on synthetic single crystals of corundum (Al_2O_3) doped with Cr^{3+}, typically with a doping level of 0.05%. The optical pumping was achieved with a flashlight (coiled around the crystal) and two silver foils at both ends of the crystal served as mirrors, one of which was somewhat thinner so that the beam could exit here after a certain amplification of the light, see also Figure 10.8. An energy level diagram of the Cr^{3+} ions in the trigonal crystal field of the corundum host and the electronic transitions of the ruby laser are shown in Figure 10.9.

Maiman's ruby laser emitted only pulsed radiation. But just a few months later, researchers at Bell Labs succeeded in producing a continuous-wave (cw) operating gas (He-Ne) laser. The research and development of lasers represents an excellent example that the separation between basic and applied research is a chimera. In the early stages of laser research, no one really knew what a laser actually could be useful for. Basic questions of laser transitions, the search for more effective four-energy-level laser systems and suitable host materials, the measurement of line widths, the improvement of spatial and temporal coherence, and the increase in intensity of the laser beam were addressed. Meanwhile, lasers have numerous applications in technology and research as well as in everyday life – from laser pointers for presentations, distance measuring

Figure 10.8: The components of Maiman's first ruby laser (photo: Guy Immega, CC0 (public domain)).

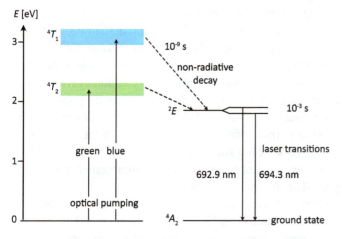

Figure 10.9: Energy level diagram of Cr^{3+} in Al_2O_3 and electronic transitions of the ruby laser. Note that the denotation of the energy states using the letters A, E, and T is not related to the orbital angular momentum, but to the symmetry properties of the electron distributions.

devices, cutting and welding tools, barcode scanners, reading optical storage media such as CDs, DVDs, and Blu-ray discs, information transmission to the applications in medicine, for instance, in ophthalmology or as a laser scalpel in surgery. Last but not least, science itself has benefited significantly from lasers: Today, researchers use optical tweezers made of laser beams to direct individual cells to a certain location. They also use lasers to hold isolated atoms captive and investigate exotic states of matter at temperatures near absolute zero. Laser radiation can even be used to fuse atomic nuclei together, at least in isolated cases, and the detection of gravitational waves is only possible with laser interferometers.

10.2.3 Classification of lasers

Typically, lasers are categorized according to the properties of the optical laser medium used. The coarsest classification is based on the state of aggregation.

Important gas lasers include the already mentioned helium-neon laser, emitting at 632.8 nm, and the carbon dioxide laser, emitting at 10,600 nm (10.6 µm). Special classes of gas lasers include excimer lasers, in which the laser medium is an excimer (excited dimer) molecule, and metal vapour lasers, in which the gaseous laser medium must first be obtained by vaporizing a metal.

Dye lasers are lasers in which dye molecules are dissolved in a liquid laser medium. These lasers are characterized by a very wide, continuous, and tuneable bandwidth of wavelengths (300–1,000 nm). The dyes used are in many cases stilbenes, coumarins, and rhodamines.

The group of solid-state lasers in the narrower sense includes lasers whose laser medium is a crystal or a glass. In a broader sense, solid-state lasers also include semiconducting laser diodes. Important examples of crystalline solid-state lasers include the already mentioned ruby laser, the titanium-doped sapphire (Ti^{3+}:sapphire) laser that is a tuneable laser which emits red and near-infrared light in the range from 650 to 1,100 nm, and the neodymium-doped yttrium aluminium garnet laser, Nd^{3+}:$Y_3Al_5O_{12}$ (Nd:YAG). The doping level is typically 1%. This four-level system is one of the most powerful laser systems and can be used both in pulsed and continuous-wave mode. For continuous-wave output, the doping is significantly lower than for pulsed lasers. As a $4f$ metal, the electronic energy levels of the Nd ion are only weakly affected by the host structure. The main laser transition occurs from the $^4F_{3/2}$ to the $^4I_{11/2}$ level by emitting electromagnetic radiation with a wavelength of 1,064 nm (i.e., in the IR region). With the help of frequency doubling by second-harmonic generation (SHG), laser beams of wavelength 532 nm (green) can be generated.

YAG crystallizes in the garnet structure. The garnets are a very large mineral group, with the general chemical formula $^{[8]}A_3{}^{[6]}B_2(^{[4]}TO_4)_3$, where

- A are predominantly divalent cations, trigondodecahedrally surrounded by eight oxygen anions, mostly Mg^{2+}, Fe^{2+}, Mn^{2+}, and Ca^{2+}, but also Y^{3+} or Na^+,
- B are predominantly trivalent cations, octahedrally surrounded by six oxygen anions, mostly Al^{3+}, Fe^{3+}, Cr^{3+}, and V^{3+}, but also Ti^{4+}, Zr^{4+}, Sn^{4+}, Sb^{5+} or Mg^{2+}, and Mn^{2+}, and
- T are predominantly tetravalent cations, tetrahedrally surrounded by four oxygen anions, mostly Si^{4+}, but also Al^{3+}, Fe^{3+}, Ti^{4+}, P^{5+}, As^{5+}, and V^{5+}.

In YAG, both the B and T sites are occupied by Al^{3+}. The crystal structure (space group $Ia\bar{3}d$, no. 230) of YAG is shown in Figure 10.10. The Y ions are surrounded by eight oxygen anions, forming a coordination polyhedron that is often described as a trigonal dodecahedron (also called snub disphenoid, also known as the Johnson solid J_{84}), but it can be also considered as a slightly deformed and twisted cube. The BO_8 coordination polyhedra are linked by common edges to form three-membered rings, whose plane is perpendicular to the body diagonal of the unit cell. Each BO_8 polyhedron belongs to two such three-membered rings. The BO_8 polyhedra are further edge-connected to the AlO_4 tetrahedra and AlO_6 octahedra. The AlO_4 tetrahedra and AlO_6 octahedra are connected by common oxygen atoms at their corners to form a framework of alternating tetrahedra and octahedra.

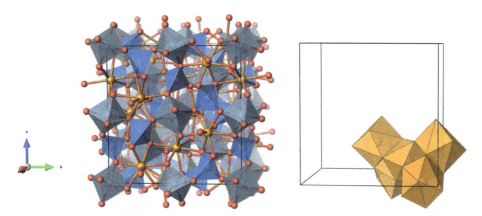

Figure 10.10: Crystal structure of YAG (left) and section of the coordination polyhedra of the yttrium ions (right); AlO_6 octahedra, light-blue; AlO_4 tetrahedra, blue; Y, orange.

Table 10.1 provides an overview of the most important laser systems.

Table 10.1: Overview of important laser systems (cw, continuous wave).

Type	Medium	Wavelength (nm)	Modus operandi	Comment
Gas	CO_2	10,600	Pulse/cw	Alongside solid-state lasers, it is one of the most powerful and most frequently used industrial lasers for materials processing
	CO	4,800–8,300	Pulse/cw	Virtually only used for research purposes and spectroscopy; requires expensive liquid N_2 cooling
	N_2	337.1	Pulse/–	Does not require a resonator
	He-Ne	632.8 / 1,152.3 / 3,392.2	–/cw	First cw-mode laser; is hardly used anymore
	Ar^+	351 / 457.9 / 488 / 514.5	Pulse/cw	Is also used in medicine (dermatology, ophthalmology and dental technology)
	Cu (gaseous)	510.6 / 578.2	Pulse/–	Operates at 1,700 °C, the partial pressure of Cu is then 0.005 mbar

Table 10.1 (continued)

Type	Medium	Wavelength (nm)	Modus operandi	Comment
Solid-state	Ruby	694.3	Pulse/cw	First laser ever; is hardly used anymore
	Ti:sapphire	670–1,100	Pulse/cw	Tuneable laser
	Nd:YAG	1,064	Pulse/cw	Most powerful solid-state laser
	Nd:YVO$_4$	914 / 1,064 / 1,342	Pulse/cw	–
	Y:YAG	1,030	Pulse/cw	Compared with Nd:YAG, the absorption bandwidth is much wider, reducing the temperature management requirements of the pumping device
	Er:YAG	2,940	Pulse/cw	Often used in dermatology; leads to a comparable low thermal load of the skin layers adjacent to the treatment zone, which is due to the short duration of the laser pulse in combination with the very high absorption in the water of the skin cells. The low heating of the tissue causes fast wound healing.
Excimer	XeCl	308	Pulse/–	Is successfully used in dermatology for all UVB-sensitive dermatoses, such as psoriasis vulgaris or neurodermatitis
Semiconductor	In$_x$Ga$_{1-x}$N laser diode	370–530	Pulse/cw	Used for Blu-Ray discs
	Al$_x$Ga$_{1-x}$As laser diode	730–850	Pulse/cw	Used for CD-ROM drives
Dye	stilbenes, coumarins, and rhodamines	300–1,000	Pulse/cw	Importance of dye lasers has declined in recent years in favour of other tuneable laser systems, in particular, easier-to-operate tuneable diode lasers

10.3 LEDs

A light-emitting diode (LED) is a semiconductor device that emits light when an electric current flows in the forward direction through it. In the opposite direction, the LED blocks the current. Thus, the electrical properties of the LED are the same as those of a diode. LEDs have become very important lighting devices and they are increasingly replacing incandescent and fluorescent lamps in households. LEDs have a much better luminous efficacy and therefore consume much lower energy.

The basic construction of an LED corresponds to that of a *p-n* semiconductor diode. A major difference is the semiconductor material used. While non-luminous diodes are made of silicon, or more rarely of germanium or selenium (all of them being indirect semiconductors), the material for LEDs is a direct semiconductor, usually a gallium compound as a III–V semiconductor compound, for instance, GaAs. When a forward voltage is applied to a semiconductor diode, electrons migrate from the *n*-doped side to the *p-n* junction. After transitioning to the *p*-doped side, electrons then move to the more energetically favourable valence band. This transition is called recombination because it can also be interpreted as a meeting of an electron in the conduction band with a defect electron (hole). The energy released during recombination is usually emitted directly as a photon. The wavelength of the emitted photon depends on the semiconductor material and the doping of the diode: The light can be visible to the human eye or in the range of infrared or ultraviolet radiation.

In Table 10.2, the most important materials used for LEDs and their wavelengths ranges are summarized.

Table 10.2: Colours, wavelengths, and materials of LEDs.

Colour	Wavelength (nm)	Material
IR	2,500–5,000	InAs/AlSb heterojunction
IR	1,400–1,600	InP
IR	>760	GaAs, $Al_xGa_{1-x}As$
Red	610–760	$Al_xGa_{1-x}As$, $Ga_xAs_{1-x}P$, AlGaInP, GaP
Orange	590–610	$Ga_xAs_{1-x}P$, AlGaInP, GaP
Yellow	570–590	$Ga_xAs_{1-x}P$, AlGaInP, GaP
Green	500–570	$In_xGa_{1-x}N$ / GaN, AlGaInP, GaP, $Al_xGa_{1-x}P$
Blue	450–500	ZnSe, $In_xGa_{1-x}N$, SiC
Purple	400–450	$In_xGa_{1-x}N$
UV	230–400	AlN, $Al_xGa_{1-x}N$

Since LEDs basically produce only monochromatic light, various additive colour mixing processes are used to produce white light. The two most frequently techniques are: (i) using RGB-(red-green-blue)-LEDs – combining them in one LED case in such a way that their light mixes well and thus appears as white from the outside when the individual LEDs are controlled accordingly, (ii) combination of a blue LED with a phosphor – here, the radiation that is emitted from the blue LED is partially used to excite a broadband emitting phosphor, whose maximum intensity range is in the yellow colour range. Such systems tend to produce suboptimal red rendering.

Further reading

V. Pawade, R. Kohale, S. Dhoble, H. Swart (Eds.), *Phosphor Handbook*. Woodhead Publishing, Cambridge, **2023**.

H. J. Eichler, J. Eichler, O. Lux, *Lasers*. Springer, Cham, **2018**.

T. Q. Khan, P. Bodrogi, Q. T. Vinh, H. Winkler (Eds.), *LED Lighting: Technology and Perception*. Wiley-VCH, Weinheim, **2015**.

11 Superconductivity

The term "superconductivity" refers to the ability of some materials to conduct electric current below a critical temperature (T_c) without any resistance. This observation was first made on April 8, 1911, by Heike Kamerlingh Onnes, a Dutch physicist (1853–1926) and pioneer of low-temperature physics. During experiments with liquid helium (which he was able to make approximately only 3 years before), he observed that the resistance of the electric current in mercury disappears when the temperature drops below the critical temperature of 4.183 K, i.e., slightly below the boiling point of helium. An article worth reading describing the circumstances of this discovery has been published in the journal *Physics Today* [69]. In 1913, Kammerlingh Onnes received the Nobel Prize in Physics "for his investigations of the properties of matter at low temperatures, which led, among other things, to the production of liquid helium".

11.1 From the metallic to the superconducting state

Metals are good-to-excellent electrical conductors. And although their electrical resistance decreases with decreasing temperature, for many metals the resistance does not fall to zero at very low temperatures. Instead, most metals show a constant residual resistance below 20 K. The concrete value of the residual resistance is strongly dependent on the purity of the material. This relationship was already recognized in the nineteenth century by the British physical chemist, Augustus Matthiessen (1831–1870). Matthiessen's empirical rule from materials science is also named after him: It states that the total resistivity caused by scattering processes is the sum of the resistivities of all individual scattering processes. The resistivity of a metal is composed of a temperature-dependent term arising from scattering of electrons by lattice vibrations, known as electron-phonon collisions, another temperature-dependent term arising from electron-electron scattering, and a temperature-independent term arising from scattering of electrons by lattice defects:

$$\rho(T) = \rho_{phonons}(T) + \rho_{electrons}(T) + \rho_{defects}$$

At room temperature the electron-phonon and electron-electron collisions are dominant, while at very low temperatures the resistivity arising from defects prevails. However, in superconducting materials, as stated above, the resistivity drops to zero below a certain critical temperature. Figure 11.1 shows a schematic comparison of the two different behaviours of the temperature dependence of the resistivity of metals and superconductors. This implies a completely different conduction mechanism, compared to the ordinary metallic state. A short explanation is given in Section 11.1.2.

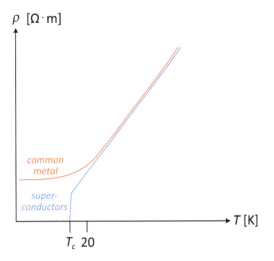

Figure 11.1: Typical temperature dependence of the resistivity for usual metallic behaviour (orange) and superconducting materials (blue).

11.1.1 The Meissner-Ochsenfeld effect and type I and type II superconductors

The superconducting state is characterized not only by the fact that the resistivity is zero, but that superconductors are perfect diamagnets, meaning that they repel (not too strong) magnetic fields inside the material *completely*. This effect is known as the Meissner-Ochsenfeld effect, named after the two German physicists Walther Meissner (1882–1974) and Robert Ochsenfeld (1901–1993), who made their discovery in 1933 [70]. This effect can be impressively demonstrated by levitating permanent magnets above superconducting materials (see Figure 11.2). The repelling of the magnetic field is independent of whether the sample was already superconducting before the magnetic field

Figure 11.2: A small permanent magnet levitates above a high-temperature superconductor cooled with liquid nitrogen (−196 °C); CC BY-SA 3.0 Mai-Linh Doan.

was switched on or is only made superconducting after the magnetic field has been switched on. Although the interior of materials in superconducting states is complete field-free, the outer magnetic field penetrates the superconductor's surface to a certain amount; it is known as the *London penetration depth* of the magnetic field and has a value of approx. 100 nm.

Maintaining the superconducting state of a certain material is immediately linked to three conditions: The material must remain below its critical temperature (T_c) and a certain critical magnetic field strength (H_c), and a certain critical electric current flux density (j_c) must not be exceeded. It is irrelevant whether the magnetic field is externally applied or whether it is generated by the electric current flowing through the material; therefore, the latter two variables are also interlinked.

Depending on how the transition from the superconducting to the normal (metallic) state occurs, superconductors can be divided into two types. In the so-called type I superconductors, an abrupt change from a superconducting to a non-superconducting state occurs at the border of the $H(T)$ curve as shown in Figure 11.3, left. Most pure elements of the periodic table are superconductors of type I and are characterized by small H_c values, which means that they lose their superconductivity in weak-to-modest magnetic fields.

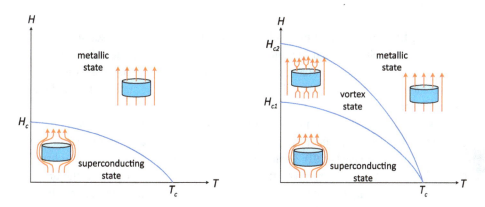

Figure 11.3: $H(T)$ diagrams for type I (left) and type II superconductors (right).

In type II superconductors, when a certain lower critical field strength H_{c1} is exceeded, a transition into the so-called vortex or Shubnikov phase takes place first. Only at an upper critical field strength H_{c2}, a transition to the normal metallic state occurs. In the vortex or mixed state, the material remains superconducting, but the magnetic field is no longer completely expelled from the interior. Rather, it penetrates the material at certain points, in the form of tubes. Inside these magnetic flux tubes, the material is normally conducting, but around these tubes, the material continues to be superconducting. These magnetic flux tubes create a circular current around them; therefore, they act like vortex generators. These vortices are also called Abrikosov vortices, because Alexei Abrikosov (a Russian physicist, 1928–2017) greatly improved the theory of type II superconductors and received the Nobel Prize (together with the Russian physicist Vitaly Ginzburg (1916–2009) and the British-American physicist Anthony James Leggett (*1938)) in Physics in 2003. These vortices can be made visible, e.g., with the help of colloidal iron on the surface of thin superconducting films. In some superconducting type II materials, the vortices form a regular triangular lattice, accordingly called an Abrikosov lattice. However, in other type II superconductors, in particular in high-temperature superconductors (see below), these vortices are not fixed; they move around when an electric current is passing through the material because of the Lorentz force. This leads to energy being dissipated and the electric current can no longer be transported without loss. However, it has been shown that, in particular, the incorporation of impurity atoms can cause the vortices to become fixed; they act as so-called pinning centres. Grain boundaries also act as pinning centres. Type II superconductors, especially those with fixed vortices, can withstand much higher magnetic fields (approx. 20 T; for some of them even much higher fields can be applied) than type I superconductors, and, therefore, also higher electric currents.

11.1.2 The BCS theory

For a long time, the phenomenon of superconductivity was an unsolved scientific mystery. This applies both to the microscopic explanation of the lossless flow of the electric current and to the macroscopic explanation of the phase transition to the superconducting state, which could not be reconciled with the state of thermodynamics developed at that time.

It was not until 1957 that a quantum mechanical explanation of superconductivity was provided by the BCS theory [71], named after its originators, the American physicists, John Bardeen (1908–1991), Leon Neil Cooper (*1930), and John Robert Schrieffer (1931–2019), for which they received the Nobel Prize in Physics in 1972. At the heart of this theory is the formation of Cooper pairs. These are two electrons with opposite spin, mediated by the lattice polarization of the positively charged atomic cores. These pairs are then able to avoid scattering or collision with the vibrating atoms. The circumstance that lattice vibrations play a crucial role in superconductivity was recognized quite early because it was observed that the critical temperature of an element exhibiting superconductivity depends on the isotope used: The heavier the isotope, the lower the frequency of the lattice vibration and the lower the critical temperature. This is also consistent with the fact that such metals that are very good ordinary electrical conductors (Cu, Ag, Au) do not exhibit superconductivity because the extremely high ordinary conductivity indicates that the electron-phonon interactions are very much reduced.

The explanation of how Cooper pairs are formed is as follows: An electron in the lattice of the surrounding positive atomic cores pulls them slightly towards it because of the opposite charge, but the electron moves on very quickly. Since the oscillation of the atomic cores happens comparatively slowly, there is a small positive charge surplus at these locations; the lattice is weakly positively polarized. As a result, a second electron is now attracted to these places and moves cooperatively in the polarization trace of the first electron (see Figure 11.4). This means that these two electrons are indirectly coupled; their bonding energy is about 10^{-3} eV. Not only is the spin state of the two electrons opposite, they also move in opposite directions, which means that Cooper pairs permanently break up, while others are newly formed. The average lifetime of a Cooper pair is approx. 10^{-12} s. However, the astonishing thing about Cooper pairs is their extent or space requirement, which is about 100–1,000 nm! Due to the fact that the average distance between two electrons is only a fraction, approx. 0.1 nm, each Cooper pair overlaps with thousands of other Cooper pairs.

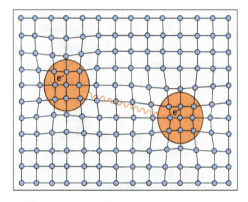

Figure 11.4: The basis of the formation of Cooper pairs is the polarization of the lattice of the atomic cores caused by the electrons. The positive excess charge attracts a second electron and forms a Cooper pair in the polarization trace of the first one.

Now, for the actual explanation of superconductivity, it is not only the fact that there are weakly bound pairs of electrons that is decisive, but that these Cooper pairs are bosons, i.e., particles with integer spin. As bosons, they are not subject to the Pauli principle and can therefore all be present in one and the same quantum state. The result is a coherent state in which the totality of all Cooper pairs can be described by a single wavefunction. Because the electrons are loosely bound to each other in the form of Cooper pairs and the bond energy is larger than any energy that can be transferred by lattice scattering, they can freely move through the material.

If a voltage is applied, the Cooper pairs are accelerated, they receive a momentum. This momentum must be exactly the same for all of them. However, when the current density becomes too high, the increased kinetic energy of the electrons causes the Cooper pairs to break up and the material then conducts like a normal metal.

The BCS theory is valid at least for low-temperature superconductors, i.e., those with a critical temperature below 77 K. Whether and to what extent it can also explain superconductivity in high-temperature superconductors ($T_c > 77$ K) is still under discussion. It seems at least that in these materials Cooper pairs form as well, even if the coupling mechanism is probably different.

In the 1950s and 1960s, the thermodynamic description of the phase change to the superconducting state was formulated in the framework of the Ginzburg-Landau theory, to which Abrikosov also contributed. Later, Gor'kov was able to derive a modified version of this theory out of the microscopic BCS theory.

11.2 Superconducting materials

11.2.1 The elements

In Figure 11.5 the current state of knowledge regarding superconducting properties of the elements is summarized. Apart from the already mentioned fact that the very good electrical conductors do not show superconductivity due to the lack of electron-phonon interactions, the figure reveals a significant amount of current lack of knowledge about the following questions: (i) what makes an element a superconductor, (ii) at which critical temperature do they become superconducting, and (iii) under which circumstances can it be transformed into a superconductor. Obvious is the fact that fabrication of thin films and the application of high pressures are good conditions for this. Nevertheless, a few trends and correlations can be inferred, see also Ref. [72]. For instance, it seems to be the case that none of the ferromagnetic elements at their normal state can be superconductors. In the high-pressure phase (above 10 GPa), Fe adopts the **hcp** structure, which is non-magnetic. Furthermore, although this is not a straight simple linear correlation but rather a general trend, the critical temperature of the elements scales with the atomic number Z: relatively high critical temperatures are observed for Li, Be, B, and C! The highest T_c value at the bulk state under normal conditions is observed for Nb (9.25 K). All elements are type I superconductors.

11.2.2 Binary compounds, alloys, and intermetallics

After superconductivity was discovered, the hunt began for materials with higher transition temperatures, higher critical magnetic field strengths, and higher critical current flow densities. Initial successes were achieved by studying alloys. In 1935, Ryabinin and Shubnikov discovered what are now called type II superconductors (see above) using alloys of lead and bismuth and lead and thallium, respectively. Then, in the 1950–1970s, the so-called A15 phases were discovered, see Section 13.5.5, all of which crystallize in the Cr_3Si structure type, of which Nb_3Ge has the highest transition temperature (T_c = 22.3 K). However, these materials are poorly processable because they are extremely brittle. Therefore, the most common material used commercially for strong magnets (e.g., for Magnetic Resonance Imaging (MRI) devices) is still an alloy of Nb and Ti, crystallizing in the **bcc** structure. NbTi has a relatively low transition temperature of only approx. 10 K but is relatively easy to process and allows very high magnetic strengths and current densities.

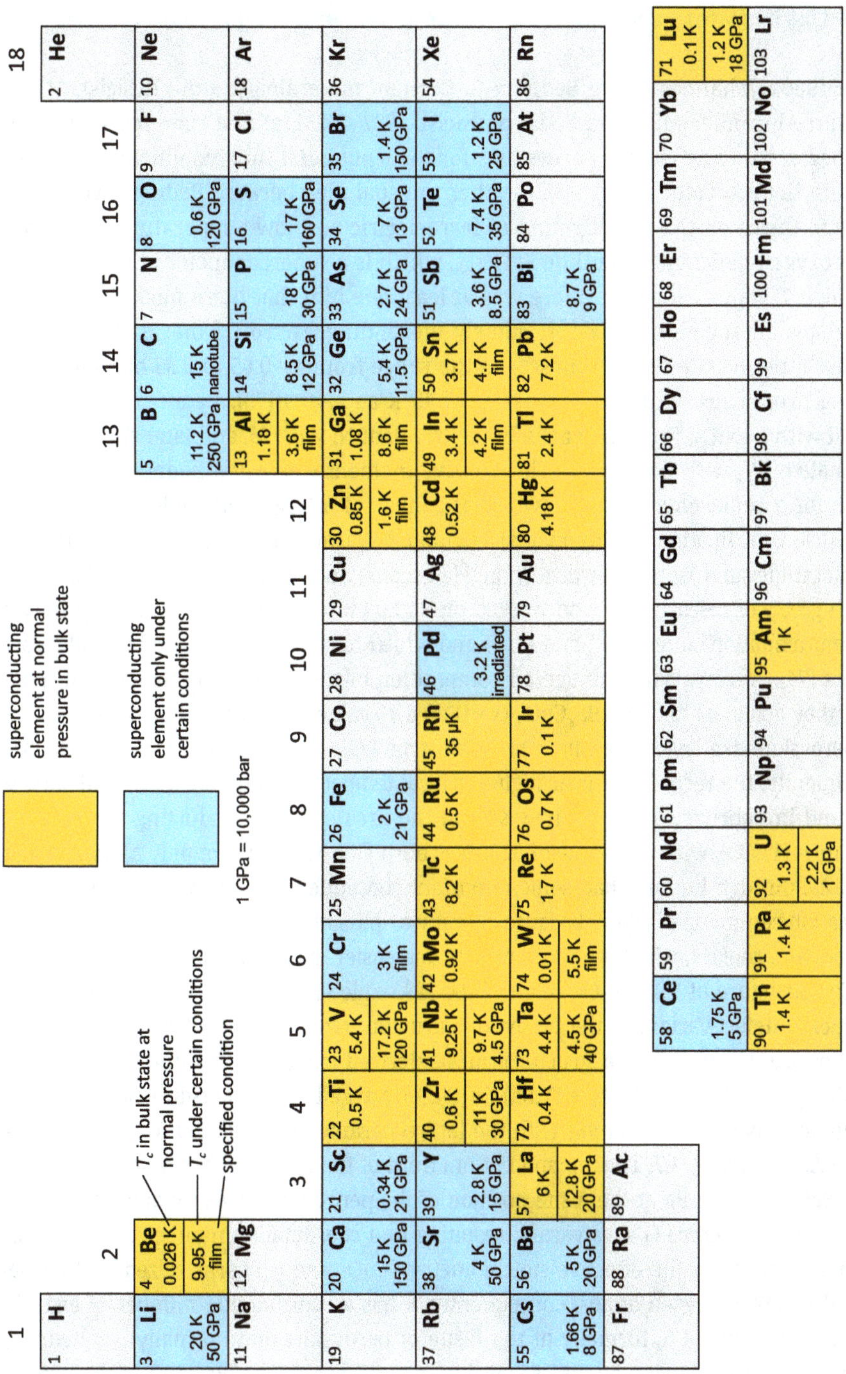

Figure 11.5: Periodic table of superconducting elements; redrawn and modified after [72].

11.2.3 Oxo cuprates – the high-temperature superconducting revolution

In the 1980s, Johannes Georg Bednorz (a German mineralogist and physicist, *1950) and Karl Alexander Müller (a Swiss physicist, 1927–2023), at that time researchers at IBM, began studying certain perovskites for their potential superconductivity properties with high T_c values. This was a rather unusual idea because Bednorz worked in his Ph.D. thesis on the investigation of ferroelectric and low-temperature properties of the oxygen-deficient perovskite $SrTiO_{3-\delta}$, which is a superconductor at 0.3 K – not a very high T_c value. However, there was at least one hint that there might be suitable candidates among certain oxides, namely the finding from 1975 that $BaPb_{1-x}Bi_xO_3$ – likewise a perovskite-related phase – in the range from $x = 0.05$ to 0.35 has superconducting properties, with the highest T_c of 13 K at $x = 0.3$ [73]. Bednorz and Müller started with $LaNiO_3$, replaced partially the trivalent Ni with Al, and later the La^{3+} with the smaller Y^{3+}, with no success. The decisive moment was when Bednorz decided to search for a replacement of Ni^{3+} (which is subject of a large Jahn-Teller distortion in octahedral coordination environments) with another transition metal, namely Cu^{3+} that does not undergo a Jahn-Teller distortion. He became aware of an article from 1985 by a French group of researchers, who worked on Ba-Lu-Cu perovskites, in which Cu has two different oxidation states (Cu^{2+} and Cu^{3+}) and which turned out to be good metallic conductors [74]. After varying the overall composition of the system and its thermal treatment, they achieved for $La_{2-x}Ba_xCuO_4$ ($x \approx 0.15$) a T_c value of 35 K in January 1986, which is a perovskite-like layer-type phase of K_2NiF_4-type [75], see Figure 11.6 and Section 5.3.6. Subsequently, the records for ever higher critical temperatures tumbled in rapid succession, and in February 1987, the new star in the area of superconducting compounds, $YBa_2Cu_3O_7$ (YBCO), with a T_c value of 92 K was born [76]. This was not only a major scientific breakthrough but also had some important consequences with regard to a possible application/commercialization, because now it was possible for the first time to use liquid nitrogen as a coolant, which is much cheaper and easier to handle than liquid helium.

The structure of $YBa_2Cu_3O_7$, often called 123 oxide ($Y_1Ba_2Cu_3$), can be considered as an ordered defect-variant of the perovskite structure type, in which 2/9th of the oxygen positions are vacant ($\square$). The structure can be derived from a tripled unit cell of the perovskite type ($3\ ABO_3 = A_3B_3O_9 \rightarrow (Y,Ba)_3Cu_3O_7\square_2$), stacked along the c-direction of the orthorhombic space group $Pmmm$ with the lattice parameters $a = 3.8209$ Å, $b = 3.8843$, $c = 11.6767$ Å, see Figure 11.7. The top and bottom third of the unit cell have a net stoichiometry of $BaCuO_{2.5}$ with Ba at the A-site position of the perovskite structure and with two of the twelve edge centre O sites vacant, resulting in a coordination number of ten for Ba. The middle unit has the effective stoichiometry $YCuO_2$ and is characterized by four vacant edge centre oxygen atoms; consequently, Y has a coordination number of eight in the form of a cube. Cu, formally at the B-site of perovskite and normally octahedrally coordinated by six oxygen atoms, is split into two different sites. Because of the oxygen deficiency, one is five-coordinated in a square-pyramidal fashion, while the second is even only fourfold coordinated in a square-planar fashion. These squares are

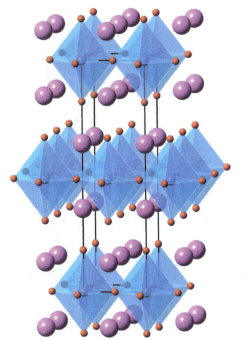

Figure 11.6: The crystal structure (tetragonal space group $I4/mmm$, $a = b = 3.8008$ Å, $c = 13.3608$ Å) of $La_{1.85}Ba_{0.15}CuO_4$; La and Ba, which share the same site, are shown in pink, Cu in blue, and oxygen atoms in red.

oriented parallel to the (b,c) plane and are corner-linked to neighbouring squares along the b-direction to form infinite chains. They are further corner-connected along the c-direction to the adjacent square-pyramids via common oxygen atoms. The square-pyramidal units are corner-connected within the (a,b) plane and form sheets. According to localized charge distribution considerations, the Cu atoms in the square-pyramidal coordination environments have oxidation state +2, while the square-planar coordinated Cu atoms are assumed to have oxidation state +3. The conductivity in YBCO in its superconducting state is anisotropic, being significantly higher within the (a,b) plane than along the c-direction.

A crucial parameter for the properties of $YBa_2Cu_3O_{7-x}$ is the oxygen content. With further increase of the oxygen deficit, the transition temperature decreases. For $x = 0$ the T_c value is 92 K, which then decreases with increasing x to about 60 K for $x = 0.25$, and at $x > 0.5$, superconductivity is no longer observed.

Subsequently, after the discovery of the superconducting properties of YBCO, more oxocuprates with increasingly complicated compositions and higher transition temperatures were synthesized. The structures of these oxocuprates can be viewed as sequences of layers of MO or M and CuO_2. The highest transition temperatures so far were found in the systems, Bi-Sr-Ca-Cu-O ($T_c \approx 110$ K), Tl-(Ba,Sr)-Ca-Cu-O ($T_c \approx 125$ K), and Hg-Ba-Ca-Cu-O ($T_c \approx 133$ K, under pressure: $T_c \approx 160$ K). The structures of two representatives in the homologous series $Tl_2Ba_2Ca_nCu_{n+1}O_{2n+6}$ for $n = 0$ and 1 are shown in Figure 11.7 (right), which are also referred to simply as 2201- and 2212-TBCCO according to their compositions.

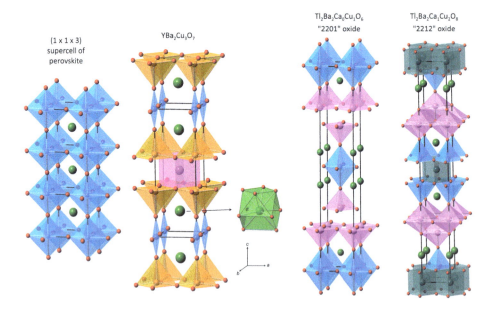

Figure 11.7: The crystal structure of YBa$_2$Cu$_3$O$_7$ in relation to the perovskite structure type (left) and the '2201' and '2212' oxocuprates of the system Tl-(Ba,Sr)-Ca-Cu-O (right); the coordination polyhedron of the Ba atom in YBCO is shown separately; A- and B-sites in perovskite are shown in blue and green, respectively, oxygen in red. In the YBCO structure, the two crystallographically distinct Cu sites are shown in blue (square-planar coordination) and orange (square-pyramidal coordination), respectively, the Y atoms in pink. In the Tl-Ba-Ca-Cu-O system, Tl is shown in pink and Ca in green-blue.

11.2.4 Further superconducting compounds

Chevrel phases

In the 1970s, an interesting class of compounds with unusual magnetic and electric properties was discovered by Chevrel *et al.* [77]. These compounds are since then known as Chevrel phases and are ternary molybdenum chalcogenides with the general formula MMo$_6$X$_y$ (y = 6–8), where M may be one of a variety of metals (Sn, Pb, Ba, Ag, Cu, La, RE, etc.) and X = S, Se, or Te. Some examples of substitution of Mo by Rh or Ru are also known. The best known Chevrel phase is PbMo$_6$S$_8$. It is superconducting below the relatively high critical temperature of T_c = 15 K. At the time of their discovery, i.e., well before the oxocuprate revolution occurred, they were even called 'high-temperature' superconductors. But the really interesting aspect of the Chevrel phases is something else: first, they can tolerate extremely high critical magnetic field strengths, which amounts to 40 T for SnMo$_6$S$_8$, 45 T for LaMo$_6$S$_8$, and 60 T for PbMo$_6$S$_8$, and second, Chevrel phases with M = RE were the first class of compounds in which the coexistence of superconductivity and magnetism was demonstrated. Usually, the exchange interaction of atoms with a magnetic moment and the conduction electrons leads to a decoupling of the Cooper pairs and thus to the loss of superconductivity. The strength of this decoupling effect is manifested by the observation that usually magnetic impurities in the order of 1 at.% leads to the loss of superconductivity. This is different in HoMo$_6$S$_8$, i.e., a compound in which as many as 6.67% of the atoms are magnetic and build a regular lattice (for the structure of Chevrel phases, see below): HoMo$_6$S$_8$ is a *paramagnetic superconductor* below the temperature of 1.8 K. Apparently, the interaction of the holmium atoms at their specific site in the crystal structure with the conduction electrons must be very small. However, the superconducting state of HoMo$_6$S$_8$ is not retained down to 0 K. At temperatures below about 0.7 K, a magnetic phase transition to the ferromagnetic state occurs and the superconductivity collapses. Such superconductors that have an upper and lower temperature limit are called *re-entrant* superconductors.

The structure of the Chevrel phases is comparatively simple, see Figure 11.8. The six molybdenum atoms form an (regular) octahedral cluster, whose atoms are located on the face centres of a surrounding cube of eight chalcogenide atoms. This inner pair of polyhedra are embedded in another, larger (slightly rhombohedrally stretched) cube of eight M atoms, but is usually rotated about the common threefold axis of rotation of all three polyhedra (the body diagonal is identical with the threefold axis of the rhombohedron). For $y = 6$ or 7, two or one of the chalcogenide atoms (located at the special position at that threefold axis of rotation) are absent. For PbMo$_6$S$_8$, the rotation angle of the inner pair of polyhedra is approx. 25°.

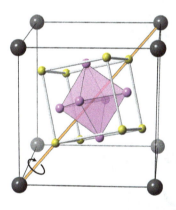

Figure 11.8: Crystal structure of PbMo$_6$S$_8$. The body diagonal of the rhombohedral structure (space group $R\bar{3}$, $a = b = c = 6.544$ Å, $\alpha = \beta = \gamma = 89.480°$) at which the inner pair of polyhedra is rotated is shown as an orange stick.

Alkali metal fullerides

After the discovery of the C$_{60}$ fullerene by Kroto et al. in 1985 [78] and the development of methods to produce fullerenes in larger quantities in the early 1990s, it was found out that some alkali metal fullerides are superconducting. Those with the composition M$_3$C$_{60}$ have the highest critical temperatures, for instance, K$_3$C$_{60}$ with T_c = 18 K, Rb$_3$C$_{60}$ with T_c = 28 K, RbCs$_2$C$_{60}$ with T_c = 33 K and Cs$_3$C$_{60}$ under pressure of 7 kbar with T_c = 38 K (at ambient pressure Cs_3C_{60} is non-superconducting). Cs_3C_{60} has the highest T_c of any *molecular* material. Interestingly, Na_3C_{60} is not superconducting at all.

Different to the fullerides that contain K and Rb, which crystallize in the space group $Fm\bar{3}m$ with the C_{60} molecules located on a **ccp** lattice and the alkali metals in the octahedral and tetrahedral voids, Cs_3C_{60} crystallizes in the A15 structure (space group $Pm\bar{3}n$, Cr_3Si structure type, compare Section 13.5.5), i.e., the fulleride anions form a "**bcc**-like" sublattice (note that the C_{60} molecules have two different orientations), while the Cs cations are at the Wyckoff position 6d to form "Cs_2 pairs" located at the face centres with their axis oriented along the x-, y-, and z-axis respectively (see Figure 11.9). Cs_3C_{60} is also special in another respect: Once the superconducting state is established, T_c initially increases with decreasing volume per C_{60} molecule (i.e., with increasing pressure), before decreasing again upon further increase in pressure, which is in complete contrast to the behaviour of the **ccp**-based compounds K_3C_{60} and Rb_3C_{60} under pressure.

Meanwhile, substantial progress has been made in order to understand the observed phenomena in the fulleride-based superconductors [79], however, a fully developed theory of C_{60}^{3-} superconductivity is still lacking.

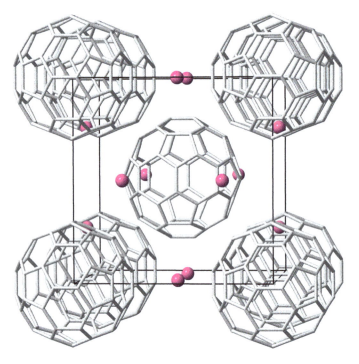

Figure 11.9: Crystal structure of Cs_3C_{60}; space group $Pm\bar{3}n$; the Cs atoms are shown as pink spheres and the C_{60} molecules are represented as stick models.

MgB$_2$

The superconducting properties of the simple metallic compound MgB$_2$, a compound known since 1953, crystallizing in the prototypical structure of AlB$_2$ (*Strukturbericht* type C32, space group *P6/mmm*, compare Figure 13.8) were discovered only in 2001 [80]. MgB$_2$ or also other simple metallic compounds can be considered in the field of superconductivity as somehow "forgotten" compounds: With the discovery of the high-temperature (type II) superconductors, researchers have been delving into the complex world of these ceramic type of materials – virtually ignoring simple metallic compounds, also because their critical temperature is usually very low. With a T_c value of 39 K – i.e., 16 K higher than any other simple metallic compound – MgB$_2$ has by far the highest critical temperature of all metallic compounds. Initially, it was thought that MgB$_2$ belongs to the type I superconductors, but meanwhile there is more and more experimental evidence that it is neither a pure type I nor pure type II superconductor. It seems that in MgB$_2$ there are two different kinds of electrons, one behaving like electrons in type-I materials, the other behaving like electrons in a type-II material, for which the term "type-1.5 superconductors" was coined. This type is characterized by the circumstance that the interaction of the vortices in the Meissner state is attractive at long range and repulsive at short range, which leads to a "semi-Meissner state": to domains of very densely packed vortices at the one hand and empty domains, domains with no vortices, no currents and no magnetic field at the other [81].

Latest developments

The commercial use of superconducting materials has so far been limited to special applications. Broader applications are hindered by two important issues: the complex cooling required even for the so-called high-temperature superconductors and the poor processability of the materials, which does not easily allow them to be formed into wires, for example. Therefore, there is still an unabated effort to explore new materials that are ideally room-temperature superconductors and easy to process. The question of whether this goal is realistic cannot be answered easily. After the discovery and optimization of the oxocuprate-based materials, the maximum achievable critical temperature remained almost the same for more than two decades. However, the elements of the BCS theory of *conventional* superconductors provide, in principle, a guide for the development of promising materials with no upper theoretical bound. Materials with a high critical temperature should exhibit high-frequency phonons, strong electron-phonon interaction, and high density of states at the Fermi level. These conditions can, in principle, be fulfilled for metallic hydrogen and covalent compounds dominated by hydrogen as hydrogen compounds provide the necessary high-frequency phonon modes as well as strong electron-phonon coupling. In this way, binary and, to some extent, ternary hydride compounds have recently come into focus, which were initially investigated theoretically and promised high transition temperatures. The highest T_c value with 250 K was recently experimentally demonstrated for the "superhydride" compound LaH$_{10}$, although only under a pressure of 170 GPa [82]. Nonetheless, this is the highest critical temperature that has been confirmed so far in a superconducting material. LaH$_{10}$ crystallizes in the space group $Fm\bar{3}m$ (a = 5.1019(5) Å) and has a very interesting clathrate-like structure, in which the hydrogen atoms form a net that is identical with the net of the zeolite framework-type AST. The La atoms occupy the Wyckoff position 4b (0, 0, 0), and the H atoms the 8c (0.25, 0.25, 0.25) and at the 32f position (0.125, 0.375, 0.125). The atoms at the 32f position constitute the characteristic H$_8$ cubes present in the structure, with their barycentres at the octahedral voids of the **ccp** packing of the La atoms. The La atoms are surrounded by a [$4^6 6^{12}$] polyhedron of 32 H atoms. The structure of LaH$_{10}$ is shown in Figure 11.10.

In addition to the renaissance taking place in the field of conventional (type I) superconductors, however, other classes of materials are also opening up new perspectives. Research is already being carried out on a completely new type of superconductors, the so-called Kagome superconductors, which, in addition to superconductivity, exhibit other extraordinary quantum phenomena, such as time-reversal symmetry breaking, as recently experimentally demonstrated with the metallic compound KV$_3$Sb$_5$ [83]. This class of material is being considered as a hot candidate for room-temperature superconductors.

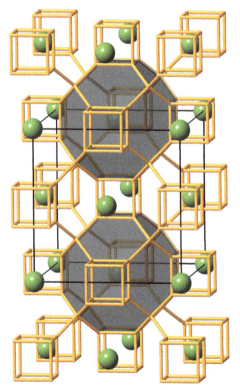

Figure 11.10: Crystal structure of LaH$_{10}$. The La atoms, shown as green spheres, are surrounded by 32 H atoms (represented as bonded orange sticks) in form of [$4^6 6^{12}$] polyhedra, highlighted in translucent black.

In Table 11.1 superconductors relevant to this chapter are listed with their critical temperatures and critical magnetic field strengths (where consolidated values exist).

Table 11.1: Overview of superconducting compounds with their critical temperatures and critical magnetic field strengths.

Compound	T_c (K)	H_c (T)
Hg	4.18	0.0415
Nb	9.25	0.2
NbTi	10	12
PbMo$_6$S$_8$	15	60
Nb$_3$Ge	23	30
Rb$_3$C$_{60}$	28	44
RbCs$_2$C$_{60}$	33	0.8
MgB$_2$	39	14 (parallel to the (a,b) planes)
YBa$_2$Cu$_3$O$_7$	92	138 (in thin films with H applied parallel to the c-axis)
Tl$_2$Ba$_2$Ca$_2$Cu$_2$O$_{10}$	125	N.A.
HgBa$_2$Ca$_2$Cu$_3$O$_8$	133	N.A.
LaH$_{10}$	250 (@ 170 GPa)	136

Further reading

S. J. Blundell, *Superconductivity: A Very Short Introduction*. Oxford University Press, USA, **2009**.

Bennemann, K. H., J. B. Ketterson (Eds.), *Superconductivity – Volume 1: Conventional and Unconventional Superconductors, Volume 2: Novel Superconductors*. Springer, Berlin Heidelberg, **2008**.

R. Kleiner, W. Buckel, R. Huebener, *Superconductivity: An Introduction*, 3rd ed. Wiley-VCH, Weinheim, Germany, **2015**.

12 Ceramics

It has already been mentioned in the introduction to this book that it is not easy to draw a clear line between the subjects of solid-state chemistry and related disciplines, such as materials science. This also applies to this chapter about ceramics. Ceramics were already invented when modern science did not exist yet. In the early times of their development, the production of ceramics had a strong empirical character. How can I treat a certain raw material in order to obtain (everyday) objects that are useful or simply beautiful? These were the questions of former times. Progress in natural sciences, specifically in the field of solid-state chemistry, has meanwhile led to developments in the area of ceramics that were never thought of before. For instance, overcoming the typical brittleness of certain ceramics has led to technical and high-performance ceramic materials. Therefore, in this chapter – after a brief introduction to ceramics – some selected aspects and materials from the field of ceramics will be presented. There are many relations to certain other chapters in this book as we will see.

12.1 Definition and classification of ceramics and the ceramic method

The question what ceramics actually are is not easy to answer. Historically, the word "ceramics" originated from an area of Athens called Kerameikos. It was the potters' quarter of the city and initially the term "ceramics" referred to the clay minerals and their shape-retaining products when they were subjected to a firing process. Today, a common definition is that ceramics are solid compounds that are formed by applying heat to certain raw materials, comprising at least two elements provided one of them is a non-metal solid and the other may be a metal or non-metallic solid element. A negative definition would be that ceramics are all solids that are neither metallic, intermetallic, nor organic compounds or polymers. The German Ceramic Society defines ceramics as materials that are inorganic, non-metallic, sparingly soluble in water, and at least 30% crystalline; as a rule, they are formed from a raw mass at room temperature and obtain their typical material properties by a temperature treatment usually above 800 °C.

The high-temperature treatment of the solid starting materials, in the course of which a *solid-solid reaction* takes place, is an essential feature of the *ceramic method*. The main process steps for the preparation of ceramics are:
- the production of the raw materials in the form of fine powders;
- mixing and processing of these powders, addition of dispersion agents (water or organic solvents) and, if necessary, other additives such as binders, deflocculants, plasticizers, etc.;
- shaping of that mixture to the so-called green body;

- baking and compacting of the green body; and finally,
- firing and sintering of the green body to the final so-called white body.

The preparation of the starting material is of decisive importance for the quality of the product, in particular for modern high-performance ceramics. To produce a uniform green body, the particle sizes of the powders must be in the submicrometre range (0.1–0.005 µm). The formation of agglomerates in the green body creates sample inhomogeneities, which after sintering appear as strength-reducing defects in the workpiece.

Ceramics consist of more or less statistically intergrown crystallites (typically in the range of 0.5–50 µm), i.e., they are polycrystalline solids. They always contain defects (compare Chapter 6), which are decisive for many properties of a ceramic material. These defects include point defects such as vacancies, incorrect atomic species (foreign atom), and the occupation of interstitial sites. Furthermore, edge or screw dislocations can occur within individual grains. Likewise, ceramics can have coherent (lattice-oriented) or incoherent (non-lattice-oriented) precipitates, or precipitates at the grain boundaries. The control over the resulting microstructure can be considered as the art of making ceramics.

A clear systematic of ceramics does not exist, because there are fluent transitions with regard to the composition of the raw materials, the firing process, and the design process. Ceramic products are therefore often differentiated according to the respective focus of attention. For instance, these can be regional aspects, meaning the classification according to the provenance of the articles.

Technical ceramics are often divided according to their chemical composition into
- silicate ceramics,
- glass-containing ceramics,
- oxide ceramics, and
- non-oxide ceramics.

Another classification scheme is based upon their intended use:
- functional ceramics,
- utility ceramics,
- building ceramics,
- sanitary ceramics, and
- structural ceramics.

High-performance ceramics can be roughly divided into structural and functional ceramics. Structural ceramics include materials that can withstand mainly mechanical loads. Functional ceramics are materials that have, for example, certain electrical, magnetic, or optical properties.

Modern ceramics have an incredibly wide range of applications and properties that the ancient Greeks could not have imagined being implemented in ceramics. In Table 12.1, an overview of selected ceramic materials and their field of application is given.

Table 12.1: Overview of fields of application of ceramic materials.

Property class	Applications and materials (examples)
Thermal	
Thermal insulation	High-temperature furnace linings (SiO_2, Al_2O_3, ZrO_2)
Fire resistance, thermal shock resistance	Furnace linings for containers of molten metals and slags, i.e. blast furnaces (SiO_2, Al_2O_3, SiC, Al_2TiO_5)
Heat resistance	Heat shields in air- and spacecraft (carbon-fibre reinforced carbon, ZrB_2, HfB_2)
Thermal conductivity	Heat sinks in integrated circuit packages (AlN)
Mechanical	
Hardness	Cutting tools for cutting silicon wafers (SiC, Si_3N_4)
Wear resistance	Bearings (Si_3N_4) and abrasives (Al_2O_3, Al_2O_3/ZrO_2, SiC, BN)
Wear resistance and biocompatibility	Prostheses (Al_2O_3, ZrO_2)
High-temperature strength retention	Turbine blades, ceramic engine parts (Si_3N_4)
Chemical	
Corrosion resistance	Heat exchangers and chemical equipment in corrosive environments (SiC)
Catalyst substrate	Synthetic cordierite ($Mg_2Al_3[AlSi_5O_{18}]$) ceramic as catalyst substrate for purification of exhaust gases
Nuclear	
Nuclear fission	Nuclear fuel (UC), fuel cladding (SiC), neutron moderators (BeO), neutron absorbers (B_4C), neutron reflectors (WC)
Nuclear fusion	Fusion reactor lining (SiC, Si_3N_4)
Optical	
Radio wave transparency	Weatherproof enclosures that protect radar antenna (radomes) (Al_2O_3, BeO)
Translucency for light and chemical inertness	Heat- and corrosion-resistant materials for discharge tubes in Na lamps (Al_2O_3, MgO)

Table 12.1 (continued)

Property class	Applications and materials (examples)
Non-linearity	Q-switches for lasers ($LiNbO_3$)
Light emitting	Green and blue LEDs (GaN)
Magnetic	
Hard magnetic	Loudspeaker (hard magnetic ferrites, $(Ba,Sr)O \cdot 6\, Fe_2O_3$)
Soft magnetic	Transformer cores (soft magnetic ferrites, $Mn_xZn_{(1-x)}Fe_2O_4$)
Electrical and dielectric	
Environmental-dependent conductivity	Gas sensors (SnO_2, ZnO)
Heat-dependent conductivity	Temperature control (thermistors) (NiO, TiO_2)
Voltage-dependent conductivity	Surge protection device (varistors) (SiC, ZnO)
Superconductivity	Wires and SQUID magnetometers ($YBa_2Cu_3O_7$)
Ferroelectricity	Capacitors ($BaTiO_3$)
Piezoelectricity	High-voltage generators, relays ($Pb(Zr,Ti)O_3$)
Ion conductivity	Solid oxide fuel cells (ZrO_2)
Dielectric	Electrical insulators with high thermal conductivity, especially in semiconducting devices (Al_2O_3, AlN)

In the following section, some types of ceramics classified according to their chemical composition will be presented in more detail.

12.2 (Alumo)silicate ceramics (traditional ceramics)

Silicate ceramics are characterized by their main component, consisting of linked $[SiO_4]^{4-}$ units, whereby in the case of aluminosilicates, the Si atom may be substituted by Al. Corresponding materials are divided into coarse and fine ceramics. Coarse ceramics include building materials such as bricks, clinker, and refractory bricks. These products tend to be coarse-grained and are often inhomogeneous and of random colouration. Fine ceramics include porcelain (tableware, dental porcelain), stoneware, or earthenware (tiles, sanitary ware). These products have a smaller grain size (< 0.05 mm) and are of defined colour (e.g., white for household ceramics, tableware, and sanitary ceramics); artistic products also belong to fine ceramics. Fine ceramics require considerably greater care in the preparation of the raw material, shaping, drying, and firing than is necessary for the production of coarse ceramics.

The main raw materials for ceramics are clay minerals, whose main components are in turn illite and montmorillonite. Illite is a group of closely related non-expanding clay minerals. They belong to the class of phylloalumosilicates (layered aluminosilicates). The characteristic of illite minerals is the 2:1 sandwich structure with the sequence of SiO_4 tetrahedra (T), AlO_6 octahedra (O), and a layer of SiO_4 tetrahedra (T) again (see Figure 12.1). The space between these three layers is occupied by poorly hydrated potassium cations which are responsible for the absence of swelling. A representative chemical formula would be $K_{0.6-0.85}Al_2(Si,Al)_4O_{10}(OH)_2 \cdot H_2O$, but the compositional space is larger and there can be also Mg or Fe present. Illite crystallizes in the monoclinic space group $C2/m$. Montmorillonite is a very similar mineral group of 2:1 clay minerals, but here essentially sodium and calcium are intercalated in the space between the three-layer T-O-T sequence. Other raw materials are quartz sand, mullite (composition from $2\,Al_2O_3 \cdot SiO_2$ to $3\,Al_2O_3 \cdot 2\,SiO_2$), and sillimanite ($Al_2SiO_5$). Mullite and sillimanite are closely related minerals, and they both belong to the group of nesosilicates that are characterized by the presence of isolated SiO_4 units. Mullite is also produced synthetically for ceramic production.

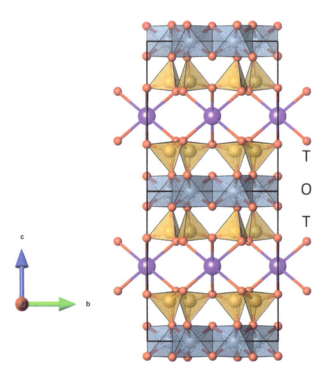

Figure 12.1: Crystal structure of the mineral illite [84] shown along the a-direction; it has a sandwich structure of a T-O-T-layer sequence between two layers of loosely hydrated potassium ions (purple); hydrogen atoms are not shown.

Porcelain is one of the most popular kinds of ceramics for kitchenware. The difference between ceramics and porcelain is the composition of the raw materials: porcelain requires the addition of kaolin, also known as China clay. Kaolin refers to rocks that are rich in the mineral kaolinite, which is a weathering product of feldspar. Kaolinite is a 1:1 (or T-O) clay mineral that consists of double layers of SiO_4 tetrahedral and AlO_6 octahedral connected sheets with the chemical formula $Al_2Si_2O_5(OH)_4$ ($Al_2O_3 \cdot 2\ SiO_2 \cdot 2\ H_2O$). These double layers are connected via hydrogen bonds to each other. In pure form, kaolinite is colourless or white, respectively. Minor constituents of kaolin are quartz, undecomposed feldspar, and mica. The sintering temperature of porcelain is typically higher (approx. 1,300–1,400 °C) compared to that of ceramics (900–1,200 °C). This causes the feldspar component of the porcelain to melt but it does not recrystallize during cooling. This special detail causes porcelain to have a slightly transparent, shimmering, noble appearance.

12.3 Binary oxide ceramics

The most important, simple (binary) oxide ceramics include α-Al_2O_3, ZrO_2, TiO_2, MgO, and BeO. In pure form, however, ceramics based on TiO_2, MgO, and BeO are used only to a very limited extent.

12.3.1 α-Al$_2$O$_3$

α-Al_2O_3 – whose structure has already been described in Section 5.4.7 – is by far the most important oxide ceramic material (approx. 85%). Sintered α-Al_2O_3 (also known as sintered corundum or sintered alumina) is chemically extremely resistant, is mechanically very stable, and tolerates extremely high temperatures (the melting point is approx. 2,053 °C). It is used as a refractory material for furnace linings. Another large field of application is abrasives of all kinds, since α-Al_2O_3 has a very high abrasion resistance. Furthermore, α-Al_2O_3 is a very good electrical insulator, but has a relatively high thermal conductivity for ceramics of approx. 25–30 $Wm^{-1}\ K^{-1}$. Due to these two properties, α-Al_2O_3 is used as an insulator for spark plugs, and in electrical engineering and semiconducting devices as a dielectric and simultaneous heat sink.

12.3.2 ZrO$_2$

ZrO_2 (zirconia) occurs in three different modifications. The most stable modification at room temperature crystallizes in the monoclinic crystal system (space group $P2_1/c$, no. 14), in which the Zr atoms have the coordination number seven. This modification also occurs in nature as the mineral baddeleyite. It transforms to the tetragonal form (space group $P4_2/nmc$, no. 137) at above 1,173 °C, and to the cubic form at above 2,350 °C, which forms a CaF_2-analogous structure (space group $Fm\bar{3}m$, no. 225) with a Zr-O distance of 2.20 Å. The tetragonal form can be considered as a distorted fluorite-type structure in which the Zr atoms have four shorter (2.07 Å) and four longer (2.46 Å) Zr-O distances. These thermal transformations are fully reversible in pure ZrO_2, which in the first instance limit the temperature range in technical applications. The large increase in volume that occurs, especially during the conversion of the tetragonal ($\rho = 6.1$ g cm^{-3}) to the monoclinic ($\rho = 5.7$ g cm^{-3}) form, easily leads to cracks in the material. However, by adding other metal oxides, the tetragonal or cubic high-temperature modification can be stabilized at low temperatures, i.e., the phase transition back to the monoclinic phase can be suppressed. Properties such as strength and translucency of the high-temperature modifications can thus be maintained at room temperature. Depending on the amount of stabilizing additives, a distinction is made between partially and fully stabilized tetragonal or cubic ZrO_2. An amount of at least 16 mol% calcium oxide (CaO) or 16 mol% magnesium oxide (MgO) or about 10 mol% Y_2O_3 is sufficient for crystallization in the cubic phase at room temperature. Smaller amounts stabilize the tetragonal phase. Partially stabilized phases increase the thermal shock resistance if, for example, the presence of mixed phases of the cubic and monoclinic phases create an internal pre-stress in the microstructure. Also, not all applications require complete suppression of phase conversion to the monoclinic phase. Conversely, the volume increase that occurs in the course of conversion to the monoclinic phase is exploited in hip joint implants made of ZrO_2. This is because this conversion to the monoclinic phase also takes place in a pressure-induced manner: When a patient with a hip implant performs an unfavourable movement (for instance when jumping) that leads to a crack in the implant, the transformation to the monoclinic phase takes place at the crack site due to the stress in the material; the volume increase thus heals the crack and is prevented from spreading further.

Zirconia is used as a refractory ceramic due to its good thermal resistance and as a technical ceramic in mechanical engineering (e.g., rolling-element bearings) due to its good mechanical properties. Due to its biocompatibility, ZrO_2 ceramics are further used as a prosthetic material in medical technology (all-ceramic dentures, component of hip joint implants as mentioned already above). Due to its good abrasion resistance and chemical resistance, zirconium oxide is also used in ball mills as grinding balls.

Yttria-stabilized ZrO_2 (YSZ) contains oxide ion vacancies which make it a good oxide ion conductor at elevated temperatures (compare Section 6.1.7). This property is exploited in the lambda (O_2) sensor, where lambda refers to air-fuel equivalence ratio, usually notated as λ. It is used to control and to optimize catalytic exhaust gas purification.

Furthermore, YSZ is used as a solid electrolyte in solid oxide fuel cells. It has been observed that, although not fully cubically stabilized, 8YSZ/8YDZ doped with 8–9 mol% has the highest conductance for oxygen ions in the Y_2O_3-ZrO_2 system almost independent of temperature in the range between 800 and 1,200 °C. In a recent work, 8–9 mol%

YSZ has been found to operate in a miscibility gap of the Y_2O_3-ZrO_2 system at typical SOFC operating temperatures, and it therefore segregates into Y-depleted and enriched regions on the nanometre scale within a few 1,000 h [85]. This chemical and microstructural segregation is directly linked to the drastic decrease in oxygen conductivity (degradation of 8YDZ) of about 40% at 950 °C within 2,500 h. Therefore, research is being conducted on ionic conductors that are permanently stable at high temperatures, which is attempted to be achieved by co-doping (e.g., Sc and Y).

12.3.3 TiO_2, MgO, and BeO

TiO_2 powder is added to various ceramic bodies to influence the sintering behaviour and grain size growth (sintering auxiliary). The proportion of TiO_2 powder for TiO_2 ceramics in the narrower sense is relatively low. Because of its relative high value of permittivity, TiO_2 ceramics are used for some capacitor materials. In this case, it is important that the firing process is done in O_2-containing atmosphere as otherwise Ti^{3+} species are generated that lead – in higher amounts – to semiconducting ceramic bodies.

Sintered MgO (MgO crystallizes in the NaCl-type structure) has a very high melting point (above 2,800 °C); therefore, the majority of MgO-based ceramic materials are used in the form of magnesia or chrome magnesia bricks as refractory materials. Particularly in the field of iron and steel metallurgy, MgO-based materials are widely used due to their very high corrosion resistance to basic slags.

BeO – which crystallizes in the wurtzite structure – has excellent properties that make it a very interesting material in the semiconductor industry: It is an electrical insulator with very good thermal conductivity. However, its use is severely limited due to its high toxicity and is increasingly being replaced by other materials (α-Al_2O_3 and AlN). Only in the field of nuclear technology, it has a certain importance as neutron moderator and adsorber.

12.4 Mixed oxide ceramics

12.4.1 Aluminium titanate

Sintered aluminium titanate (Al_2TiO_5, alternatively written as $Al_2O_3 \cdot TiO_2$), also known as tialite and in the ceramic community only referred to as ATI, is a material with a very low thermal expansion coefficient, high thermal shock resistance, high refractoriness, and good corrosion resistance. Usually, the production is carried out by preparing a stoichiometric mixture of fine-grained corundum (Al_2O_3) and rutile (TiO_2) in the form of a pellet and subsequent sintering. Aluminium titanate can decompose into its parent materials in the temperature range between 900 and 1,300 °C. This is prevented by the addition of various additives (e.g., MgO).

The excellent thermal shock resistance of the final product results from a high amount of residual porosity and microcracks in the structure. This is caused by the highly anisotropic thermal expansion coefficient of the single ATI crystals. This coefficient is positive in the *a*- and *b*-direction of the orthorhombic structure (see Figure 12.2) and negative in the *c*-direction. During production, and especially during cooling, this leads to stresses in the microscopic range, which result in the formation of these microcracks. With regard to the macroscopic expansion behaviour, the material exhibits very low thermal expansion due to the anisotropic thermal expansion of the individual crystals, as this initially occurs in the interior of the material into the microcracks, meaning that these cracks close.

Due to its good thermal shock resistance as well as low thermal conductivity, aluminium titanate is used in the refractory industry, especially where non-ferrous metal melts occur, because ATI is characterized by a very low wettability by such melts.

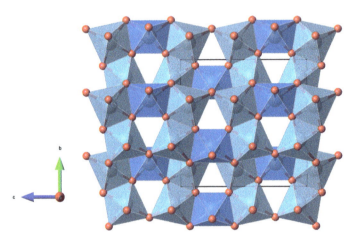

Figure 12.2: The crystal structure (space group *Cmcm*, no. 63; $a = 3.605$ Å, $b = 9.445$ Å, $c = 9.653$ Å) of Al_2TiO_5 viewed along the *a*-direction according to Ref. [86]; distorted MO_6 octahedra (M = Al, Ti) are present, which are edge-shared in the (*b*,*c*)-plane and further corner-connected to neighbouring octahedra along the *a*-direction. The metal ions occupy the Wyckoff positions 4*c* and 8*f* (the respective coordination octahedra are shown in darker and lighter blue). There is a slight preference of the 4*c* position for the Ti ions.

12.4.2 Barium titanate and lead zirconate titanate

Barium titanate ($BaTiO_3$) and lead zirconate titanate ($Pb(Zr_xTi_{1-x})O_3$, abbreviated as PZT) belong to a special class of dielectric ceramics, namely those with ferroelectric properties. Ferroelectrics are characterized by the fact that (a) they have very high values of permittivity ($\varepsilon_r = 1{,}000\text{--}10{,}000$), (b) that the polarization is maintained after application of an electric field, whereas a distinction is made between soft and hard ferroelectrics according to the value of remanence and the ease of switching the direction of polarization, and (c) that the direction of polarization is switchable. All relevant processes and terms are completely analogous to the field of ferromagnetic materials (see Section 9.2.1), the only difference being that in ferroelectrics electrical dipole moments are involved instead of magnetic moments.

The emergence of electric dipole moments is best illustrated on the basis of the phase transition of barium titanate. Above 120 °C BaTiO$_3$ is present in the cubic perovskite structure with the space group $Pm\bar{3}m$ (no. 221). In this phase, called the paraelectric phase, there is no electric dipole present. However, if BaTiO$_3$ is cooled down below 120 °C, a phase transitions towards the ferroelectric tetragonal phase (space group $P4mm$, no. 99) takes place. This happens in such a way that the titanium ions move from the octahedral centre in one of the six possible directions towards an oxide ion (Figure 12.3). The displacement of the titanium species is only about 0.1 Å. Nevertheless, this tiny movement is sufficient to induce a rather strong dipole moment. In areas with diameters of several 10 to over 100 Å, the titanium ions move from the centre of the octahedron in the same direction, corresponding to one of six possibilities (along the negative or positive x-, y-, or z-axis). Therefore, a uniformly aligned polarization occurs in these coherent domains. But overall, the material consisting of many such domains is initially unpolarized since domains with all six possible polarization directions occur with equal probability. The behaviour of ferroelectric materials in the electric field can be described completely analogous to that of ferromagnetic materials in the magnetic field.

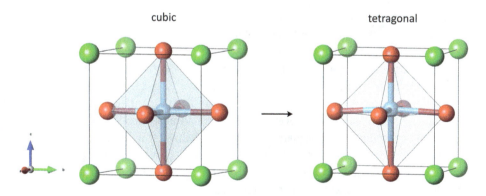

Figure 12.3: Phase transition from the cubic paraelectric phase of BaTiO$_3$ to the ferroelectric tetragonal phase: Ba, green; Ti, light blue; O, red.

While in the case of BaTiO$_3$ practically only the cubic and tetragonal phases occur at normal pressure and in the relevant temperature range, the phase diagram of Pb(Zr$_x$Ti$_{1-x}$)O$_3$ is considerably more complicated and, of course, depends in particular on the extent of substitution of Ti by Zr. Below about 50% titanium, a so-called morphotropic phase boundary (MPB) to a rhombohedral ferroelectric phase occurs. Originally, the term "morphotropic" referred to any phase transition due to changes in composition, but it is now mainly used for the specific phase transition from tetragonal to rhombohedral symmetry in ferroelectrics by changing the composition. Interestingly, pure $PbZrO_3$ is antiferroelectric below 230 °C and paraelectric above 230 °C. A simplified version of the phase diagram of PZT in dependence on the composition is shown in Figure 12.4; the hitherto most complete phase diagram can be found in Ref. [87].

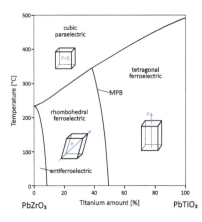

Figure 12.4: Simplified phase diagram of PZT. The direction of polarization is shown with a blue arrow; MPB, morphotropic phase boundary.

Barium titanate and PZT are the most important materials used for ceramic capacitors, with $BaTiO_3$ being the main material used here. Initially, $BaTiO_3$ was also used for ferroelectric RAM modules (abbreviated as FRAM or FeRAM; RAM = random-access memory), but these have now been almost completely displaced by PZT-based ones, as they are more powerful and cheaper.

Not only the ferroelectric properties of $BaTiO_3$ and PZT but also their simultaneously existing piezoelectric properties are used in applications. PZT represents the most important piezoelectric ceramic. Piezoceramic components are used as transducers in telecommunications, acoustics, hydroacoustics, materials testing, ultrasonic processing, liquid atomization, flow measurement, level measurement, distance measurement, and medical technology. In the form of actuators, they can be found, for instance, in atomic force microscopy (AFM) devices, micropumps, optical systems, gas valves, inkjet printers, and textile machines. As sensors, they respond to force, pressure, and acceleration and enable the monitoring of a wide variety of processes.

Two main types of synthesis are used in industrial-scale production: the conventional PZT powder preparation by one-stage solid-state reactions in a mixture of PbO, ZrO_2, and TiO_2 powders, and the wet co-precipitation route from solution.

12.5 Boride ceramics

Of technical importance of boron ceramics are mainly the diborides of some transition metals, such as TiB_2, which crystallizes in an AlB_2-analogous structure. TiB_2 has the highest hardness of all known borides and is an electrically conductive compound that melts at about 3,225 °C. Ceramics made of transition metal borides are used for the production of moulded parts (electrodes, wear parts in engine construction). Hexaborides of lanthanides are among the best electron emitters. LaB_6 is used as a high-performance electrode in electron microscopes. LaB_6 crystallizes in the space group $Pm\bar{3}m$ (no. 221) and belongs to the structure type of CaB_6, comprising B_6 octahedra in the centre of the primitive cell with longer B–B distances (1.7784 Å) within the octahedron and shorter distances (1.6421 Å) between neighbouring octahedra (see Figure 12.5).

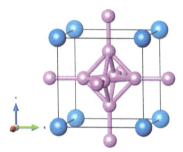

Figure 12.5: Crystal structure of LaB_6; La, blue; B, pink.

The ternary boride $Nd_2Fe_{14}B$ is one of the most important permanent magnetic materials due to its hard magnetic properties.

12.6 Carbide ceramics

The most important carbide ceramics are those of boron and silicon. In addition, titanium carbide (TiC), tungsten carbide (WC), and tantalum carbide (TaC) are also important, all of which are typical hard materials and are used in the corresponding manufacturing processes where hardness, abrasion resistance, etc. are required (cutting, drilling, milling). TiC and TaC crystallize in a NaCl-analogous structure, while WC forms its own structure type (*Strukturbericht* type B_h), in which, along the c-axis, linear strands of face-shared trigonal-prismatic WC_6 units (or alternatively described CW_6 units) are present, which are further edge-linked in the (a,b) plane (see Figure 12.6).

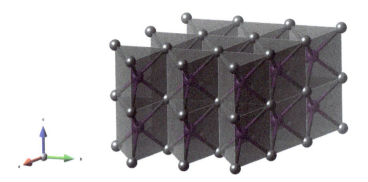

Figure 12.6: Crystal structure of tungsten carbide; space group $P\bar{6}m2$, no. 187; $a = b = 2.9017$ Å, $c = 2.8486$ Å; W, purple; C, grey.

12.6.1 Boron carbide

Boron carbide, with a chemical formula of approx. B_4C (see below), can be synthesized by reduction of boron trioxide either with (1) carbon (then course, shiny black crystals are obtained) or (2) magnesium in the presence of carbon (then a black powder is obtained) in an electric arc furnace at temperatures of ~ 2,400 °C:

$$(1) \quad 2\,B_2O_3 + 7\,C \rightarrow B_4C + 6\,CO$$
$$(2) \quad 2\,B_2O_3 + 6\,Mg + C \rightarrow B_4C + 6\,MgO$$

Boron carbide is the third hardest (hardness on the Mohs scale is ~ 9.5) substance known, after diamond and cubic boron nitride; therefore, boron carbide is also known as "black diamond".

Boron carbide is not a stoichiometric line compound but has a phase range between B_4C and approx. $B_{10.4}C$. The characteristic structural units of the carbon-deficient $B_{13}C_2$ phase are B_{12} icosahedra and linear C-B-C chains, where the carbon atoms connect at both ends of these chain elements three further B_{12} icosahedra (see Figure 12.7). The structure of the boron-poor phase B_4C can be described by the formula $(B_{11}C)CBC$, in which one boron atom of the B_{12} icosahedra is replaced by a carbon atom.

Boron carbide is used as an abrasive, in cutting tools, for nozzles in sandblast devices, bulletproof vests, and as a neutron absorber in nuclear reactors.

12.6.2 Silicon carbide

Silicon carbide (SiC), also known as carborundum, is by far the most important and versatile carbide ceramic, with about 1 million tonnes produced in 2020. It is a typical hard

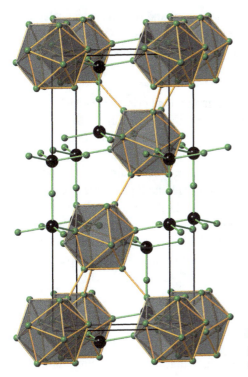

Figure 12.7: Crystal structure of $B_{13}C_2$, space group $R\bar{3}m$ (no. 166) with $a = b = 5.586$ Å and $c = 12.048$ Å; B, green; C, black.

material which is used, among other things, as an abrasive and in cutting tools (e.g., for cutting wafers).

SiC is technically produced by the Acheson process, for which a patent was filed as early as 1893 by Edward Goodrich Acheson (1856–1931). In this process, long carbon mouldings embedded in pulverized coke and covered with sand are heated to 2,200–2,400 °C by electric current flow through a graphite heater rod in large basins (12–18 m), at which the carbothermic reduction of the sand yields the SiC. The material formed in the Acheson furnace varies in purity, according to its distance from the graphite resistor heat source. Colourless, pale yellow, and green crystals have the highest purity and are found closest to the resistor. The colour changes to blue and black at greater distance from the resistor, and these darker crystals are less pure. Nitrogen and aluminium are common impurities, and they affect the electrical conductivity of SiC.

The result of the Acheson process is powdery SiC with a typical particle size of 5–10 mm. This is then ground and prepared and fed into different, typical ceramic processes, resulting in a whole range of open pore and also dense SiC ceramics, e.g., sintered silicon carbide (SSiC), silicon infiltrated (SiSiC) silicon carbide, liquid-phase SSiC, and hot-pressed silicon carbide, in which the typical properties of the SiC are realized to varying degrees.

SiC occurs in two modifications. Below 1,700 °C, the cubic β-SiC is formed, which has a sphalerite-analogous structure and is also known as the 3C polytype due to the layer sequence *ABC*. Above temperatures of about 2,000 °C, β-SiC transforms into α-

SiC, which is generally a mixture of different hexagonal and rhombohedral polytypes. The simplest hexagonal structure is wurtzite analogous and is also called 2H due to the layer sequence AB. Technologically more significant are the polytypes 4H and 6H (layer sequences ABCB and ABCACB), which are a mixture of the purely hexagonal and the purely cubic polytype.

Because of its very high hardness, which is maintained even at very high temperatures, one of the main applications of SiC is to use it as an abrasive (carborundum). It is also used in the optical field as an abrasive for lenses and mirrors. Due to its high stiffness, low weight, and low thermal expansion, SiC is also used as the basis for space telescope mirrors. Furthermore, SiC-based ceramic sand filter systems are used in oil and gas production. They are extremely durable and highly resistant to corrosion, acids, high temperatures, and various downhole fluids. SiC ceramics are also used as a component for refractory applications due to its good oxidation and corrosion resistance as well as its resistance to temperature changes. Finally, as being a semiconductor with band gaps between 2.36 eV (β-SiC) and 3.23 eV (6H polytype), it is also used in power electronic devices.

12.7 Silicide ceramics

The only technically significant silicide ceramic is that of molybdenum.[13] Molybdenum disilicide ($MoSi_2$) is synthesized directly from the elements at temperatures of approx. 1,400 °C: $Mo + 2\,Si \rightarrow MoSi_2$. The isostructural compounds $MoSi_2$, WSi_2, and $ReSi_2$ crystallize in the tetragonal space group $I4/mmm$ (no. 139) and belong to the *Strukturbericht* type $C11_b$. Their structures are related to the structure of tetragonal CaC_2 (*Strukturbericht* type $C11_a$), although the silicon atoms are not associated in pairs like C_2^{2-} in CaC_2. $MoSi_2$ may be considered as constructed from closest-packed pseudohexagonal layers consisting of Mo and Si atoms, stacked in AB sequence (with 2 + 2 rather than three atoms of the next layer in contact with any atom in the previous layer) perpendicular to the [110] direction of the body-centred tetragonal unit cell. This type of arrangement is regarded as derived from the **bcc** packing. However, instead of coordination number 8 + 6, the tetragonal distortion induces coordination numbers 10 + 4 for both components. Molybdenum atoms are surrounded by 10 silicon and 4 molybdenum atoms, while the silicon atoms are surrounded by 5 Si and 5 Mo atoms, with 4 silicon atoms at larger distances (see Figure 12.8).

---

[13] The author had the pleasure of visiting many places of mineralogical interest during his several months' trip to the USA in 2019, including the largest molybdenum mine in the world, the Climax mine in Climax, Colorado. In general, as a chemist, it is a real pleasure to travel through the mining interesting regions of the USA, because you travel through very many places and regions that are named after an element, mineral, or rock: Leadville, Telluride, Antimony, Silver City, Sulphur Springs, Platina, Carbondale, Basalt, Agate, Aragonite, Beryl, Chloride, Quartzsite, and many more.

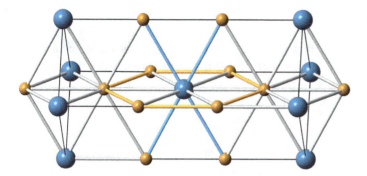

Figure 12.8: Tetragonal body-centred ($a = b = 3.211$ Å, $c = 7.825$ Å) crystal structure of $MoSi_2$; approx. viewed along the [110] direction; Mo, blue; Si, yellow. The hexagonal environment of the Mo atoms is emphasized by orange stick bonds, and the 2 + 2 coordination of Mo atoms above and below the hexagonal layers are emphasized by blue stick bonds (see text).

$MoSi_2$ is a grey solid with a density of ~ 6.2 g cm^{-3}. It has a high melting point of >2,000 °C and is a metallic conductor. Molybdenum silicide ceramics are used as high-temperature heating elements up to 1,700 °C and for lining combustion chambers and gas turbines.

12.8 Nitride ceramics

Among the nitride ceramics, those of boron, aluminium, and silicon are of greater relevance. The production volume of titanium nitride (TiN) and zirconium nitride (ZrN), on the other hand, is relatively low, which is also due to the fact that these are usually only used as ultra-thin coatings on cutting, drilling, or punching tools (see also Figure 12.9). TiN and ZrN crystallize in a rock salt-analogue structure (space group $Fm\bar{3}m$, no. 225, $a = 4.235$ Å for TiN and $a = 4.559$ Å for ZrN) and have a golden yellow colour as a thin layer. They are extremely hard and corrosion resistant and can thus increase the lifetime of the tools mentioned.

Figure 12.9: Titanium nitride-coated drill, 70 mm (photo: Peter Robert Binter, CC BY-SA 3.0).

12.8.1 Boron nitride

Of the three known modifications of boron nitride, two are of technical importance: the hexagonal α-BN and the cubic β-BN. The third modification, γ-BN, which has a wurtzite-analogous structure, is stable only under elevated pressures.

Technically, α-BN is synthesized by ammonolysis of boron oxide (1) or boron halogenides (2) at temperatures between 800 and 1,200 °C:

(1) $B_2O_3 + 2\,NH_3 \rightarrow 2\,BN + 3\,H_2O$
(2) $BCl_3 + 4\,NH_3 \rightarrow BN + 3\,NH_4Cl$

In the structure of α-BN (see Figure 12.10), the hexagonal BN layers lie, unlike the graphene layers in graphite, congruently on top of each other, with one N atom below and above each B atom (and vice versa). The interplanar distance is with 3.33 Å only slightly smaller than those in graphite (3.35 Å).

α-BN is also referred to as "inorganic" or "white graphite". It is used in engineering as an abherent and lubricant. In contrast to other solid lubricants (graphite, MoS_2), α-BN can be used even at temperatures up to about 1,000 °C, even in the presence of oxygen; graphite would already oxidize at these temperatures. α-BN is characterized by excellent non-sticking properties, especially towards molten aluminium, and the friction coefficient remains very low at high temperatures. This prevents molten metal from sticking to mould walls and promotes sliding of the molten metal on melt linings. In particular, the material protects the mould walls from corrosion and improves the overall service life of all components that come into contact with the molten metal.

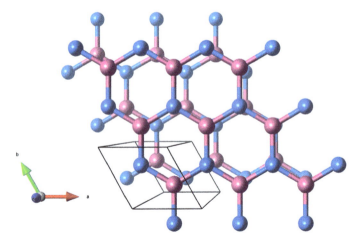

Figure 12.10: Perspective view along the c-axis of the crystal structure (space group $P6_3/mmc$, no. 194) of α-BN. Boron, pink; nitrogen, blue.

In powder form, α-BN is also used in the cosmetic industry as an inorganic filler (up to 3%) for make-up that improves some properties of the products, for instance the intensity of the colour or the covering power.

β-BN is produced from α-BN by applying high pressures of about 50–90 kbar and temperatures of about 1,500–2,200 °C. β-BN crystallizes in a sphalerite-(ZnS)-analogous structure (space group $F\bar{4}3m$, no. 216, = 3.615 Å).

In 1969, β-BN, also known by the abbreviation CBN (cubic boron nitride), was introduced to the market by the General Electric company under the name BORAZON. However, commercial production in higher quantities has only been established since the early 1990s. β-BN is one of the hardest materials available and is the second hardest cutting material after diamond. The material therefore has a high abrasion resistance, combined with very good thermal conductivity and chemical resistance. Machining tools made of CBN are therefore significantly more wear-resistant than conventional cutting materials made of corundum or silicon carbide. CBN is particularly suitable for cutting steel because – unlike diamond – it does not release any carbon to the steel, even when exposed to high temperature. Furthermore, the hardness remains almost unchanged even at temperatures of up to 1,000 °C, whereas diamond already loses its hardness significantly at temperatures of approx. 700 °C. CBN is therefore also suitable for grinding diamond under the influence of heat.

12.8.2 Aluminium nitride

Aluminium nitride powder is usually produced in two ways: firstly from aluminium oxide, nitrogen or ammonia, and carbon in excess at a temperature of ~1,600 °C in a carbothermic reaction (1) and secondly via direct nitridation in which metallic aluminium or aluminium oxide powder reacted with N_2 or NH_3 at temperatures of ~900 °C to form AlN (2):

(1) (a) $Al_2O_3 + 3C + N_2 \rightarrow 2AlN + 3CO$
    (b) $2Al_2O_3 + 9C + 4NH_3 \rightarrow 4AlN + 3CH_4 + 6CO$
(2) (a) $2Al + N_2 \rightarrow 2AlN$
    (b) $Al_2O_3 + 2NH_3 \rightarrow 2AlN + 3H_2O$

Aluminium nitride ceramics are usually sintered pressure-less at temperatures of approx. 1,800 °C. With the aid of suitable sintering additives, liquid-phase sintering is employed, which makes it possible to obtain a virtually pore-free, dense material. In practice, doping with calcium and yttrium oxide has become widely accepted as the standard process. For preparing thin films of AlN, usually physical vapour deposition or metal-organic chemical vapour deposition methods are used.

AlN crystallizes in a wurtzite-analogous structure (space group $P6_3mc$, no. 186, with a = 3.115 Å and c = 4.988 Å). Aluminium nitride powder exhibits a high sensitivity to hydrolysis. In water, incomplete cleavage of aluminium nitride into aluminium hydroxide and ammonia can be observed. Sintered ceramics are not sensitive to hydrolysis. In sodium hydroxide solution, aluminium nitride decomposes both as powder and as sintered ceramic to ammonia and aluminate.

Concerning the field of applications, AlN ceramics are characterized by their extreme high electric insulating properties ($\sigma \sim 10^{-12}$ S m^{-1}) and simultaneously very high thermal conductivity; single crystals of AlN reach values of about 320 W m^{-1} K^{-1}, while polycrystalline AlN samples still reach values of about 200 W m^{-1} K^{-1}. AlN ceramics are therefore used as dielectric layers in optical storage media and as electronic substrates and chip carriers where high thermal conductivity is essential.

12.8.3 Silicon nitride and SiALONs

Silicon nitride

Silicon nitride can be prepared by three different routes: (1) by heating powdered silicon between 1,300 and 1,400 °C in a nitrogen atmosphere, by carbothermal reduction of silicon dioxide in a nitrogen atmosphere at 1,400–1,450 °C (2), which is considered as the most-cost-effective industrial route to high-purity silicon nitride powder, and also the "low-temperature" or "polymer" route via the diimide species can be employed (3):

(1) $3\,Si + 2\,N_2 \rightarrow Si_3N_4$
(2) $3\,SiO_2 + 6\,C + 2\,N_2 \rightarrow Si_3N_4 + 6\,CO$
(3) (a) $SiCl_4 + 6\,NH_3 \rightarrow Si(NH)_2 + 4\,NH_4Cl_{(s)}$ at 0 °C
   (b) $3\,Si(NH)_2 \rightarrow Si_3N_4 + N_{2(g)} + 3\,H_{2(g)}$

In (3), as in other reactions via molecular precursors, X-ray amorphous products are initially formed, which can crystallize at higher temperatures. At temperatures of 1,300–1,500 °C, α-Si$_3$N$_4$ is formed, which irreversibly transforms above 1,700 °C into the high-temperature form β-Si$_3$N$_4$.

In both the α- and β-modification of Si$_3$N$_4$, the silicon and nitrogen atoms build a three-periodic network (see Figure 12.11) consisting of SiN$_4$ tetrahedra and three-coordinating N atoms. While in the hexagonal β-modification (space group $P6_3$, no. 173), only trigonal-planar NSi$_3$ units are present, in the trigonal α-modification (space group $P31c$, no. 159) also trigonal-pyramidal NSi$_3$ units exist.

With regard to applications of Si$_3$N$_4$ ceramics, the main issue is not of technical nature, but the costs. The costs dropped in the recent two decades significantly, but at the moment of writing this book, the price is still a factor of ~ 20 higher than for corundum ceramics. As the cost has come down, the number of production applications is accelerating. Silicon nitride has a high fracture toughness, a very high strength, low coefficient of thermal expansion, and a relatively large Young's modulus, a combination making it also very thermal shock resistant. Therefore, Si$_3$N$_4$ is used, for instance, in aluminium casting processes. Another field of application is specialized bearings.

Silicon nitride ball bearings can be found in high-end automotive bearings, industrial bearings, wind turbines, motorsports, bicycles, rollerblades, and skateboards. Silicon nitride bearings are especially useful in applications where corrosion or electric or magnetic fields prohibit the use of metals, for example, in tidal flow metres, where seawater attack is a problem. Furthermore, in semiconductor technology, silicon nitride is used as an insulation or passivation material in the manufacture of integrated circuits.

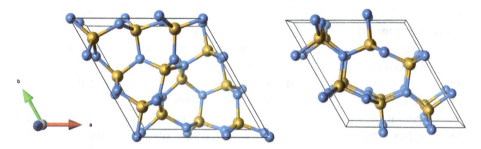

Figure 12.11: Perspective view along the c-axis of the crystal structure of α-Si$_3$N$_4$ (left) and β-Si$_3$N4. Si, orange; N, blue.

SiALONs

The name "SiALON" is derived from the four constituent elements contained in this group of ceramic materials. They can be regarded as substitutional variants of the Si$_3$N$_4$ structure. They are formed by sintering Si$_3$N$_4$ at about 1,900 °C with the addition of Al$_2$O$_3$. In the solid solutions of SiALONs, Si^{4+} is partially substituted by Al^{3+}, and N^{3-} is partially substituted by O^{2-}. In the resulting compounds of the type (Si$_{3-x}$Al$_x$)(N$_{4-x}$O$_x$), charge balance is maintained.

SiALON ceramics are high-temperature refractory materials, with high strength at ambient and high temperatures, good thermal shock resistance, and exceptional resistance to wetting or corrosion by molten non-ferrous metals, compared to other refractory materials such as alumina. A typical use is with handling of molten aluminium. They are also exceptionally corrosion resistant and hence used in the chemical industry. SiALONs also have high wear resistance, low thermal expansion, and good oxidation resistance up to above ~1,000 °C. In metal forming, SiALON is used as a cutting tool for machining chill cast iron and as brazing and welding fixtures and pins, particularly for resistance welding. Some rare-earth-activated SiALONs are photoluminescent and can serve as phosphors. They are considered to have a great potential for white LEDs [88].

12.8.4 Aluminium oxynitride

Aluminium oxynitride with a chemical formula of $Al_{23}O_{27}N_5$, which is known as γ-AlON, is a relatively new ceramic material that is characterized primarily by its high mechanical strength and simultaneously optical transparency, in particular in the visible range of the electromagnetic spectrum. Aluminium oxynitride is marketed under the trade name ALON^{TM} of the company Surmet Corp. γ-AlON can be fabricated to transparent windows, plates, domes, rods, tubes, and other forms using conventional ceramic powder processing techniques to give in its sintered form a complete dense and pore-free material. Full transparency in polycrystalline materials can be obtained only in isotropic materials that do not contain any secondary phases and are pore-free. The transparency range of γ-AlON lies approx. between 220 and 5,000 nm; however, the quality of the optical transparency is also highly dependent on the details of the polishing procedure after the material was sintered.

Because of its relatively low weight, distinctive optical and mechanical properties – it is four times harder than glass based on fumed silica and currently the hardest polycrystalline transparent ceramic commercially available – and resistance to oxidation or radiation, it shows promise for applications such as bulletproof, blast-resistant, and optoelectronic windows.

The two most common processing routes to γ-AlON are (1) reacting alumina and aluminium nitride powders in nitrogen atmosphere above 1,650 °C and (2) carbothermal reduction of alumina in the presence of carbon and nitrogen above 1,700 °C:

(1) $9\ Al_2O_3 + 5\ AlN \rightarrow Al_{23}O_{27}N_5$
(2) $23\ Al_2O_3 + 15\ C + 5\ N_2 \rightarrow 2\ Al_{23}O_{27}N_5 + 15\ CO$

In route (1), the sintering can be performed either in one step (reaction sintering), or in two steps by first reacting the powders to form γ-AlON followed by densification of the product. One of the problems is obtaining a high-purity AlN powder.

The γ-AlON phase has an inverse spinel-type crystal structure (space group $Fd\bar{3}m$, no. 227). As a phase of variable composition $Al_{(64+x)/3}O_{(32-x)}N_x$ ($0 \le x \le 8$, Al_2O_3 for $x = 0$ and Al_3O_3N for $x = 8$), it is frequently considered in the so-called constant or permanent anion model, whose composition (at $x = 5$) is characterized by the formula $Al_{23}O_{27}N_5$. In this model, it is assumed that the anion positions in the crystal lattice of the spinel phase are completely occupied by oxygen and nitrogen, whereas its cation positions are occupied by aluminium and contain vacancies. The spinel phase was found to be stable only when the disordered vacancy is distributed over the octahedral position (Wyckoff position 16d).

12.9 Glass-ceramics

Glass-ceramics consist of a glassy, amorphous phase and a polycrystalline, ceramic-like phase, i.e., they are materials in which both material properties are combined to a certain extent. The amount of the crystalline phase usually lies between 30 and 90 wt%. Compared to glasses, glass-ceramics are characterized by their high dimensional stability at high temperatures, high thermal shock resistance (they have a very low or even negative thermal expansion coefficient), mechanical hardness, and, if desired, transparency – i.e., the typical property of glasses per se. Compared to other ceramics, they are easier to shape and have excellent resistance to corrosion. As other ceramics too, many glass-ceramics are also biocompatible.[14]

There are a number of different glass-ceramic systems, the most important of which are the so-called LAS ($Li_2O \cdot Al_2O_3 \cdot n\ SiO_2$), MAS ($MgO \cdot Al_2O_3 \cdot n\ SiO_2$), and ZAS ($ZnO \cdot Al_2O_3 \cdot n\ SiO_2$) systems, in which these substances are the main components, but to which further glass-phase-forming agents such as Na_2O, K_2O, CaO, or carbonates are usually added. After forming a typical glass melt, nucleation agents are added, most commonly zirconium(IV) oxide in combination with titanium(IV) oxide is used, and the melt is cooled down to a certain temperature (typically 500–700 °C), the nucleation temperature, and maintained at that temperature at which the glass object is then annealed in a controlled manner, until the desired amount of crystal nuclei have formed in the glass phase ("controlled devitrification"). In the LAS system, the dominant crystalline phase is typically a solid solution based on β-quartz (also known as "high-quartz"); if the glass-ceramic is subjected to a more intense heat treatment, this β-quartz solid solution transforms into a keatite-based solid solution. While both β-quartz and keatite have a negative coefficient of thermal expansion, the remaining glass phase has a positive one; adjusting the proportion of these two phases offers a wide range of possible thermal expansion behaviours in finished material, even a material with zero expansion.

When the crystallites that were formed are significantly smaller (about 50 nm) than the wavelength of visible light, transparent glass-ceramics are formed.

Originally developed for use as mirrors and mirror mounts of astronomical telescopes, LAS glass-ceramics have become popular through its use as cooktops (a well-known brand name is Ceran produced by the company Schott), as well as cookware and bakeware, and as high-performance reflectors for digital projectors.

---

[14] The IUPAC defines biocompatibility as "the ability to be in contact with a living system without producing an adverse effect".

Further reading

J. Hojo (Ed.), *Materials Chemistry of Ceramics*. Springer Nature, Singapore, **2019**.
M. Barsoum, *Fundamentals of Ceramics*. 2nd ed. CRC Press, Boca Raton, **2019**.
F. Baino, M. Tomalino, D. Tulyaganov (Eds.), *Ceramics, Glass and Glass-Ceramics – From Early Manufacturing Steps Towards Modern Frontiers*. Springer Nature, Cham, **2021**.

13 Intermetallic phases

According to the common definition, which goes back to Schulze [89], intermetallic phases, or intermetallics for short, are homogeneous, *ordered*, solid phases containing two or more metallic elements, with optionally one or more non-metallic elements, whose crystal structure is not isotypic to that of the constituents. Since ordered alloys (cf. Section 6.1.5) also fall under this definition, the question arises, what is the difference between alloys and intermetallic phases? In addition to ordered, homogeneous intermetallics consisting of a single phase, alloys also include disordered phases, which can be present as substitutional or interstitial solid solutions as well as heterogeneous, multi-phase systems. The term *phase* is used to express that many intermetallics have a certain phase width. If this is not the case and the phase is exactly stoichiometric, one speaks of intermetallic *compounds* instead.

While alloys are characterized by their metallic bonds and their typical metallic properties, in many intermetallic phases a certain degree of covalency and/or ionic bond character is prevalent. This is the reason why usually intermetallics have higher melting points, significantly increased strength, and high wear resistance, but at the expense of ductility: most of the intermetallics are very brittle. The limited ductility hampered a wider industrial use of intermetallic materials for a long time. However, as techniques and processes in the field of materials science continue to advance, remarkable progress is being made in this area as well. One example is the intermetallic phase TiAl, which has been considered as a potential material for construction work since the 1970s, in particular because of its low density and high oxidation resistance; however, the poor processability of TiAl precluded its application. These difficulties have now been overcome and TiAl is used, for example, in aircraft construction for the design of turbines and is increasingly replacing the Ni-based alloys that were used previously.

Except for polar intermetallic compounds (Zintl phases), there are no universally applicable models to explain their compositions and structures. Thus, even the occurrence of stoichiometric compounds cannot always be explained by the chemical valence of the bonding partners. The structures of certain subtypes of intermetallic phases are based on geometric or topological criteria, i.e., structures that can be derived from the size of the species involved and by space-filling aspects (Laves and Frank-Kasper phases).

The following chapter presents basic features of intermetallic phases and some of their structures, which can be quite complex on the one hand but are, on the other hand, also fascinating because they are still periodically ordered. For those readers who want to dive deeper into the topic of intermetallic phases, there are at least three comprehensive monographs already available [90–92].

13.1 Classification scheme

Although it is not possible to clearly distinguish all types of intermetallic phases from each other, attempts have been made to divide the metals that are involved into different groups and then to consider certain group combinations that lead preferably to a certain type of intermetallic phases. A common grouping that leads to a more or less satisfactory system of classification of intermetallic phases is as follows (see also Figure 13.1):

- A – alkali and alkaline earth metals, i.e., pronounced electropositive metals, some of them with very large metal radii.
- B – d-block elements of groups 3–11 (i.e., except Zn, Cd, and Hg); they are characterized by their similar metal radii and electronegativities, but differ, of course, in their valence electrons.
- C – d-block elements of the group 12, plus elements of group 13 except B, plus Sn and Pb of group 14; these metals are more electronegative and many of them crystallize in special structures.
- D – Si, Ge (group 14), plus As, Sb, Bi (group 15), plus Se, Te, and Po (group 16); these elements are at the border between typical metals and typical non-metals.

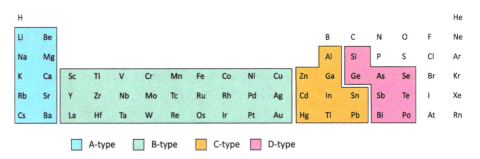

Figure 13.1: Grouping of metals into different blocks that allow for a classification scheme of different types of intermetallic phases.

Before we will further explore the structures of the different types of intermetallics, a short overview will highlight which combinations of the metal groups will preferably lead to which type of intermetallic phases.

– **A + A**: If the metals have the same number of valence electrons and if they have similar radii (the difference should not exceed approx. 15%), they often form disordered (substitutional) solid solutions over the entire range of compositions, i.e., without a miscibility gap. These types of solid solutions were already discussed in Section 6.1.7. Examples for combinations within the A-type group that do form such solid solution are K-Rb ($\Delta r_M = 6\%$), K-Cs ($\Delta r_M = 13\%$), Rb-Cs ($\Delta r_M = 8\%$), and Ca-Sr ($\Delta r_M = 9\%$).

If, on the other hand, two **A**-type metals with a different number of valence electrons or with a larger difference of their radii are combined, then either packing-dominated

(Laves) phases (see Section 13.5) are obtained or the metals cannot be combined anymore and they are immiscible over a large fraction of or even the entire composition range; this is, for instance, the case for Na-Mg.

– **B + B**: Combinations of metals within the **B**-type block usually lead, similar as in the case for A + A, to (disordered) solutions with a wide phase width. For some combination ordered superstructures (see Section 13.2) are observed at lower temperatures, e.g., for the system Cu-Au.

The solubility of two **B**-type metals within each other is essentially dependent on the radius difference of the two metals. Almost all technically important alloys belong to this group including the interesting group of so-called shape-memory alloys [93].

– **C + C and D + D**: Combinations of elements of either the **C** or **D** block among each other generally lead to solid solutions, if the metals are of the same group of the periodic table. If they belong to different groups of the periodic table, they exclusively form stoichiometric compounds with a highly covalent bond character and obey the Grimm–Sommerfeld rule [94]. The Grimm–Sommerfeld rule states that binary compounds with covalent character that have an average of four electrons per atom will have structures where both atoms are tetrahedrally coordinated, i.e., have the wurtzite/zincblende structure. The technical relevant II–VI and III–V semiconductor compounds belong to this class.

– **A + B**: The combinations of metals from the **A** block with a metal from the **B** block tend to form again packing-dominated phases (see Section 13.5), some selected pairs form Hume-Rothery phases (Section 13.3), or they do not build a common phase.

– **A + C**: Typically, combinations of these blocks lead to Laves phases (Section 13.5.2), and some combinations also form Zintl phases (Section 13.4). The difference of the metal radii and that of their electronegativities is relatively large, leading most frequently to stoichiometric compounds.

– **A + D**: Combinations between these types of metals lead to Zintl phases (Section 13.4), which are stoichiometric compounds with predominantly polyanionic fragments; some polycationic species are also known.

– **B + C**: Mixing **B**- and **C**-type metals leads to Hume-Rothery phases (Section 13.3), whose structures are almost exclusively derived from the concept of valence electron concentration (VEC). Hume-Rothery phases have small-to-intermediate phase widths. The two most important systems are Cu-Zn (brass) and Cu-Sn (bronze).

– **B + D**: The phases of these systems are either stoichiometric or they have narrow phase widths. Frequently, the **D** component forms a densest packing in which the voids are occupied by the **B** component. Many phases crystallize either in the NiAs, CdI_2, or Ni_2Ge structure type. Compounds that are isostructural to marcasite (FeS_2), pyrite (FeS_2), or MoS_2 always form stoichiometric compounds. Various other important structure types also belong to these systems, e.g., Cr_3As and Cr_3Si.

13.2 Ordered solid solutions – superstructures

As already explained in Section 6.1.7, the miscibility of metals in the solid state is linked to four prerequisites, which will be repeated here:
1. The metals have the same crystal structure.
2. The atomic sizes of the two metals differ not more than by 15%.
3. The metals have roughly the same electronegativity.
4. The metals have a similar number of valence electrons.

If these conditions are not met, the metals are either only miscible to a limited extent (or only at higher temperatures) or are even completely immiscible. This is illustrated by some examples in Table 13.1.

Table 13.1: Examples of metal combinations that are either only miscible to a limited extent or do not form any solid solution.

System	Structure	Block according to classification scheme in Figure 13.1	Δr_M (%)	Solid solution?
Na–K	bcc–bcc	A–A	25	No
Ca–Al	ccp–ccp	A–C	38	No
Pb–Sn	ccp–special	C–C	10	Limited
Cr–Ni	bcc–ccp	B–B	3	Limited
Ag–Al	ccp–ccp	B–C	1	Limited
Mg–Pb	hcp–ccp	A–C	9	Limited
Cu–Zn	ccp–hcp	B–C	7	Limited

However, even if the four criteria are fulfilled, there are some pairs that constitute exceptions: in the Ag–Pt system, a wide miscibility gap occurs although both metals crystallize in the **ccp** structure and Δr_M is only 4%, while the system Ag–Pd with Δr_M = 5%, on the other hand, forms a solid solution over the entire composition range. To the best of the author's knowledge the deeper reasons for these exceptions are mainly unknown.

If the two metals are at the border of obeying the four criteria mentioned above, for some of such pairs a transition from *disordered* to *ordered* solid solutions at special compositions (cf. Section 6.1.5) is observed, if the cooling rate is low enough (otherwise a frozen solid state with still statistically distributed metals is obtained). These metal pairs form structures that can be directly derived from their parent structure but – as they are crystallographically ordered – form *superstructures*. Often such superstructures are formed in which the metal atoms of one type are only surrounded by atoms of the other metal type. In the following the most important superstructures are presented.

13.2.1 Superstructures of the bcc packing

There are two main types of superstructures that are derivatives of the **bcc** packing. The first one leads to a 1:1 composition of the metals and results in structures analogous to the CsCl structure type (*Strukturbericht* type B2). The most prominent example is the β'-brass phase, i.e., CuZn (see Figure 6.5), already discussed in Section 6.1.5. Another example is TiFe.

The second important structure type is based on a (2 × 2 × 2) supercell of the **bcc** structure with composition 3:1 for binary systems and 2:1:1 for ternary systems. The prototypical structure types are α-BiF$_3$ (*Strukturbericht* type D0$_3$, space group $Fm\bar{3}m$), which is, for instance, realized in Fe$_3$Al and Cu$_2$MnAl (*Strukturbericht* type L2$_1$, space group $Fm\bar{3}m$). In Fe$_3$Al the centres of the eight octants are occupied by one Fe with Wyckoff position 8c at (¼, ¼, ¼) while Al and the remaining Fe occupy the Wyckoff positions 4a (0,0,0) and 4b (½, ½, ½), respectively, building an interwoven **ccp** NaCl-type structure; in Cu$_2$MnAl the 8c site is occupied by Cu. The structures of Fe$_3$Al and Cu$_2$MnAl are shown in Figure 13.2, left and middle.

Intermetallic phases of the Cu$_2$MnAl type are known as Heusler compounds or Heusler alloys (after Friedrich Heusler, 1866–1947, a German mining engineer and chemist, who investigated Cu$_2$MnAl in 1903), specifically they belong to the so-called full Heusler compounds of general composition X$_2$YZ, where X and Y are transition metals and Z an element of the *p*-block of the periodic table, usually from group 13 to 15. Many of the Heusler compounds have properties relevant to spintronics (for a review, see Ref. [95]) and magnetism. For instance, Cu$_2$MnAl is a ferromagnetic material, although none of the constituting elements are ferromagnetic! Relevant full Heusler compounds comprise:

- Cu$_2$MnZ, with Z = Al, In, and Sn,
- Ni$_2$MnZ, with Z = Al, In, Sn, Sb, and Ga,
- Co$_2$MnZ, with Z = Al, Si, Ge, and Ga,
- Co$_2$FeZ, with Z = Al, Si, and Ge,
- Co$_2$NiGa,
- Mn$_2$VGa,
- Fe$_2$VAl, and
- Pd$_2$MnZ, with Z = Al, In, Sn, and Sb.

The other class of Heusler compounds is the so-called half-Heusler compounds of the form XYZ, where one sublattice site of the X sites, i.e., half of the centres of the octants remain unoccupied, leading to non-centrosymmetric compounds with space group $F\bar{4}3m$; the prototypical compound is AgMgAs (*Strukturbericht* type C1$_b$), see Figure 13.2, right. Some of these half-Heusler compounds are relevant with respect of a new quantum state of matter – so-called topological insulators – that was predicted in 2006 by Shoucheng Zhang from the Stanford University for the compound HgTe and soon

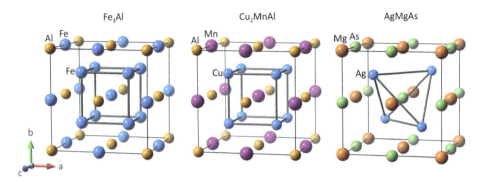

Figure 13.2: Crystal structures of Fe$_3$Al (left), the full Heusler compound Cu$_2$MnAl (middle), and the half-Heusler compound AgMgAs (right); the grey lines connecting the Fe, Cu, or Ag atoms, respectively, are only a visual aid.

afterwards experimentally confirmed. Topological insulators consist of compounds that are insulators or semiconductors in the bulk but have metallic properties at their surface, for a review see Ref. [96]. One of the most attractive aspects of half-Heusler compounds is the tunability of the band gap over a wide energy range from about 4 eV (e.g., LiMgN) to values more in the classical regime of semiconducting compounds (for instance, AsLiMg with 2.3 eV) to compounds with a narrow gap (for instance, SbScPt with 0.7 eV) down to approx. 0 eV (BiScPt).

13.2.2 Superstructures derived from densest packings (ccp or hcp)

By selective substitution of subsets of atoms forming close-packed structures and/or by the selective occupying of the available tetrahedral or octahedral voids, families of substitutionally ordered structures as well as of other derivative structures can be obtained.

A rich family of binary intermetallic phases is derived from the **ccp** packing (Cu, *Strukturbericht* type A1) by substituting a subset of the atoms with another kind of atom as shown in Figure 13.3, leading to two derivative phases, one with stoichiometry 1:1 and one with 3:1 (or 1:3). These are realized in the binary system Cu–Au, which shows full miscibility of the two components above 683 K. Below this temperature, several ordered phases with broad stability ranges around the compositions Au$_3$Cu, CuAu, and Cu$_3$Au are formed. The latter two are the prototypical structures of the respective structure types CuAu (*Strukturbericht* type L1$_0$, space group $P4/mmm$, Cu at Wyckoff position 2e at (0, ½, ½) and Au at Wyckoff positions 1a (0,0,0) and 1c (½, ½, 0)) and Cu$_3$Au (*Strukturbericht* type L1$_2$, space group $Pm\bar{3}m$, Cu at Wyckoff position 3c (0, ½, ½), and Au at Wyckoff position 1a (0,0,0)). The densest layers in Cu$_3$Au can be described as a 3.6.3.6 kagome net[15] [97] (**kgm** net, see http://rcsr.net/layers/kgm) constituted by the Cu atoms, while the Au atoms are located in the centres of the hexagons (Figure 13.3, bottom-left). Due to the fact that Au is larger than Cu, the structure of CuAu is tetragonally distorted. In that regard, CuAu can also be described as a tetragonally distorted CsCl analogue structure, being derived from the prototypical **bcc** structure of W (*Strukturbericht* type A2). In terms of layers CuAu can either be described by a stacking of alternating 4^4 layers (a **sql** net, see http://rcsr.net/layers/sql) of Au and Cu atoms along the *c*-direction (this is shown in Figure 13.3, bottom-right) or as a *ABC* stacking sequence of densest layers along [111].

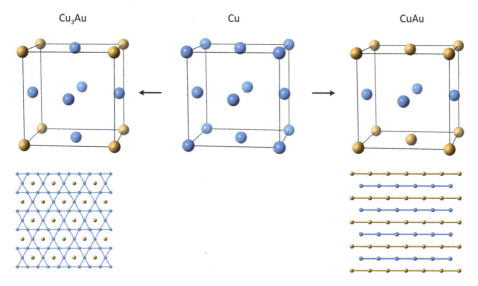

Figure 13.3: Derivatives of the Cu structure type by substituting a certain subset of Cu with Au atoms leading to the prototypical structure types CuAu and Cu$_3$Au (top); in the bottom part the atom configurations within one dense(st) layer are shown: In Cu$_3$Au the Cu atoms form a 3.6.3.6 kagome net and the Au atoms are located in the centre of the hexagons; in CuAu alternating 4^4 layers of Cu and Au atoms are stacked along the *c*-direction of the tetragonal cell; Cu: blue, Au: orange.

The hexagonal analogue of Cu$_3$Au with a stacking sequence of *AB* of the decorated 3.6.3.6 kagome-like layers (along the *c*-direction of the hexagonal unit cell) is Ni$_3$Sn (*Strukturbericht* type D0$_{19}$), which crystallizes in the space group *P*6$_3$/*mmc* and for which more than 200 representatives are known.

There are two further ordered superstructures that can be derived from the **ccp** packing, leading again to 1:3 compounds. The first one can be derived from a (1 × 1 × 2)

---

15 In Japanese, kagome means a bamboo-basket (kago) woven pattern (me) that is composed of interlaced triangles whose lattice points each have four neighbouring points. How the word came into scientific terminology is explained in Ref. [97].

supercell of the cubic cell of Cu, present in the TiAl$_3$ family of structures (*Strukturbericht* type D0$_{22}$, space group $I4/mmm$), and the second one is an even more complex superstructure derived from a (1 × 1 × 4) supercell, present in structures of the type ZrAl$_3$ (*Strukturbericht* type D0$_{23}$, space group $I4/mmm$).

The densest layers in TiAl$_3$ have the same orientation as in the pseudo cubic subcells, i.e., these are (112) planes in the tetragonal supercell. The sequence of the 4^4 layers along [001] is equal to that in two unit cells of Cu$_3$Au, except that the layer in the middle is shifted by (½, ½, 0), see Figure 13.4, left. If only connections between the major component are considered (i.e., Al-Al), then the densest layers form a **bew** net (see http://rcsr.net/layers/bew) with two kinds of vertices and their vertex symbols are 3.6.3.6 and 3^2.6^2, respectively; again the minor component is located at the centres of the hexagons. From a structural point of view TiAl$_3$ is interesting because the Ti-Al distances within the 4^4 layers are shorter (2.725 Å) than the distance calculated by their atomic radii, and also the Al-Al distances (2.725 Å) within these layers are shorter than in bulk Al (2.864 Å). This indicates some covalent character, and therefore, it would also be appropriate to categorize TiAl$_3$ into intermetallic phases with covalent character.

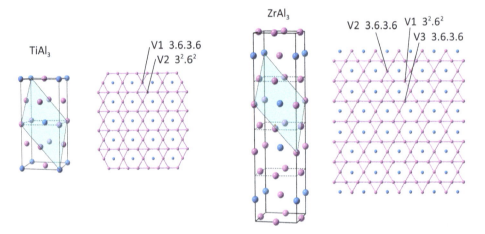

Figure 13.4: Crystal structures and atom configuration of the densest layers in TiAl$_3$ (left) and ZrAl$_3$ (right). The densest planes are drawn as semi-transparent, turquoise rhombi; the former cubic subcells are indicated by dashed squares; Al, pink; Ti/Zr, blue.

In ZrAl$_3$ the densest layers are (114) planes in the tetragonal cell. The subnet formed by the Al atoms is a hitherto unrecognized trinodal two-periodic net, in which two of the vertices have the same vertex symbol but are not symmetry-equivalent (see Figure 13.4, right).

An ordered intermetallic phase with composition 1:4 and based on densest packings is realized in MoNi$_4$, which crystallizes in the space group $I4/m$. The densest layers are (2$\bar{1}$1) planes in the tetragonal cell (see Figure 13.5). The Ni subnet forms a krh net (http://rcsr.net/layers/krh), a binodal (4,5)-c two-periodic net. The hexagonal counterpart of MoNi$_4$ with a stacking sequence of AB of the densest layers is ZrAu$_4$.

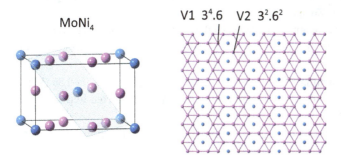

Figure 13.5: Crystal structure of the intermetallic phase MoNi$_4$ (left) and atomic configuration of the densest planes (right) in MoNi$_4$.

13.3 Hume-Rothery phases

Many combinations of the **B**-type (preferably from groups 8–11) with **C**-type metals form Hume-Rothery phases (brass-like phases), named after William Hume-Rothery (1899–1968), an English metallurgist and material scientist who studied the constitution of alloys. Some **A/B** (AuMg, BePt) and **B/D** (Cu$_3$Si) combinations also lead to Hume-Rothery phases. Hume-Rothery phases are binary alloys of non-stoichiometric composition whose structures are determined by their VEC, i.e., sum of their valence electrons per atom. The classical example is that of the Cu-Zn (brass) system, which passes through five (or six, if the high temperature δ-phase is included) distinct phases, which are called – according to an increasing VEC – α-, β-, γ-, δ-, ε-, and η-phases. They have small-to-intermediate phase widths (see Figure 13.6), and their structures are explained below. Not all Hume-Rothery phases pass through all of these five phases.
- The α-phase is a disordered solid solution with a **ccp** structure. For the Cu$_{1-x}$Zn$_x$ system it is stable from $x = 0$ to 0.38 and the VEC ranges from 1 to 1.38 as Cu has one and Zn has two valence electrons.
- The β-phase crystallizes in the **bcc** structure and is stable from $x = 0.45$–0.49 (VEC = 1.45–1.49). It is often formulated as CuZn, ignoring the fact that this phase has a certain phase width. Above temperatures of approx. 470 °C it is again a completely disordered solid solution. Below that temperature the β'-phase forms, which is the ordered variant with a CsCl analogue structure (compare Section 6.1.5). Other Hume-Rothery phases that have VEC in this regime crystallize not in the β-brass structure but in the β-Mn type.

- The γ-phase, stable from $x = 0.58$–0.66 (VEC = 1.58–1.66), is a special-ordered, defect-containing superstructure of the **bcc** structure that is further explained below. It is often formulated as Cu_5Zn_8 although the phase width corresponds to compositions from $Cu_5Zn_{6.9}$ to $Cu_5Zn_{9.7}$.
- The δ-phase is an insufficiently characterized, cubic high-temperature phase – an entry in the ICSD database does not exist, a literature survey in order to find atomic coordinates failed.
- The ε-phase, often formulated as $CuZn_3$, actually stable from $x = 0.78$–0.86 (VEC = 1.78–1.86), is a disordered solid solution with a regular, undistorted **hcp** structure (structure type of Mg).
- The η-phase from $x = 0.98$–1 crystallizes again as a disordered solid solution in the **hcp** structure.

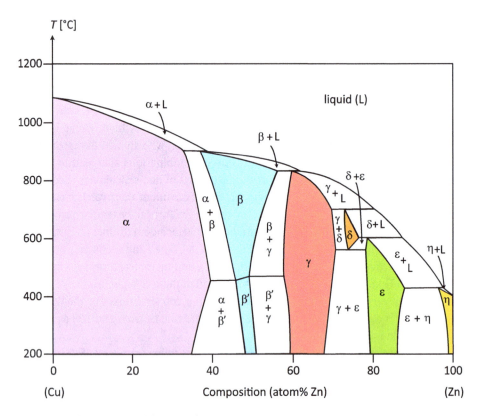

Figure 13.6: Phase diagram of the Cu-Zn (brass) system.

The concept of VEC, according to which their number determines the predominant structure type of Hume-Rothery phases, does have a few shortcomings. First, there are overlapping regions so that no sharply separated VEC ranges can be assigned that lead unambiguously to a certain structure type. For instance, some systems with a VEC up to 1.59 still form the β-brass phase, while others already adopt the γ-brass structure with VEC numbers larger than 1.54. Second, there is the ambiguity that some systems with a VEC of about 1.5 form the β-brass structure, while others crystallize in the β-Mn type; these are amongst others: Ag_3Al, Au_3Al, Cu_5Si, and $CoZn_3$. And third, for the concept to be reasonably applicable at all, the number of valence electrons of the group 3–10 metals must be set to zero, for which there is initially no good justification. This means that the VEC can only give an indication, which structure the respective Hume-Rothery phase will adopt. In Table 13.2 an overview of some binary Hume-Rothery phases together with their VEC is given.

Table 13.2: Examples of binary Hume-Rothery phases and their VEC.

β-Brass phase, VEC ≈ 1.5	γ-Brass phase, VEC ≈ 1.6	ε-Brass phase, VEC ≈ 1.75
CuZn (3/2)	Cu_5Zn_8 (21/13)	$CuZn_3$ (7/4)
Cu_3Al (6/4)	Cu_9Al_4 (21/13)	$CuCd_3$ (7/4)
Cu_5Sn (9/6)	$Cu_{31}Si_8$ (63/39)	Cu_3Si (7/4)
AgZn (3/2)	$Cu_{31}Sn_8$ (63/39)	Cu_3Sn (7/4)
AgCd (3/2)	Ag_5Zn_8 (21/13)	$AgZn_3$ (7/4)
AuZn (3/2)	Ag_5Cd_8 (21/13)	$AgCd_3$ (7/4)
AuMg (3/2)	Au_5Zn_8 (21/13)	$AuZn_3$ (7/4)
FeAl (3/2)	Au_5Cd_8 (21/13)	Au_5Al_3 (14/8)
CoAl (3/2)	Fe_5Zn_{21} (42/26)	Ag_3Sn (7/4)
NiAl (3/2)	Co_5Zn_{21} (42/26)	
	Pt_5Be_{21} (42/26)	

A first theoretical explanation of the observed stability sequence of the phase sequence (**ccp** – **bcc** – γ-phase – **hcp**) of Hume-Rothery phases with increasing VEC was provided by Mott and Jones in 1936 [98], but this theory had to be further adjusted over the following decades because the original explanations gave wrong values for the stability regions of the phases. The key point of the explanation is related to the relation of the Fermi edge to the boundary of the first Brillouin zone. When in a certain structure type this limit is reached by increasing occupation of higher energy levels, a new structure type is realized which leads to a lower overall total energy. The interested reader is referred to Ref. [99], in particular Chapter 9, "Electronic structures of alloys".

13.3.1 The γ-brass structure

The γ-brass structure can be described in several ways. The most traditional and in the view of the author also the most convenient way is the following: (1) we will start with a (3 × 3 × 3) supercell of the **bcc** structure. This leads to a unit cell with 54 atoms;

(2) now two atoms are removed, namely the one in the centre and the (8 × ⅛) atoms at the corners of this supercell, leading to only 52 atoms per supercell. Due to these missing sites the atoms relax and occupy slightly different positions in comparison with the parent structure; (3) around these empty sites the following polyhedra, all of which are distorted, are now partly nested beginning from the inside to the outside: (a) an inner tetrahedron (IT) of Zn atoms, (b) an outer tetrahedron (OT) of Cu atoms; in the parent structure these eight atoms form a regular cube; (c) an octahedron (O) of Cu atoms, and finally (d) a cuboctahedron of Zn atoms. These nested polyhedra comprise 4 + 4 + 6 + 12 = 26 atoms; (5) these 26-atom clusters are then packed in a body-centred cubic fashion (see Figure 13.7); there are no shared atoms between neighbouring cuboctahedra. The distortion around the empty sites leads to a symmetry reduction from the parent structure with space group $Im\bar{3}m$ to $I\bar{4}3m$. Since the structure of the cubic γ-phase consists of a total of 52 atoms in the unit cell, the molecular formulae of these phases often contain 13 or 26 atoms, which occur with $Z = 4$ or 2 formula units in the unit cell.

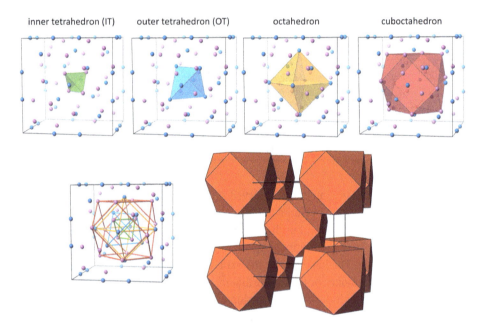

Figure 13.7: The structure of the γ-brass phase is based on a (3 × 3 × 3) supercell of the **bcc** unit cell. Four partly nested polyhedra are enclosing an empty site: a Zn_4 inner tetrahedron (green), followed by a Cu_4 outer tetrahedron (blue), a Cu_6 octahedron (orange), and finally a Zn_{12} cuboctahedron (red) (top row). These polyhedra are shown together at the bottom left. This four-shell arrangement is packed in a body-centred cubic fashion (bottom right).

13.4 Zintl phases

Intermetallic compounds that are constituted from more-electropositive elements from groups 1 and 2 of the periodic table on the one hand (type **A** metals) and of more electronegative elements of groups 13–15 (**C**- or **D**-type metals), on the other hand, are called Zintl phases after Eduard Zintl (1898–1941), a German Chemist and the pioneer in that field. According to the Zintl-Klemm concept (for an interesting historical review of this concept, see Ref. [100]), the structures of such phases can be understood on the basis of a generalized (8–N) rule (where N is the number of valence electrons, and the valency is then 8–N) by assuming that the more electropositive partner completely transfers its valence electrons to the more electronegative partner. Depending on the number of electrons transferred, either isolated (less common) or polyatomic (more common) anions are then formed (see Table 13.3). The atoms in the polyanions are bonded covalently, and in most cases, they consist of larger subunits such as clusters, layers, or 3D networks. According to the Zintl-Klemm concept extended later by Busmann, they should form structures of their corresponding isoelectronic elements. In the case where clusters or cages are formed, the Wade rules apply. Some compounds containing also transition metals and rare-earth elements as the more electropositive partner can also be considered as Zintl phases.

The classical example of Zintl phases is that of the binary compound NaTl (*Strukturbericht* type B32, space group $Fd\bar{3}m$). Na donates one electron to Tl, a group 13 element. The Tl anion is then isoelectronic to a group 14 element and adopts, accordingly, a covalently bonded diamond network with tetravalent Tl atoms. The Na$^+$ ions fill the space in-between and adopt a second interpenetrating diamond-like structure (but of course not covalently bonded), see Figure 13.8, left. The covalent character of the polyanionic Tl substructure is also evident in the short Tl-Tl distance in NaTl (3.231 Å) that is shorter than in the element Tl itself (3.420 Å).

The concept and the way in which the number of valence electrons are counted may gain clarity with another example: The more electronegative component in BaSi$_2$ is silicon with four valence electrons in the elemental state. If we assume that Ba transfers its two valence electrons completely to Si, we have altogether 10 electrons in the anionic part (Si$_2^{2-}$) or 5 per atom. Accordingly, the valency is (8–N) = 3.

One weakness of the Zintl-Klemm concept might be that it does not allow for a prediction what allotropic form of the isoelectronic element the polyanionic substructure adopts. While Tl$^-$ constitutes a diamond network, the Ga$^-$ ions in SrGa$_2$ form a graphene-like layered structure; but unlike in graphite the hexagonal layers are stacked in a primitive manner, see Figure 13.8, right. The prototypical structure is that of AlB$_2$, *Strukturbericht* type C32, space group $P6/mmm$. Furthermore, the range of structures of polyanions in Zintl phases is much wider than for pure elements. For instance, in SrSi$_2$, the Si atoms, which are trivalent, form crosslinked spiral chains according to a 4$_1$ screw axis.

In Table 13.3 some compounds are listed that obey the Zintl-Klemm-Busmann concept.

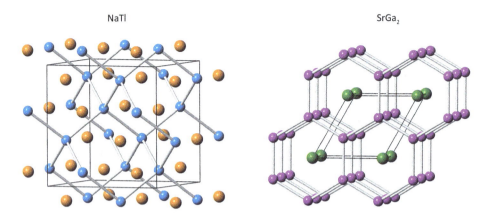

Figure 13.8: The crystal structure of the Zintl phases NaTl (left) and SrGa$_2$ (right); Na: orange, Tl: blue, Sr: green, Ga: purple.

Table 13.3: Zintl phases that obey the Zintl-Klemm-Busmann concept; the space group of the compound, the formal charge per atom of the more electronegative component, the number N of their valence electrons, their valency (8–N), and a short description of the structure of the anionic component is given.

Compound	Space group	Formal charge	N	Valency (8–N)	Anionic substructure
NaTl	$Fd\bar{3}m$	−1	4	4	Diamond network
CaIn$_2$	$P6_3/mmc$	−1	4	4	Distorted diamond structure (In-In = 2.92–3.13 Å)
KGe	$P\bar{4}3n$	−1	5	3	Isolated Ge$_4$ tetrahedra like in white phosphorous
BaSi$_2$	$Pnma$	−1	5	3	Isolated Si$_4$ tetrahedra like in white phosphorous
CaSi$_2$	$R\bar{3}m$	−1	5	3	Layers of puckered six-membered rings like in α-As
SrSi$_2$	$P4_132$	−1	5	3	Network of crosslinked spiral chains
LiAs	$P2_1/c$	−1	6	2	Helical chains like in Se and Te
CaSi	$Cmcm$	−2	6	2	Planar zigzag chains
CoAs$_3$	$Im\bar{3}$	−1	6	2	Rectangular four-membered rings
CaAs	$P\bar{6}2m$	−2	7	1	Dimers like in Cl$_2$
Na$_3$As	$P6_3cm$	−3	8	0	Isolated anions

Zintl phases do not have any technical relevance but from a chemist's point of view they are extremely interesting due to the fact that the bonding situation is very special: The anionic partial structure can be characterized as covalently bonded. The bond between the electropositive and electronegative partners can be described more as salt-like. This is expressed, for example, in the comparatively high melting temperatures of Zintl phases and also makes it partly understandable why Zintl phases are usually quite brittle. On the other hand, Zintl phases often have a metallic luster or are deeply coloured; however, their electric conductivity is rather low and increases with increasing temperature. Many theoretical studies confirm that Zintl phases belong to the class of semiconductors. In this sense, then, one could say that Zintl phases resemble a kind of trinity of metal, salt, and semiconductor.

13.5 Packing-dominated phases (Frank-Kasper and Laves phases)

In the case of metals and their underlying sphere packings (cf. Chapter 4), we have already become acquainted with two types of densest packings: the cubic-closest packing (**ccp**) and the hexagonal-closest packing (**hcp**), both of which have a space filling of about 74%. Since these packings are called *closest* or *densest*, one could be led to the erroneous assumption that no denser sphere packings exist. This is of course not the case since we only have to drop the premise that we consider *equally sized* spheres. The space filling degree or packing fraction depends on the composition/stoichiometry of *differently* sized spheres and their size ratio(s). It is easy to infer that much higher packing densities than 74% can be achieved if we start from a **ccp** and add a second and third kind of sphere, one of which fits into the octahedral and the other into the tetrahedral voids.

When two metals are mixed together that obey the first four Hume-Rothery rules, i.e., do not differ too much chemically and do not differ in size by more than 15%, then there is a pronounced tendency to form substitutional solid solutions (cf. Section 6.1.7). However, if the size difference is more than 20%, then often phases are formed, in which exclusively tetrahedral structural units are present (as opposed to the densest sphere packings of equally sized spheres, in which tetrahedral and octahedral voids are present), i.e., they form tetrahedrally close packings (**tcp**), sometimes also called topologically close packings. As regular tetrahedra do not fill the space completely – opposed to the proclamation of Aristotle – the tetrahedra in **tcp** phases must be necessarily (slightly) distorted ones. To avoid misunderstandings, please note that "complete space filling" refers here to the overall packing of the (distorted) *tetrahedra* and not to the sum of the volumes of the spheres that are located at the corners of a packing of tetrahedra.

In 1958, Frank and Kasper described the geometrical principles behind such phases in their two-part paper "Complex Alloy Structures Regarded as Sphere Packings" [101, 102], which is why these phases are also called Frank-Kasper phases (FK phases). Laves phases, which are described in Section 13.5.2, are a subgroup of FK phases.

In the following decades after their important work, it turned out that not only complex alloys can be described by the principles derived by Frank and Kasper. A broad range of other physical entities also form structures that belong to the same (or similar) structure family of FK phases: colloidal dispersions, soft matter systems like micellar phases of surfactants and lipids, copolymers, nanoparticle superlattices and foams (see the **BOX** "The Kelvin Conjecture, Foams, Zeolites, the Weaire-Phelan Structure, and Gas Hydrates" below).

Although the Frank-Kasper phases are referred to as "packing-dominated phases", they are also influenced by electronic factors. The formation of certain types of structures cannot be explained by packing theory alone. This influence of electronic factors, especially the VEC, similar to the Hume-Rothery phases, has been documented many times and the tendency to form one of the structure types can be understood on the basis of band structure calculations, but there is no generally valid concept that would allow a prediction.

13.5.1 The geometrical principles in FK phases and the FK coordination polyhedra

FK phases are characterized by two geometrical rules that are related to the way that the tetrahedral building units are organized: (1) there are clusters of either five or six tetrahedra that share one common edge (if spheres are put at the vertices of the tetrahedra, these clusters correspond to (compressed and empty) pentagonal and hexagonal bipyramidal coordination polyhedra, respectively, in which the line between the apical vertices builds the common edge, see Figure 13.9) and (2) the six-connected vertices are never linked by a common edge. As a consequence of these two rules, there can be only four different coordination polyhedra with coordination numbers of 12, 14, 15, and 16 (see Figure 13.10), all of which have only triangular faces, which is why they belong to the polyhedra class of deltahedra:

- the icosahedron with a coordination number of 12 (12 vertices, all five-connected, 20 faces)
- a two-fold hexagon-face-capped hexagonal antiprism with CN = 14 (14 vertices, 12 of them five-connected, two of them six-connected, 24 faces)
- a three-fold hexagon-face-capped vertex-truncated trigonal prism with CN = 15, also called the µ-phase polyhedron (15 vertices, 12 of them five-connected, three of them six-connected, 26 faces)
- a three-fold hexagon-face-capped vertex-truncated tetrahedron with CN = 16, also called the Friauf polyhedron (after James Byron Friauf, 1896–1972, an American professor of physics, who solved the crystal structure of $MgCu_2$ and Al_2Cu in 1927) (16 vertices, 12 of them five-connected, four of them six-connected, 28 faces).

Since these coordination polyhedra partially interpenetrate each other in the structures of FK phases, it is in some cases necessary to choose a different representation method that provides a view on the structure that is easier to comprehend.

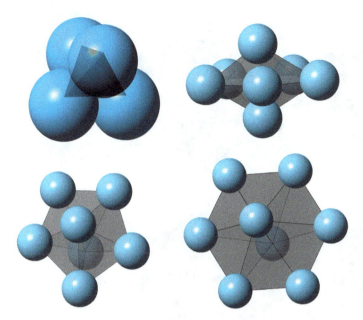

Figure 13.9: Tetrahedral close packing of four identical spheres with their centres at the vertices of a regular tetrahedron (top-left), two views of a cluster of five (slightly distorted) tetrahedra sharing a common edge (top-right and bottom-left), and a cluster of six (slightly distorted) tetrahedra sharing a common edge (bottom-right). The spheres of the clusters were scaled down in order to visualize the underlying tetrahedral arrangements. The five- and six-cluster arrangement of the spheres correspond to (compressed) pentagonal and hexagonal bipyramids, respectively.

13.5.2 Laves phases

Laves phases (after Fritz Laves, a German mineralogist and crystallographer, 1907–1978) are a huge family of binary (or pseudo-ternary) intermetallic compounds of the stoichiometry AB_2 with thousands of representatives, which crystallize in three different, but closely related structure types. In all Laves phases only 12- and 16-vertex FK polyhedra are present. The three prototypical compounds are:
- $MgCu_2$, *Strukturbericht* type C15, space group $Fd\bar{3}m$,
- $MgZn_2$, *Strukturbericht* type C14, space group $P6_3/mmc$,
- $MgNi_2$, *Strukturbericht* type C36, space group $P6_3/mmc$.

For a composition of AB_2, the packing fraction can be theoretically as high as 77.9% [103] if the radius ratio of the spheres is $r_A/r_B = 1.730$. However, in the **tcp** packing that the Laves phases adopt, the largest packing density is 0.71, for which the two kinds of atoms have to be an ideal radius ratio of $r_A/r_B = (3/2)^{1/2} \approx 1.225$. However, the atomic radius ratio of known structures of this type varies from 1.672 to 1.050.

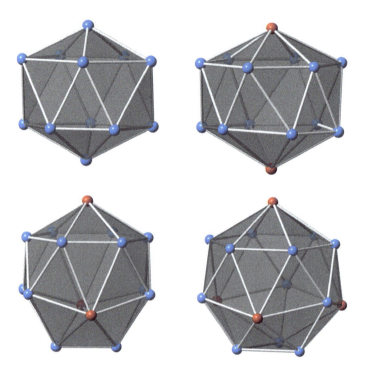

Figure 13.10: The Frank-Kasper polyhedra with CN = 12 (top-left), 14 (top-right), 15 (bottom-left), and 16 (bottom-right). The six-connected vertices are coloured in red.

The structure of the Laves phases can be described in several ways, the two most common explanations are given in the following section. The first description is based on atomic layers and nets, and the second one uses a more conventional approach based on (coordination) polyhedra.

In MgCu$_2$ a sequence of the following layers is stacked on top of each other (see also Figure 13.11):
- In the first flat layer, on (111) planes, there are Cu atoms (the smaller component) present, forming a 3.6.3.6 kagome net, i.e., a **hcp** layer in which one quarter of the spheres are removed; let us call this kagome layer the A layer,
- On one half of the Cu$_3$ triangles of this kagome net, we place further Cu atoms, building Cu$_4$ tetrahedra with the Cu atoms of the kagome net below; let us call this an intermediate Cu layer.
- The larger Mg atoms are now placed (a) on top of the other half of the Cu$_3$ triangles and (b) on top of the centres of the hexagons of the first kagome A layer. The Mg atoms form a net of corrugated six-membered rings, in which the barycentres of half of the Mg atoms are below the barycentres of the Cu atoms of the intermediate layer and the other half above them.

– Finally, another Cu kagome layer (*B* layer), which is shifted within the (111) plane with respect to the first *A* layer and rotated by 180°, is stacked on top of these two intermediate Cu/Mg layers. After two further intermediate Cu/Mg layers, a third Cu kagome layer, in position *C*, is stacked on top on these.

The other two Laves phases differ only in the stacking sequence of the kagome layers. In MgZn$_2$ the layer sequence is *AB*... and in MgNi$_2$ it is *ABAC*... .

Alternatively, MgCu$_2$ can be described as a two-fold interpenetrated diamond-like structure (see Figure 13.12): the Mg atoms form the first diamond net and the centres of the (empty!) Cu$_4$ building units form the second diamond net. The Cu atoms are then located at the midpoints of the connection lines between the centres of the tetrahedra. The Cu$_4$ tetrahedra are exclusively corner-connected. The Cu atoms are surrounded icosahedrally by six (3 + 3) Cu atoms and further six (in the form of a corrugated six-membered ring) Mg atoms. The Mg atoms are surrounded by a 16-vertex FK polyhedron, which can be decomposed to a Cu$_{12}$ truncated tetrahedron and an Mg$_4$ tetrahedron.

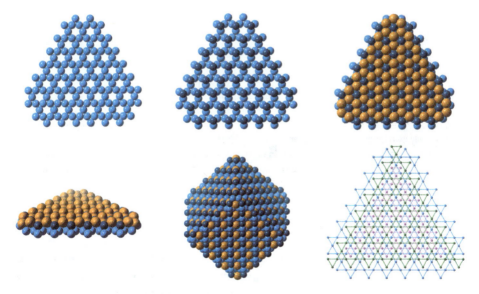

Figure 13.11: Illustration of the crystal structure of MgCu$_2$ in terms of atomic layers. The start point is a kagome net of Cu atoms shown in light-blue (top-left). On half of the Cu triangles a further Cu layer shown in dark blue is stacked (top-middle). On the other half of the Cu triangles as well as on the centres of the hexagons of the first kagome layer the Mg atoms are placed shown in orange (top-right). The Mg atoms form a corrugated six-membered ring structure. A perspective view of these three layers is shown in the bottom-left. (2 × 2 × 2) unit cells viewed parallel to the (111) planes are shown in the bottom-middle panel. The three kagome nets of Cu atoms (*ABC*) that are shifted within the (111) plane against each other are shown at the bottom-right, *A* = blue, *B* = green, *C* = pink.

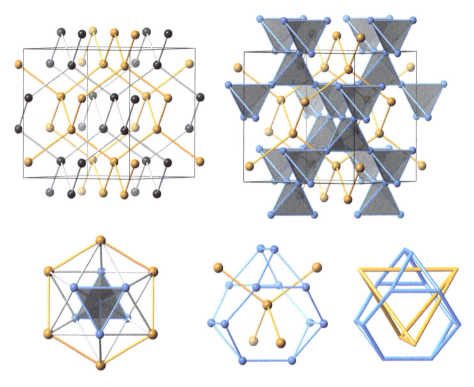

Figure 13.12: Alternative description of the crystal structure of MgCu$_2$, which can be described as two interpenetrating diamond networks of Mg atoms and corner-connected Cu$_4$ tetrahedra (top); the centres of the Cu$_4$ tetrahedra are shown as black spheres in the top-left panel. The Cu atoms are surrounded icosahedrally by 6 Cu and 6 Mg atoms (bottom-left). The Mg atoms are surrounded by 4 Mg atoms in a tetrahedral fashion and further 12 Cu atoms in form of a vertex-truncated tetrahedron; this is shown in two different styles in the bottom-middle and bottom-right panel. The connections between the atoms are only a visual aid.

In MgZn$_2$ the (empty) Zn$_4$ tetrahedra build columns running along the z-axis of the unit cell and are alternately corner- and face-connected; thus, the latter pairs are forming (empty) trigonal bipyramids (see Figure 13.13). The basal faces of two trigonal bipyramids lying on top of each other are maximally rotated against each other. There are two crystallographically distinct Zn atoms: The first kind of Zn atoms is at the tip of the trigonal bipyramids, and they are surrounded by six Zn atoms in the form of a trigonal antiprism and by six Mg atoms that build a six-membered ring in chair conformation. The other kind of Zn atoms belongs to the basal face of the trigonal bipyramids, and they are surrounded by six Zn atoms in the form of an extremely distorted tetragonal bipyramid and by six Mg atoms building a six-membered ring in boat conformation. If the Mg atoms are considered separately, then they form a structure analogous to lonsdaleite, i.e., the hexagonal modification of diamond. Each Mg atom is surrounded by 4 Mg atoms and 12 Zn atoms forming a 16-vertex FK polyhedron.

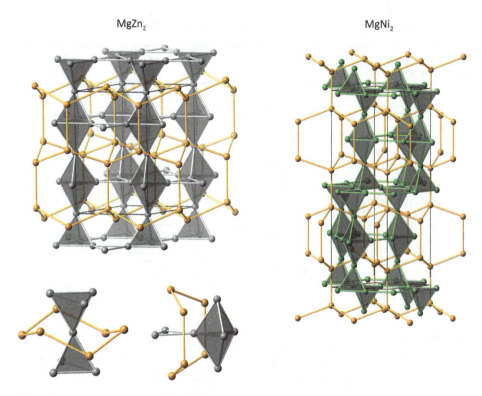

Figure 13.13: Crystal structure of MgZn$_2$ (left) and MgNi$_2$ (right) as well as the surrounding of the two kinds of Zn atoms in MgZn$_2$ (bottom left). Mg: orange, Zn: grey, Ni: green.

Finally, MgNi$_2$ can be described as a mixture between the MgCu$_2$ and MgZn$_2$ type. There are three crystallographically distinct Ni sites, two of them have a surrounding analogue to the first kind of the Zn atoms in MgZn$_2$ and the third kind has an analogue surrounding of the second kind of Zn atoms in MgZn$_2$. The Mg atoms are again surrounded by a 16-vertex FK polyhedron.

Typical representatives of binary compounds crystallizing in the MgCu$_2$, MgZn$_2$, and MgNi$_2$ structure type are gathered in Table 13.4.

13.5.3 Variants of Laves phases

In addition to the classical Laves phases of composition AB$_2$, there are variants characterized by a different stacking of the kagome nets and having composition AB$_3$, AB$_5$, and A$_2$B$_{17}$. These phases also occur rather frequently. The prototypical structure of CaCu$_5$ is realized for some binary intermetallic compounds which occurs when there are significant radius differences between the two partners. It is characterized by a primitive stacking (AA...) of the 3.6.3.6 Cu kagome nets. The (empty) Cu$_4$ tetrahedra

Table 13.4: Examples of binary Laves phases of the type $MgCu_2$, $MgZn_2$, and $MgNi_2$.

Major component belonging to group	$MgCu_2$ type	$MgZn_2$ type	$MgNi_2$ type
1		$CaLi_2$, KNa_2	
2	(Ti,Nb,Ta,Cu,Ag)Be_2, $REMg_2$	(V,Cr,Mo,W,Mn,Re,Fe)Be_2, (Ca,Sr,Ba,Er)Mg_2	$ThMg_2$
5	(Zr,Hf,Ta)V_2		
6	(Ti,Zr,Hf,Nb,Ta)Cr_2, (Zr,Hf)Mo_2, (Zr,Hf)W_2	(Ti,Zr,Hf,Nb,Ta)Cr_2	(Ti,Zr,Hf)Cr_2, $HfMo_2$
7	(Y,RE)Mn_2	(Ti,Zr,Hf,RE)Mn_2, (Zr,Hf,RE)Re_2	$HfMn_2$
8	(Y,RE,Zr,Hf)Fe_2, $RERu_2$, $REOs_2$	(Sc,Ti,Nb,Ta,Mo,W)Fe_2, (Sc,Y,RE,Zr)Ru_2, (Zr,Hf)Os_2	(Sc,Zr,Hf)Fe_2
9	(Y,RE,Zr,Hf,Nb,Ta)Co_2, (Ca,Sr,Ba,Y,RE)Rh_2, (Ca,Sr,RE,Zr)Ir_2	$TaCo_2$, $ZrIr_2$	$TaCo_2$
10	(Sc,Y,RE,)Ni_2, (Ca,Sr,Ba)Pd_2, (Li,Na,Ca,Sr,Ba,Y,RE)Pt_2	$ZrNi_2$	UPt_2
11	$NaAg_2$, (Na,Pb,Bi)Au_2	$CdCu_2$	$CdCu_2$
12	(Zr,Hf)Zn_2	(Ti,Ta)Zn_2, $CaCd_2$	(Hf,Nb)Zn_2
13	(Ca,Sc,Y,RE,Nb)Al_2	(Zr,Hf)Al_2	
15	(K,Rb,Cs)Bi_2		

RE, rare-earth metal. Note that some compounds show polymorphism and crystallize in more than one type.

are alternately corner- and face-connected along the c-axis of the hexagonal unit cell (space group $P6/mmm$). The Cu atoms are icosahedrally surrounded by nine Cu and three Ca toms, while the Ca atoms form a (18 + 2)-vertex coordination polyhedron (six Cu atoms at a distance of 2.948 Å, 12 Cu atoms at a distance of 3.259 Å, and two Ca atoms at a distance 4.051 Å), see Figure 13.14. Some important hard magnetic materials like $SmCo_5$, YMn_5, and $HoMn_5$ (compare Section 9.2.1) belong to this structure type.

Two further structure types with stoichiometry AB_3 are obtained, if the stacking sequence of the kagome nets formed by the smaller metal proceeds in the way AABB... ($CeNi_3$, space group $P6_3/mmc$) or AABBCC... ($PuNi_3$, space group $R\bar{3}m$). Their structures are shown in Figure 13.15.

Further stoichiometric variants are obtained, if one-third of the larger atoms are substituted by pairs of the smaller atoms, leading to compounds with composition A_2B_{17}. The two prototypical compounds are Th_2Zn_{17} (space group $R\bar{3}m$) and Th_2Ni_{17} (space group $P6_3/mmc$). The partial Zn and Ni structures can be decomposed into empty tetrahedra and empty hexagonal bipyramids as shown in Figure 13.16.

13.5 Packing-dominated phases (Frank-Kasper and Laves phases) — 287

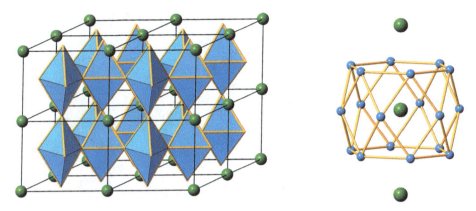

Figure 13.14: The crystal structure of CaCu$_5$ shown as (2 × 2 × 2) super cell (left) and the (18 + 2)-vertex coordination environment of Ca in CaCu$_5$ (right); Ca: green, Cu: blue.

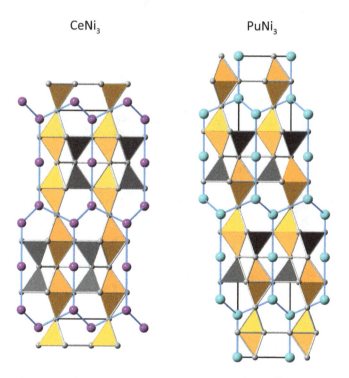

Figure 13.15: The crystal structures of CeNi$_3$ (left) and PuNi$_3$ (right); view along the a-axis; Ce, purple; Pu, cyan; Ni, grey.

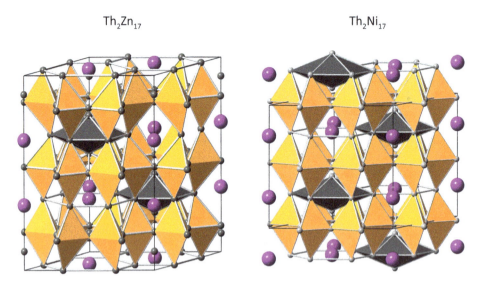

Figure 13.16: Crystal structure of Th$_2$Zn$_{17}$ (left) and Th$_2$Ni$_{17}$ (right); Th: purple, Zn: dark-grey, Ni: light-grey; empty Zn/Ni tetrahedra: orange, empty Zn/Ni hexagonal bipyramids: dark-grey.

Another ordered intermetallic phase, in which even higher coordination numbers are realized, occurs in the prototypical structure of NaZn$_{13}$, which crystallizes in the cubic space group $Fm\bar{3}c$. There are two crystallographically independent Zn sites. Both are icosahedrally surrounded by the other type of Zn atoms, one of them in an almost regular fashion, and the second one is significantly distorted due to the presence of the Na atoms. The Na atoms are surrounded by 24(!) Zn atoms (all at the same distance of 3.565 Å) forming a so-called snub cube (also called snub cuboctahedron) as coordination polyhedron, a polyhedron with 38 faces (6 squares and 32 equilateral triangles), see Figure 13.17. If we disregard the different orientations of the icosahedra and snub cubes, or to express this in an alternative way, if we only consider the barycentres of these two types of polyhedra, they build an CsCl analogous structure, in which each type of polyhedron is surrounded by eight polyhedra of the other type.

13.5.4 σ-, μ-, M, P, and R phases

While in Laves phases only FK polyhedra with CN = 12 and 16 are present, in the phases mentioned in the title of this subsection,[16] other compositions of different types of FK

---

[16] Even after intensive research efforts, the author has not succeeded in finding out what the Greek and Latin letters stand for, in short, why the phases are called in such a way or if there is systematics – presumably none. Probably they have their origin in the field of metallurgy and foundry.

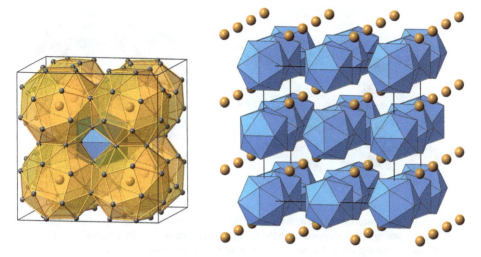

Figure 13.17: Two views of the crystal structure of NaZn$_{13}$; a Zn-Zn$_{12}$ icosahedron (blue) is surrounded by eight NaZn$_{24}$ snub cubes (orange) (left); a slightly extended view of the unit cell emphasizing the Zn-Zn$_{12}$ icosahedra while the sodium atoms are shown as orange balls.

polyhedra constitute the structure. An overview together with prototypical compounds is given in Table 13.5.

Table 13.5: Structure types of the σ-, μ-, M, P, and R Frank-Kasper phases together with the relative frequencies of the structure constituting FK polyhedra.

Structure type	Name	Space group	Relative frequency of polyhedra with			
			CN = 12	CN = 14	CN = 15	CN = 16
Cr$_{49}$Fe$_{51}$	σ-phase	$P4_2/mnm$	0.33	0.54	0.13	0
W$_7$Fe$_6$	μ-phase	$R\bar{3}m$	0.55	0.15	0.15	0.15
Nb$_{10}$Ni$_9$Al$_3$	M phase	$Pnma$	0.55	0.15	0.15	0.15
Mo$_{21}$Cr$_9$Ni$_{20}$	P phase	$Pnma$	0.43	0.36	0.14	0.07
Mo$_3$Cr$_2$Co$_5$	R phase	$R\bar{3}$	0.51	0.23	0.11	0.15

The σ-phase can be described in several ways, and the most convenient is to decompose the structure into a packing of (empty) hexagonal bipyramids, which are vertex-connected along the c-direction of the tetragonal unit cell (see Figure 13.18, left). There are two crystallographically distinct hexagonal bipyramids. The connection scheme of the hexagonal bipyramids within the (a,b) plane at height z = 0 is shown in Figure 13.18, right.

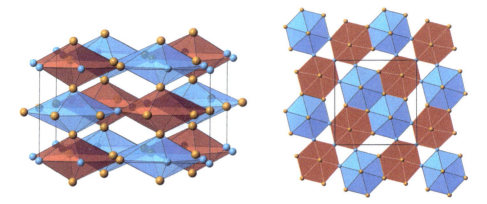

Figure 13.18: Crystal structure of the σ-phase exemplified on the structure of $Al_{40}Ta_{60}$ which can be described by stacking corner-connected (empty) hexagonal bipyramids along the c-direction (left); Al: blue, Ta: orange. The two kinds of (empty) hexagonal bipyramids are shown in blue and red, respectively. The connection scheme of the hexagonal bipyramids within the (a,b) plane is shown on the right.

The structure of the μ-phase (W_7Fe_6) can be described as layers of W-centred Fe_{12} truncated tetrahedra, W@Fe_{12}, alternating with layers of edge-sharing (empty) hexagonal W_8 bipyramids (see Figure 13.19). The M phase ($Nb_{10}Ni_9Al_3$) can be described as a network of pairs of hexagonal face-sharing truncated tetrahedra (the corners are occupied by the Al and Ni atoms in a disordered manner) in which pairs of edge-sharing hexagonal bipyramids (corners occupied by the Nb atoms) are integrated in the voids of this network (Figure 13.19). In the P phase ($Mo_{21}Cr_9Ni_{20}$) there are two crystallographically distinct types of (empty) hexagonal bipyramids and metal-centred icosahedra present. One type of the hexagonal bipyramids (shown in blue in Figure 13.19) is corner-connected to six other hexagonal bipyramids of the same type within the (a,c) plane and are further edge-connected to the icosahedra. The other type of hexagonal bipyramids (shown in green) shares common edges and forms rods along the c-axis. They are corner-connected to the icosahedra and share a common corner with the other type of hexagonal bipyramid. The R phase, $Mo_3Cr_2Co_5$, is built from close-packed chains of face-sharing FK polyhedra with the sequence CN16-(CN12)$_3$-CN16 (two different views are provided in Figure 13.19).

13.5.5 The Cr₃Si structure (*Strukturbericht* type A15)

A relatively simple structure is formed by the classical FK phase of Cr_3Si. This structure is also known as the A15 phase, designated to the structure of the (metastable) β-allotrope of the element tungsten, which was first reported in 1931 [104]. For a long time, it was, however, not clear if the samples that were identified as β-W consists of

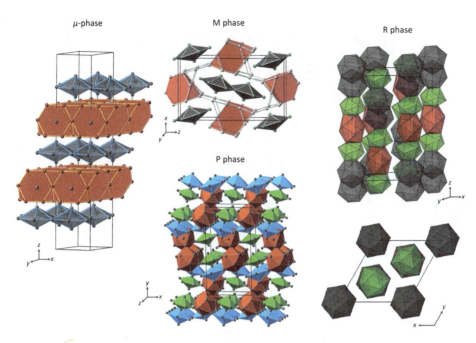

Figure 13.19: The structure of the μ-, M, P, and R phases (see text).

pure W or if they constitute a suboxide of W, W_xO with x = 3–20. Today, β-W is considered as an interesting material for spintronics [105].

In β-W, which crystallizes in the space group $Pm\bar{3}n$, there are two crystallographically distinct W sites: one at the Wyckoff position 2a (0,0,0) and one at the Wyckoff position 6c (0.25, 0, 0.5). The atoms at the position 2a build a **bcc**-like sublattice, while the atoms at 6c form "W_2 pairs" located at the face centres with their axis oriented along the x-, y-, and z-axis, respectively, which extends into mutually perpendicular, infinite rows of W atoms, taken the neighbouring cells into account. The β-W structure is also known as the Weaire-Phelan structure, see BOX below.

As Cr_3Si is a binary compound, it builds, of course, its own structure type. The silicon is at Wyckoff position 2a and the chromium atoms occupy the Wyckoff position 6c; the (slightly extended) unit cell of Cr_3Si is shown in Figure 13.20. The silicon atoms are surrounded by 12 Cr atoms in the form of a (slightly distorted) icosahedron and the Cr atoms are surrounded by an FK-14 polyhedron, build from 4 Si atoms (at a distance of 2.549 Å) and 12 Cr atoms (at a distance of 2.792 Å).

The structure of Cr_3Si is of relevance in the field of superconducting compounds (see also Chapter 11): While Cr_3Si itself is not a superconductor, several other binary or pseudo-ternary isostructural intermetallic compounds are superconductors of type II with relatively high critical temperatures (and high critical magnet fields) including

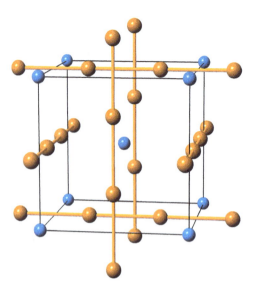

Figure 13.20: A perspective view of the crystal structure of Cr_3Si; Cr = orange, Si = blue.

V_3Si (T_c = 16.8 K), Nb_3Sn (T_c = 18.3 K), $Nb_3(Al,Ge)$ (T_c = 21.0 K), and Nb_3Ge (T_c = 22.3 K). Nb_3Ge was discovered to be a superconductor in 1973 and for 13 years (until the discovery in 1986 of the cuprate superconductors) it held the record as having the highest critical temperature.

Box 13.1 The Kelvin conjecture, foams, zeolites, the Weaire-Phelan structure, and gas hydrates

In the beginning of this section, it was already mentioned that the structure of packing-dominated phases is of relevance in very different areas. These structures also touch upon interesting mathematical questions, for instance, the Kelvin problem or conjecture. The Kelvin conjecture (it is named Kelvin conjecture, although, he himself never called it a conjecture) is the 3D analogue of the 2D honeycomb conjecture, which states: any partition of the plane into regions of equal area has perimeter at least that of the regular hexagonal honeycomb tiling. Interestingly, the proof of the honeycomb conjecture could only be provided around the turn of the millennium by Thomas Callister Hales [106]; Hales also proved the Kepler conjecture, which states that "no packing of congruent balls in Euclidean three space has density greater than that of the face-centred cubic packing" (see Chapter 4).

In 1887, Lord Kelvin (Sir William Thomson) asked the question: how can space be partitioned into cells of equal volume with the least area of surface between them? Or, in short: what is the most efficient soap bubble foam? [107] His answer was that this is a convex uniform honeycomb (a uniform tessellation which fills three-dimensional Euclidean space with non-overlapping convex uniform polyhedral cells) based on truncated octahedra, see Figure 13.21. A truncated octahedron is a polyhedron with 14 faces, made of eight regular hexagons and six squares. However, a packing with these polyhedra does not satisfy Plateau's laws, formulated by Joseph Plateau in the 19th century, according to which minimal foam surfaces meet at 120° angles at their edges, and that these edges meet each other in sets of four with angles of 109.47°, i.e., the tetrahedral angle. Therefore, Kelvin proposed a structure with slightly bent edges and slightly warped surfaces for its faces, obeying Plateau's laws and reducing the area of the structure by 0.2% compared with the corresponding polyhedral structure. Interestingly, the honeycomb of truncated octahedra resembles the structure of the zeolite sodalite or to be precise the Al/Si atoms are at the vertices of the truncated octahedra; in the area of zeolites this truncated octahedron cage is called the sodalite or β-cage (cf. Chapter 14).

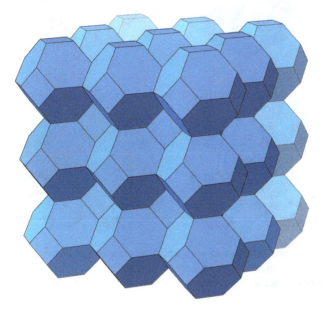

Figure 13.21: A honeycomb of truncated octahedra; it is the variant of the Kelvin foam structure with linear edges and flat surfaces and resembles the structure of the zeolite sodalite.

For more than 100 years no counter-example of the Kelvin foam structure was found. But in 1994 the physicist Denis Weaire and his student Robert Phelan discovered a structure with a surface area that is 0.3% smaller than that of the Kelvin structure [108]. In contrast to the Kelvin structure, the Weaire-Phelan structure is composed of two kinds of cells, both having equal volumes. One is a pyritohedron, an irregular version of the dodecahedron with 12 pentagonal faces (5^{12}), possessing tetrahedral symmetry. The second has a form of a truncated hexagonal trapezohedron, a species of a tetrakaidecahedron with 2 hexagonal and 12 pentagonal faces ($5^{12}.6^2$), see Figure 13.22. Like the hexagons in the Kelvin structure, the pentagons in both types of cells are slightly curved. The primitive cubic Weaire-Phelan structure with space group $Pm\bar{3}n$ consists of columns of tetrakaidecahedra parallel to the cube edges; they form pairs connected by the hexagonal faces. The pyritohedra are at the centre and the corners of the unit cell and build a pseudo **bcc** structure; pseudo, since the pyritohedron in the centre is rotated by 90° with respect to the pyritohedra at the corners. The pyritohedra have no common corners or edges among themselves. If only the centres of the polyhedra are considered, this corresponds exactly to the structure of Cr_3Si or the β-tungsten structure!

The polyhedral honeycomb associated with the Weaire-Phelan structure (obtained by flattening the faces and straightening the edges) is also loosely referred to as the Weaire-Phelan structure. It was

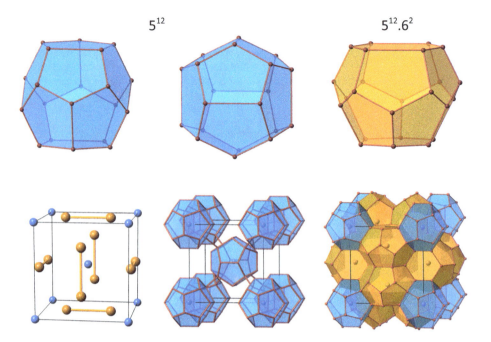

Figure 13.22: A pyritohedron (shown in two different views) and a tetrakaidecahedron (top) are the constituents of the Weaire-Phelan structure, in which the pyritohedra are at the centre and corners of the unit cell and pairs of the tetrakaidecahedra at the centre of the faces oriented parallel to the x-, y-, and z-axis, respectively (bottom).

known well before the Weaire-Phelan structure was discovered, but the application to the Kelvin problem was overlooked. Already in 1954, the structure was described as the structure of the SO_2-in-ice gas hydrate, in which the oxygen atoms of the water molecules are at the vertices of the Weaire-Phelan cages [109]. This is also the structure of one of the naturally occurring methane hydrate structures, precisely methane hydrate with "structure-I".

14 Porous crystals, reticular chemistry, and the net approach

So far, almost exclusively dense materials have been discussed in this book. Until three decades ago, they also made up the bulk of all known materials. There is a good reason for this because nature tends to avoid too large gaps in structures because this is energetically unfavourable; dense or compact structures have a lower interfacial energy. However, the principle of minimizing the interfacial area does not necessarily mean that porous structures cannot occur in nature. In many cases, these are kinetically frozen systems that cannot overcome the energy barrier to reach a more compact state; porous volcanic rocks may serve as an example.

However, certain techniques are required to produce porous substances in the laboratory (or in large-scale plants) in a desired way. One principle that is frequently used is to first produce a composite material from at least two components and, after synthesis, to selectively remove one of them (the template), leaving gaps or cavities behind. This technique is, for instance, applied for synthesizing mesoporous carbons or (organo)silica phases. Apart from the fact that often the remaining gaps form a regular lattice, the compact matter around the gaps, in most cases, is in an amorphous or semi-crystalline state.

However, there are also some classes of substances that can be produced as true, at least potentially porous crystals. Two of the most important classes will be presented first. On the one hand zeolites, which belong to the class of framework (tecto) aluminosilicates; they are of great importance in industrial applications. And on the other hand, metal-organic framework (MOF) compounds (and related structures), which have developed into a very important class of materials in the last 20 years and now comprise several 10,000 representatives. Finally, there will be a short introduction to the topological description of such framework or network-like structures.

14.1 Zeolites and zeotypes

The naturally occurring zeolites (gr. *zeein* = boiling or to boil, gr. *lithos* = stone) were discovered in 1756 by the Swedish mineralogist Axel Fredrik Cronstedt. He observed, upon heating, the evaporation of water from a mineral specimen, probably largely composed of the zeolite stellerite, so that the impression was given that the "stone" was boiling. Subsequently, many more natural zeolites were discovered, and more than 60 different ones are known today. Most typically, they are formed as alteration products of volcanic rocks (mostly basalts) in areas with hydrothermal water veins. Despite the relatively large number of representatives of zeolites and although interesting properties of zeolites were successively recognized, for instance their potential to act as *molecular sieves*, they remained for a long time a rather insignificant class of minerals. This was also because it was believed that they only occur in trace amounts but not as large deposits. However, in the middle of the twentieth century – about 200 years after the discovery of zeolites – things began to change. The first successful zeolite syntheses were reported in 1948 by Richard M. Barrer, applying a hydrothermal synthesis route to mimic geological conditions. One of these zeolites was a structural analogue of the naturally occurring mineral mordenite, others had a structure that is not found in any naturally occurring zeolite. This laid the foundation for the almost meteoric rise of zeolites, which are now one of the most important industrial chemicals. A nice overview about the historical development of zeolites can be found in Ref. [110]. Interestingly, in the late 1950s also huge beds of zeolite-rich sediments were also discovered in the western part of the USA and some other places in the world. These beds were generally flat-lying and easily mined by surface methods. The most common zeolites that are found in these large-scale deposits are clinoptilolite, mordenite, chabazite, and phillipsite. Since natural zeolites are often accompanied by other mineral phases (quartz, feldspars, mica, etc.) and as it is very costly to separate the zeolites from these phases, mined zeolites tend to be used rather as zeolite-rich rocks in applications where high purity is not required. Nevertheless, the production volume of natural zeolites is comparable to that of synthetic zeolites (even if reliable exact figures are difficult to obtain in the zeolite business).

Today, more than 250 unique zeolite topologies (see below), i.e., framework types, have been recognized by the Structure Commission of the International Zeolite Association (IZA), each of which is given a unique three-letter (in capitals) code. Only about 40 of these topologies occur in natural zeolites.

The property par excellence that makes zeolites so interesting and versatile is that they have *accessible cavities*. This property distinguishes them from other framework compounds of corner-linked tetrahedral building units and is also part of the definition of zeolites [111] according to the Subcommittee on Zeolites of the International Mineralogical Association (IMA):

> Zeolites are crystalline materials whose structure consists of a framework of corner-sharing tetrahedra. Their structures contain cavities (cages, channels), which are usually occupied by extra-framework species like H_2O molecules, inorganic cations, or organic species. These guest species are able to diffuse through the channels, allowing for an exchange of the extra-framework content. Dehydration of hydrated zeolites typically occurs at temperatures below about 400 °C, it is largely reversible. [...]

As the maximum cavity size of zeolites is about 18 Å, they all belong to the class of *microporous* materials.

The spectrum of applications of zeolites is extremely broad and the production volume of synthetic zeolites is in the order of 2 million tonnes per year. Zeolites are used among others:
- as catalysts in the hydrocracking and fluid catalytic cracking (FCC) process of medium- and heavy-weight raw oil fractions; in this process the high-boiling oil

fractions that are mostly useless are converted into lighter hydrocarbons that can be used either as fuels (gasoline, diesel, jet fuel) or as platform chemicals in chemical industry;
- as catalyst in the methanol to gasoline and methanol to olefin (MTO) process;
- as ion-exchange agent and detergent additive for water softening;
- for separation of gases, e.g., oxygen from air;
- as lightweight aggregate and pozzolans in cements and concretes and as filler materials in paper; here mainly naturally occurring zeolite-rich rocks are used;
- for the removal of ammonia from municipal, industrial, and agricultural waste and drinking waters;
- for the removal of certain radionuclides ($^{137}Cs^+$, $^{90}Sr^{2+}$) from radioactively contaminated wastewaters;
- in the flue gas desulfurization (FGD) process;
- as abrasives in toothpaste;
- for the regeneration of dialysis fluids;
- in the freeze-drying process of food;
- as adsorption agent in cigarette filters;
- for reconditioning of used engine oil;
- as desiccant in the manufacturing process of insulating windows and as drying component of refrigeration systems to absorb water carried by the refrigerant,
- as additives in animal feed and, finally, to name a niche application,
- for the construction of self-cooling beer kegs.

Approximately 70% of the production volume of synthetic zeolites are used for ion-exchange processes, further 17% are used in the field of catalysis (however, these 17% make about 55% of the market value), and 10% are used as adsorbents, in particular for separation processes. Detailed information about the industrial use in catalysis and separation can be found in the monograph edited by Kulprathipanja [112].

14.1.1 Structural chemistry and building units of zeolites

The large majority of zeolite minerals and the most common synthetic zeolites are aluminosilicates, with silicon and aluminium occupying the tetrahedral sites (T sites). The Si/AlO_4 tetrahedra which are exclusively corner-connected via the oxygen atoms constitute the so-called primary building units of zeolites. The negative charge of the framework is balanced through positively charged extra-framework species (metal cations of groups 1 and 2, typically sodium, potassium, magnesium and calcium, or organic cations, or through framework protons bonded to oxygen atoms). The range of Si/Al ratios, this ratio is also known as module, extends from 1 to infinity. Zeolites with a module of infinity are called *all-silica* zeolites and have a pure SiO_2 composition and, hence, a neutral framework. They do not occur in nature but can be synthesized in the laboratory. Si/Al ratios below 1 would violate Löwenstein's rule, which states that corner-sharing AlO$_4$ tetrahedra are unstable. As the Si/Al ratio increases, zeolites become more hydrophobic and the framework charge decreases so that fewer cations are needed for charge balance. Furthermore, the thermal stability is increasing with higher silicon content; zeolites with a Si/Al ratio of 1 are approximately stable up to 700 °C, while all-silica zeolites decompose only at temperatures of ~1,300 °C. With an increasing silicon content the number of catalytic centres also decreases, however, the activity of each catalytic active site is increasing. This is why in catalysis zeolites are favourably used that have a medium-to-high Si/Al ratio.

The general empirical formula of zeolites can be written as

$$M_{2/n}O \cdot Al_2O_3 \cdot y\,SiO_2 \cdot w\,H_2O$$

where n is the valency of the cation M, w the number of formula units of water in the voids of the zeolite, and y is typically in the range from 2 to 200. One typical example of a zeolite structure is shown in Figure 14.1, which depicts the structure of the naturally occurring zeolite natrolite with the chemical formula Na$_2$Al$_2$Si$_3$O$_{10}$ · 2 H$_2$O, i.e., with regard to the general formula, $n = 1$, $y = 3$, and $w = 2$.

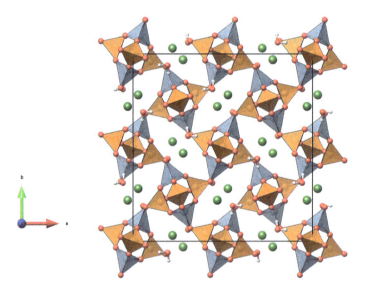

Figure 14.1: Crystal structure of the zeolite natrolite, Na$_2$Al$_2$Si$_3$O$_{10}$ · 2 H$_2$O, space group *Fdd*2 (no. 43) (a = 18.326 Å, b = 18.652 Å, c = 6.601 Å; $\alpha = \beta = \gamma = 90°$); SiO$_4$ tetrahedra, orange; AlO$_4$ tetrahedra, blue-grey; oxygen, red; hydrogen, white; sodium, green.

For the sake of completeness, it should be mentioned that since the late 1970s, various materials with zeolite-like structures, but other compositions, have been synthesized. This includes neutral-framework materials like silicogermanates, aluminophosphates (AlPOs), and gallophosphates (GaPOs) as well as charged-framework materials like silicoaluminophosphates (SAPOs), metal AlPOs (MeAPOs), and borosilicates. Usually, the term *zeotype* is used to designate non-aluminosilicate materials that have a zeolite-like structure.

As mentioned in the introduction, the structural-topological classification of zeolites is done by the IZA, in which a unique three-letter framework-type code (FTC) is assigned for each principally different linkage type of the tetrahedral building units, for instance LTA (Linde type A), MOR (mordenite), or FAU (faujasite). The structure is abstracted to the T-atoms only, i.e., the chemical composition, possible counterions, etc., are not taken into account. Structural information about these framework types is provided in the IZA Database of Zeolite Structures, which is also available online (www.iza-structures.org). Each framework type can be topologically unambiguously characterized by a combination of two topological descriptors, namely the vertex symbol and the coordination sequence (CS) of all non-equivalent T-atoms in the structure. These descriptors will be further explained in Section 14.3. Apart from this topological description, several approaches to identify larger building units than the primary building units, i.e., the TO_4 units, have been proposed.

Secondary building units

Zeolite frameworks can be thought of as consisting of finite or infinite (i.e., chain- or layer-like) component units, which are called secondary building units (SBUs) and which consist of up to 16 T-atoms. By definition, SBUs should be derived assuming that the entire framework is made up of only one single type of SBU. A unit cell always contains an integral number of SBUs. However, the choice of an SBU for a framework is not unique. For instance, the LTA framework can be built up from six different types of SBUs including the SBUs 4, 6, and 8 that correspond to four-, six-, and eight-membered rings. The choice of SBUs is not unique in another sense, namely that different frameworks can be built up from the same type of SBU. For example, the three frameworks FAU, LTA, and SOD can all be constructed using only six-membered rings. It should be noted that SBUs are a theoretical construct and should not be considered to be or equated with species that may be in the solution/gel during the crystallization of a zeolitic material.

The number of identified SBUs increased from 16 in 1992 to 23 in 2007 (see Figure 14.2). With the recognition of an increasing number of very complex framework types that cannot be decomposed into a single type of SBU (this is the case for the frameworks DDR, DOH, IHW, ITH, MEP, MTN, SFG, and UTL), hence violating the definition of SBUs, the listing of SBUs for new FTCs was discontinued in 2007.

Composite building units

Composite building units (CBUs) are typically larger units than SBUs, although there is some overlap between SBUs and CBUs as the double-four ring (*d4r*), double-six ring

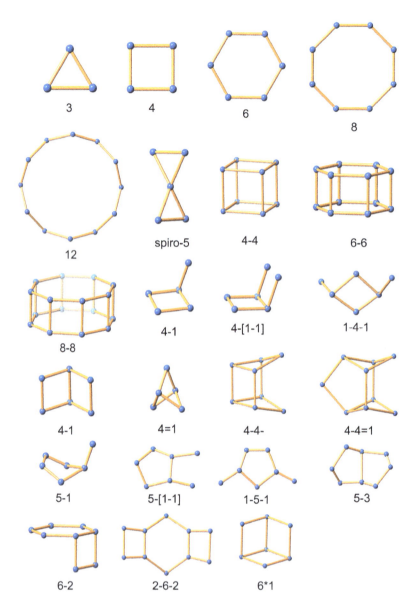

Figure 14.2: SBUs of zeolites that were recognized by the IZA until 2007.

(*d6r*), and the double-eight ring (*d8r*) constitute SBUs and CBUs as well. CBUs are different from SBUs that they cannot necessarily be used to build the entire framework. With the exception of the mentioned CBUs *d4r*, *d6r*, and *d8r*, each CBU has been assigned a lowercase italic three-character designation. CBUs correspond usually to more intuitively defined building units in zeolites and are sometimes identical to cages or cavities (see below) that are present in their structure. For instance, the LTA framework can be constructed by assembling the *d4r* unit, the sodalite (*sod*) cage, and the alpha (*lta*) cavity; the latter two are shown in Figure 14.3. The definition of CBUs lacks strict definition rules and they are therefore also kind of arbitrary. This problem of arbitrariness can be solved by the topological concept of *natural tilings* [113].

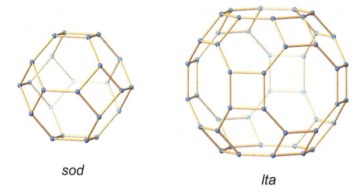

Figure 14.3: Two typical CBUs that occur in zeolites; left: *sod* cage, right: *lta* cage/cavity. Together with the *d4r* unit they can be used to assemble the LTA framework.

Rings, pores, cages, cavities, and channels

The faces of polyhedral CBUs are composed of *rings* of T-atoms of different sizes. These *n*-rings with $n = 3$–12 are called pore windows or simply pores. They are the entries to larger cages and cavities: polyhedra with faces that are not larger than six-rings are called *cages*. These faces are too narrow to let molecules pass through that are larger than water. Polyhedra that consist at least of one face that is larger than a six-ring are called cavities (although some authors do not distinguish between cages and cavities and call them equally cages). Cages and cavities that are infinitely extended in one direction and whose windows are large enough to allow diffusion of guest species (i.e., larger than six-rings) are called channels. The number of *intersecting* channels gives rise to the *dimensionality* of the channel system. In a one-dimensional system of channels, the channels are not connected to each other. In zeolites with a two-dimensional channel system, the channels are interconnected to form a layered system, and finally, in a three-dimensional system the channels form a 3D-interconnected system. In the database of zeolite structures of the IZA two different specifications for the dimensionality of channels are given:
(1) the topological dimensionality and
(2) the dimensionality with respect to the sorption of an organic molecule, the "sorption dimensionality".

The first specification is identical with the six-ring criterion of the pore opening given above, i.e., is irrespective of the actual geometric pore opening. In contrast to this, in the "sorption dimensionality" only channel directions that have a pore opening larger than 3.4 Å are counted. Thus the "sorption dimensionality" will provide a guide whether a small organic molecule might be able to diffuse along a channel direction.

The dimensionality of the channel system and the width of the pore openings are technologically important characteristics of a zeolite species. The more important zeolites usually have 3D channel systems, allowing for an effective diffusion of the guest species through the zeolite material. The pore opening or the effective width of the channels is another important characteristic of zeolites, which determines the accessibility of the channels to guest species. The effective width is limited by the smallest free aperture along the channel. The aperture is calculated by subtracting the diameter of an oxygen ion in silicate structures (0.27 nm) from the interatomic distance of two opposing oxygen atoms across the ring. Technologically zeolites are classified based on pore sizes into narrow- (free aperture 0.30–0.45 nm), medium- (0.45–0.60), large- (0.60–0.80), and extra-large-pore zeolites (ring size > 12, e.g., UTD-1). The free apertures for channels limited by several ring sizes are shown in Table 14.1 [112].

Table 14.1: Maximum free and typical free apertures for different ring sizes of T-atoms.

Ring size	Max. free aperture (nm)	Typical free aperture (nm)
4	0.16	
5	0.15	
6	0.28	
8	0.43	0.30–0.45
10	0.63	0.45–0.60
12	0.80	0.60–0.80

The maximum free aperture is given for non-distorted, symmetrical rings.

In mineralogy, zeolites are also classified by their external morphology of the crystals. According to this classification, zeolites belong to three main groups:
- fibrous structures formed by chains of tetrahedra weakly linked laterally as in natrolite and scolecite,
- lamellar structures formed by tetrahedra strongly linked in one plane with fewer bonds at right angle to this plane, as in heulandite and clinoptilolite, and
- block- or cube-like structures as in phillipsite and chabazite.

Although there is some degree of correlation between the dimensionality of the morphology and the dimensionality of the channel system in zeolites, there are also some major exceptions, for instance, natrolite, which has a fibrous morphology (needle-like) but a 3D-channel system.

14.1.2 Technically important zeolites

The technically most important zeolites include the synthetic zeolite A (LTA framework), zeolite X and Y (both FAU), zeolite L (LTL), synthetic mordenite (MOR), ZSM-5 (MFI), zeolite beta (*BEA/BEC), MCM-22 (MWW), zeolite F (EDI), and W (MER) as well as the natural zeolites mordenite, chabazite (CHA), erionite (ERI), and clinoptilolite (HEU). Some of them are presented in more detail in this section.

Zeolite A (Linde type A, LTA)

The LTA framework (see Figure 14.4) can be built by assembling *sod* cages via *d4r* units in a cubic primitive fashion, leading to an *lta* cavity in the centre of the cubic cell, which is accessible for molecules larger than water via the eight-rings with a diameter of ~4.1 Å. LTA has a 3D-channel system. The type material – zeolite A in its hydrated sodium variant – has the chemical formula $|Na_{12}(H_2O)_{27}|_8 [Al_{12}Si_{12}O_{48}]_8$ and is related to the unit cell of the T-atom framework by doubling the cubic cell in all directions ($a' = 2a$) leading to a face-centred cubic cell (space group $Fm\bar{3}c$, no. 221) with $a' = 24.61$ Å. Variants with other cations such as potassium or calcium and also the completely dehydrated forms are collectively called zeolite A.

Figure 14.4: The framework of T-atoms of zeolite A (LTA) is composed of *sod* cages linked via *d4r* units leading to a *lta* cavity in the centre of the cell that is accessible via eight-rings.

Usually, zeolite A is synthesized with a Si/Al ratio of 1, but also forms with a ratio up to 6 are produced; even an all-silica form exists, which is known as ITQ-29. The main application for the hydrated sodium variant of zeolite A (NaA) is in detergents, where it acts as a water softener, i.e., during the washing process the sodium ions are exchanged for the calcium ions of the (hard) water; zeolites can account for up to 30 wt% of detergents. The dehydrated form of NaA is used for the dehydration of solvents or in drying processes for gases, for instance air or natural gas.

Water-free forms of zeolite A are also used as molecular sieves for several separation processes. The effective size of the eight-ring pore windows to the *lta* cavity depends on the type of the extra-framework cations: The available aperture in Ångström is eponymous for the molecular sieves 5A (calcium form), 4A (sodium form), and 3A (potassium form). In particular the 5A form is used in the petrochemical industry for the separation of certain linear (*n*) alkanes (which are able to enter the zeolite) from branched (*iso*) isomers (which are too bulky to enter the pores) and aromatic hydrocarbons.

Other synthetic forms of LTA zeolites and zeotypes, besides those already mentioned, are zeolite Alpha, LZ-215, N-A, SAPO-42, ZK-21, ZK-22, ZK-4, and UZM-9.

Faujasite (FAU)

The framework of zeolites of the FAU type can be constructed by assembling *sod* cages via *d6r* units leading to a large cavity which is also known as *supercage* (according to the definition concerning cages and cavities, it should be called supercavity), see Figure 14.5. The supercavity with an inner diameter of approx. 12 Å is accessible via 12 rings with a diameter of approx. 7.4 Å. FAU has a 3D-channel system. The unit cell of the type material has a composition of $|$ $(Ca,Mg,Na_2)_{29}$ $(H_2O)_{240}$ $|$ $[Al_{58}Si_{134}O_{384}]$. Forms in which a certain cation is prevalent are called FAU-Na, FAU-Mg, etc.

The number of Al atoms per cell can vary from 96 to less than 4 (Si/Al ratios of 1 to more than 50). The two most technically important forms are zeolite X (Linde X), referring to synthetic FAUs with an Al content between 96 and 77 atoms per cell (Si/Al ratios between 1 and 1.5), and zeolite Y (Linde Y) with less than 76 Al atoms per cell (Si/Al ratio higher than 1.5).

The most important application of FAU-type zeolites is as a catalyst in the FCC and hydrocracking process in the refinery for the production of light-boiling fractions (gasoline and diesel) from petroleum distillation residues. Initially, zeolite X was used for this purpose. However, this has the disadvantage that it is not permanently stable due to the low Si/Al ratio, especially in humid atmosphere. Efforts have therefore been made to synthesize FAUs with a higher Si/Al ratio. Zeolite Y has been a major advance in this regard, but it cannot, without major effort, be produced with an Si/Al ratio larger than 2.5 directly by crystallization. The preferred technical variant to further increase the Si/Al ratio consists of a downstream hydrothermal dealumination step (in which Al is gradually removed from the lattice first and subsequently

Figure 14.5: The framework of T-atoms of FAU is composed of *sod* cages linked via *d6r* units leading to a supercavity in the centre that is accessible via 12-rings.

replaced by Si) followed by recalcination. This is also how the variant known as USY (ultra-stable zeolite Y) is produced.

In some cases, also rare-earth-ion-exchanged (mainly lanthanum) Y-type zeolites are used in the FCC process. Those with a high content of rare earth metals are mainly used to maximize gasoline yield, and those with lower content are mainly used to reduce coking and formation of cracked gases.

Nickel-containing zeolite Y can be used as a catalyst in hydrogenation reactions, e.g., in the hydrogenation of carbon monoxide in the Fischer-Tropsch synthesis.

Other synthetic forms of FAU zeolites and zeotypes, besides those already mentioned, are Beryllophosphate X, Li-LSX, LZ-210, SAPO-37, siliceous Na-Y, and zincophospate X.

Mordenite (MOR)

The framework of zeolites of the mordenite type (MOR) is characterized by chains along the x-direction composed of alternating units of two fused five-rings and one four-ring that are undulated in the y- as well as the z-direction. These chains are further laterally linked by four- and five-rings to give the overall framework as shown in Figure 14.6. This gives rise to a topologically 2D channel system consisting of a 1D channel of slightly elliptical shape running along the z-direction with limiting pore windows of 12-rings (approx. 7.0 × 6.5 Å) alternating with channels running in the same direction limited by windows of eight-shaped eight-rings (approx. 5.7 × 2.6 Å in size), see Figure 14.6. The channels limited by 12-rings are further connected by secondary 1D channels limited by eight-rings. Because of the small aperture of the eight-rings MOR is essentially regarded as a zeolite with a 1D channel system. The chemical composition of the type material is [Na$_8$(H$_2$O)$_{24}$] [Al$_8$Si$_{40}$O$_{96}$] and it crystallizes in the orthorhombic space group *Cmcm* (no. 63) with a = 18.1 Å, b = 20.5 Å, and c = 7.5 Å.

Natural mordenite is highly siliceous and exists with a nearly constant Si/Al ratio of 5, while synthetic mordenites have been made with Si/Al ratios from about 4 to 12. MOR zeolites were some of the first commercialized for use as solid-acid catalysts in the petrochemical industry for isomerization of alkanes and aromatics. They are also used in industry for the removal of volatile organic compounds (VOCs).

Mordenite-rich zeolitic tuff is used as a building stone and in its crushed and dried form in the agricultural area, for instance as animal feed supplement.

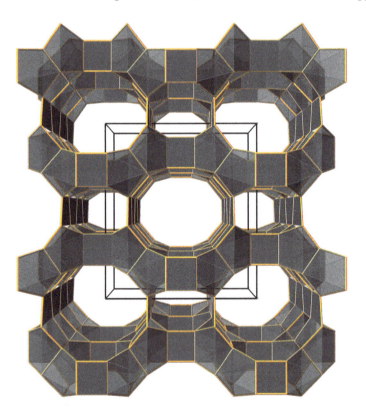

Figure 14.6: The framework of T-atoms of MOR viewed along the *z*-direction. It is characterized by chains along the *x*-direction composed of alternating units of two fused five-rings and one four-ring that are undulated in the *y*- as well as the *z*-direction. These chains are further laterally linked by four- and five-rings to give the overall framework.

Chabazite (CHA)

The framework of CHA zeolites and zeotypes can be built by linking *d6r* units through tilted four-rings (see Figure 14.7). The specific linkage of these units in CHA defines a large cavity, called the *cha* cavity, that is accessible through a 3D-channel system limited by pore windows of eight-rings (approx. 3.8 Å in diameter). The type material has the composition $[Ca_6(H_2O)_{40}]_{1/3}$ $[Al_{12}Si_{24}O_{72}]_{1/3}$ and crystallizes in the trigonal space group $R\bar{3}m$ (no. 166).

The CHA framework is a member of a larger family of zeolites which is known as the ABC-6 family. All of them consist of a hexagonal array of (non-connected) planar six-rings that are related by pure translations along the *a*- and *b*-direction of the hexagonal cell. These six-rings are centred at (0,0) in the (*a,b*) plane and this position is called the A position. Neighbouring layers of six-rings that are connected by the tilted four-rings along the *c*-direction can now be located at three different positions:

- the second layer is shifted by +⅔ along the *a*- and by +⅓ along the *b*-direction; this is the B position,
- the second layer is shifted by −⅔ along the *a*- and +⅓ along the *b*-direction; this is called the C position,
- the second layer can have a zero lateral shift in the (*a,b*) plane, leading to *d6r* units (AA or BB or CC).

The concrete stacking sequence of layers along the *c*-direction now determines the framework type. An A layer can be followed by an A, B, or C layer, but a sequence of three or more identical layers (*e.g.*, AAA) is not likely because of the severe strain that would be imposed on bond angles. The most simple member of the ABC-6 family with only two layers in the repeating sequence (AB) is the CAN framework (as in cancrinite). Chabazite has six layers in the repeating sequence (AABBCC). The (ordered) members with highest number of layers in the repeating sequence are the frameworks SAT (as in STA-2) and AFT (as in AlPO-52). While AlPO-52 can be essentially synthesized without any stacking faults, other frameworks are more prone to stacking faults.

Chabazite has only a moderate thermal stability. To withstand certain industrial processes, it was necessary to synthesize forms with a higher Si/Al ratio, for instance the high-silica form SSZ-13. Cu-containing high-silica variants are used to remove the majority of harmful nitrogen oxides from diesel engine exhaust gases by selective catalytical reduction. Furthermore, the zeotype SAPO-34 is used in the methanol-to-olefine (MTO) process.

Further synthetic forms of zeolites and zeotypes of the CHA framework include AlPO-34, CoAPO-44, CoAPO-47, DAF-5, GaPO-34, Linde D, Linde R, LZ-218, MeAPO-47, MeAPSO-47, (Ni(deta)$_2$)-UT-6, Phi, SAPO-47, UiO-21, ZK-14, and ZYT-6.

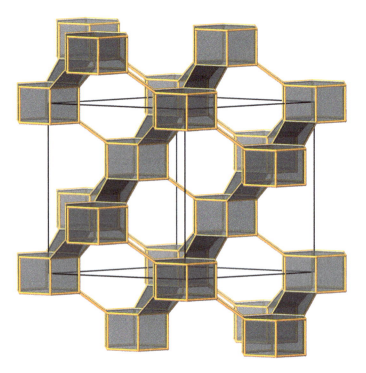

Figure 14.7: The framework of T-atoms in chabazite (CHA) is composed of *d6r* units linked by tilted four-rings, defining a *cha* cavity that is accessible via eight-rings. The sequence of layers of *single* six-rings is AABBCC, see text.

ZSM-5 (MFI)

The MFI framework (Figure 14.8) can be built-up by assembling five-rings, specifically 5–1 SBUs (compare Figure 14.2) and is a representative of the pentasil class of zeolitic frameworks. The term pentasil is used with two slightly different meanings. In mineralogy the term is used for any framework (alumo)silicate that contains five-rings as a characteristic building unit. In the field of zeolites, the term is defined narrower and is used for such frameworks that contain the pentasil subunit, a building block that consists of eight five-rings as shown in Figure 14.8, right.

The framework contains an interesting channel system consisting of straight channels running along the y-direction limited by 10-rings of almost circular shape (5.3 × 5.5 Å) and a further sinusoidal-like or zig-zag channel system perpendicular to the first one, limited also by 10-rings with a slightly more elliptical shape (5.1 × 5.5 Å). The interconnected 3D channel system is shown in Figure 14.9.

ZSM-5 (Zeolite Socony Mobil-Five) is a proprietary, synthetic zeolite invented and patented as early as 1969 by the Mobil Oil Corporation (now part of the Exxon Mobil Corporation), however, the structure was solved only in 1978 by Kokotailo et al. [114].

Figure 14.8: The framework of T-atoms of MFI that is largely consisting of five-rings (left) and that contains specifically the so-called pentasil subunit (right).

The reference material has the chemical formula $[Na_x(H_2O)_{16}]$ $[Al_xSi_{96-x}O_{192}]$ with $x < 27$ and it crystallizes in the orthorhombic space group *Pnma* (no. 62) with lattice constants of $a = 20.07$ Å, $b = 19.92$ Å, and $c = 13.42$ Å. The Si/Al ratio is relatively high, making the zeolite rather hydrophobic and useful for removing organics from water streams and stable for separations and catalysis in the presence of water. In particular, the *all-silica* variant, silicalite, which is a synthetic SiO_2 polymorph, selectively adsorbs organic molecules over water because the hydrophobicity is even further increased [115].

Because of the very regular channel system and its highly active acidic sites ZSM-5 is used in a number of acid-catalysed reactions such as hydrocarbon isomerization and the alkylation of hydrocarbons. One prominent example is the production of *para*-xylene (an important chemical feedstock for polyethylene terephthalate production) either via isomerization of *ortho*- or *meta*-xylene or by reaction of toluene with methanol. While *para*-xylene (less bulky) is able to diffuse very quickly through the channel system it can be easily separated from *ortho*- or *meta*-xylene (bulkier). This is a typical example of a product selectivity of a shape-selective catalyst. ZSM-5 is also used in catalytic cracking processes for propene production and in MTO processes that converts methanol to light olefins such as ethene and propene. Yet another field of application of ZSM-5 is the dewaxing process of lubricants and fuels in which long-chain linear alkanes undergo cracking and/or isomerization to form branched species.

Titanium-doped silicalite (TS-1) is an important catalyst for the production of propylene oxide from the reaction of hydrogen peroxide with propylene. Unlike in polymorphs of titanium dioxide, the Ti centres in titanium silicalite have tetrahedral coordination geometry.

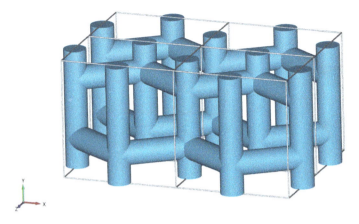

Figure 14.9: The channel system of the MFI framework, consisting of straight channels along the y-direction and zig-zag-channels in the (x,z) plane. The image was generated with the software package iRASPA [116].

Further synthetic forms of zeolites and zeotypes with MFI framework are AMS-1B, AZ-1, Bor-C, Boralite C, FZ-1, LZ-105, NU-4, NU-5, TSZ, TSZ-III, TZ-01, USC-4, USI-108, ZBH, ZKQ-1B, and ZMQ-TB. A calcium containing mineral that possesses the MFI framework is mutinaite.

14.2 Metal-organic frameworks

MOFs are a relatively new class of (potential) porous crystals. They consist of two main components: an inorganic building block consisting of metal atoms or metal-oxygen clusters (often referred to as the *secondary building unit*, SBU) and an organic building block (called *linker*) bridging the inorganic building blocks. Typical examples of linkers are di-, tri-, or tetracarboxylic acids, usually with a rigid backbone, such as terephthalic acid (benzene-1,4-dicarboxylic acid, bdc), biphenyl-1,4-dicarboxylic acid (bpdc), trimesic acid (benzene-1,3,5-tricarboxylic acid, btc), or biphenyl-3,3',5,5'-tetracarboxylic acid (bptc), see also Figure 14.10. Because MOFs are composed of these two different units, they belong to the larger class of hybrid organic-inorganic materials. The two components form infinitely extended networks, which generally represent a relatively rigid framework.

The chemistry of MOFs has its roots in the field of coordination chemistry with bridging (ditopic) ligands. But unlike in coordination chemistry, where usually only single atoms are coordinated, like for instance in the famous three-periodic infinite extended coordination compound Prussian blue (iron(II,III) hexacyanoferrate(II,III)), the special feature of MOFs is that the inorganic part consists of a *polynuclear cluster*. This leads to becoming less dependent on the preferred coordination number and geometry of individual atoms and also explains the enormous structural diversity of

14.2 Metal-organic frameworks — 311

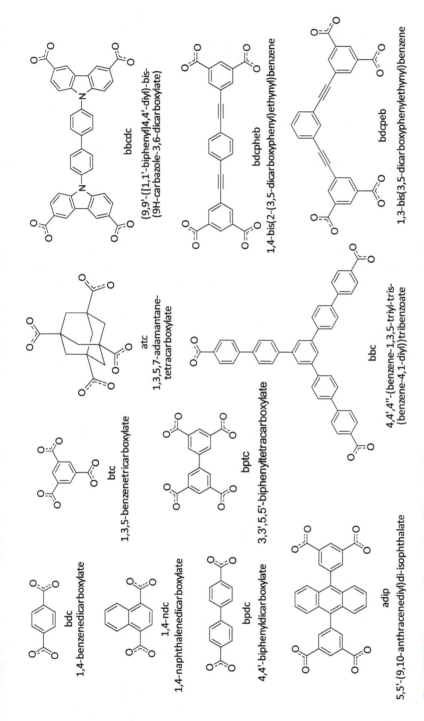

Figure 14.10: A representative selection of linkers in MOFs.

MOFs. Although the class of MOFs has been developed only since the late 1990s, meanwhile more than 70,000 different structures are known. And also the number of different network types, which describe the way the SBUs and linkers are connected, has grown enormously, especially in the last decade, to approx. 3,500 (see below).

The combination of inorganic SBUs with certain predefined coordination geometries with organic building blocks with a certain topicity in a rational way in order to obtain a crystalline framework with desired properties and topology – this is the art of reticular chemistry, a term coined by Omar Yaghi; reticular is derived from Latin and means "net-like". He defines reticular chemistry as:

> In essence, reticular synthesis can be described as the process of assembling judiciously designed rigid molecular building blocks into predetermined ordered structures (networks), which are held together by strong bonds.

In order to get more familiar to the Aufbau principle of MOFs, let's look more closely at one of the most famous MOFs, the prototypical MOF-5, and its assembly of the individual components, which consist of a Zn_4O tetrahedron and the dianion of terephthalic acid as a linker (see Figure 14.11). The terephthalic acid dianion coordinates with both oxygen atoms, thus bridging two zinc atoms each. Overall, six linkers

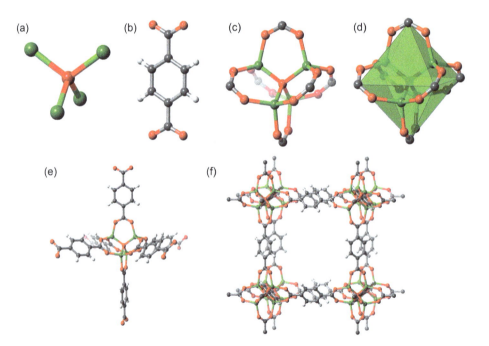

Figure 14.11: Structure of MOF-5: (a) Inorganic tetrahedral Zn_4O unit, (b) terephthalate dianion, organic bridging ligand (linker), (c) coordination sphere around the Zn_4O unit, (d) green octahedron representing the shape of SBU, (e) six linkers bind to a Zn_4O unit, and (f) section of the primitive cubic network of MOF-5 (reproduced from [33] with kind permission from SpringerNature © 2020).

coordinate on a Zn_4O tetrahedron (a tetrahedron has six edges) in octahedral fashion. By connecting the carbon atoms of the six linkers that coordinate to the Zn_4O unit, they define the coordination polyhedron of the SBU. These transition points from the inorganic SBU to the actual bridging moiety are also called *points of extension*. A short form to note the number of points of extension is "n-c", where n is the number of such points and "c" stands for connected or coordinated. In MOF-5 the SBUs are, thus, 6-c. Overall, an infinitely extended, primitive cubic framework forms that is abbreviated as **pcu** (see below). MOF-5 is considered to be the starting point of the whole field of MOFs. It was synthesized by Omar Yaghi and coworkers in 1999 [117].

In the meantime, a whole construction kit is available from which an inorganic SBU with a desired number of points of extensions in different geometries can be selected (see Figure 14.12).

MOFs are typically synthesized via a solvothermal synthesis, mixing together one or more metal salts with one or more type of linkers solved in an appropriate solvent or mixture of solvents. The specific synthesis conditions (concentration of the reactants, reaction temperature, duration, pH value, etc.) are of detrimental importance not only for whether an MOF is formed at all, but possibly also which structure it has, because another peculiarity of MOFs is that different structures can sometimes be built from the same reactants.

Typically, after the synthesis of MOFs, the voids of the framework still contain the solvent, which must be removed to create the permanent porosity of MOFs. This process, which often can be accomplished simply by evaporating the solvent (optionally under reduced pressure, optionally after the initial reaction solvent was exchanged with a more volatile solvent), is also called *activation*. Most of the MOFs are microporous materials, i.e., have pores with diameters up to 2 nm, but there are also some mesoporous MOFs [119]. Compared to zeolites, MOFs are characterized by the fact that the relative void volume can assume much larger values (70% are not uncommon) and that also the internal specific surface area is usually much higher. The world record holder is currently a MOF called DUT-60, which has a specific surface area of 7,840 square meters per gram [120], comparable to the area of a soccer field!

The large void volume and high specific surface area make MOFs suitable for a number of applications, particularly for gas storage and gas separation. In the following, a few selected areas of active research in the field of MOFs are presented. Because of the incredible high level of activity in this field – approx. 60 paper per day(!) are currently being published – it is impossible to give a whole overview and the reader is referred to the section Further Reading.

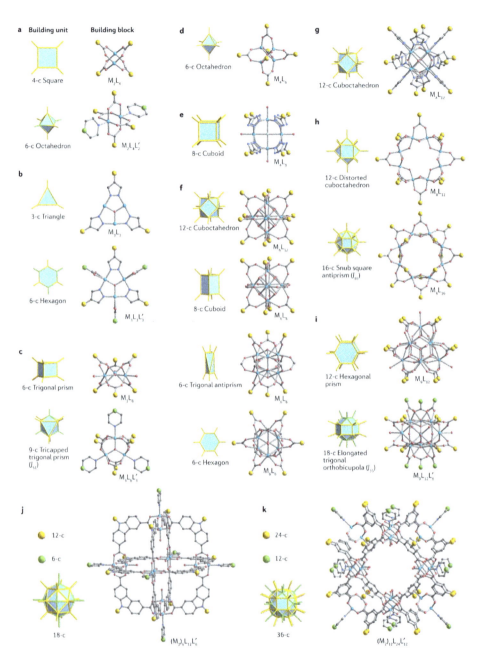

Figure 14.12: Overview of prominent inorganic building units. Colour scheme: metals (light blue), carbon (grey), oxygen (red), nitrogen (dark blue), halogens (cyan), basic points of extension (yellow), extra points of extension (coordinatively unsaturated metal sites to which additional linkers can be added) (green). L, linker; M, metal (reproduced from [118] with kind permission from SpringerNature © 2021).

14.2.1 Carbon capture and sequestration

As concern over the rising concentration of atmospheric CO_2 and its negative impact on the earth's climate has increased, recent research efforts have focused on capturing CO_2 from large anthropogenic point sources, like, for instance, coal-fired power plants. Due to their small pore sizes and high void fractions MOFs are promising materials to provide more efficient alternatives to the traditional amine solvent-based methods (also known as amine scrubbing) of CO_2 capture from flue gas. A major drawback of this traditional technology for post-combustion capture of CO_2 is that it causes an energy penalty of approximately 30% of the output of the power plant. This energy loss is associated with the large energy input required to liberate captured CO_2 from the capture medium, which is mainly water that has a comparable high heat capacity.

Post-combustion flue gas mainly consists of N_2 (about 77%) and CO_2 (about 16%) and some other minor components such as H_2O (about 6%), O_2, CO, NO_x, and SO_2. SO_2 is typically removed before the CO_2 capture process takes place. For using MOFs for post-combustion CO_2 capture the flue gas (that has a temperature between 40 and 60 °C) would be fed through a MOF in a packed-bed reactor setup. To separate the main components of the gas mixture (i.e., CO_2 and N_2), it is of importance to provide adsorption sites that interact more strongly with CO_2 than with N_2. Once the MOF is saturated, the CO_2 is extracted from the MOF through either a temperature swing or a pressure swing regeneration procedure.

In principle, MOFs with so-called "open" metal sites, i.e., coordinatively unsaturated metal sites (CUS), are suitable for CO_2 capture as such Lewis acidic sites provide a partial positive charge on the pore surface and are therefore capable of interacting strongly with CO_2. This strong interaction is typically accompanied by high values of the isosteric heat of adsorption (Q_{st}) values at low pressures, a high selectivity (in the absence of other molecules with a strong dipole or quadrupole moment), and a high CO_2 uptake. One of the most studied systems in this context is the MOF-74 series, M-MOF-74, with M being, for instance, Mg^{2+}, Ni^{2+}, Fe^{2+}, or Zn^{2+}. The linker is 2,5-dihydroxyterephthalic acid. The hydroxyl groups are being deprotonated during synthesis and the oxygen atoms also coordinate to the metal, resulting in an overall rod-shaped (1D) SBU. Every open metal site can bind one molecule CO_2 in the hexagonal channels (see Figure 14.13) leading to an uptake of up to ~28 wt% in the case of Mg-MOF-74 at 298 K and 1 bar with a high selectivity for CO_2 in CO_2/N_2 mixtures. However, as flue gas contains water that also binds very strongly to open metal sites the CO_2 uptake capacity of M-MOF-74 materials drops dramatically in the presence of water. To circumvent the water problem other polarizable adsorption sites such as Lewis bases that cannot interact with water must be introduced.

Another strategy for capturing CO_2 with the help of MOFs is to use amine-functionalized linkers. Here, the CO_2 molecules are bound covalently to the amine group. It was shown that Mg-IRMOF-74-III-CH_2NH_2 in which the linker 2′,5′-dimethyl-3,3″-dihydroxy-[1,1′:4′,1″-terphenyl]-4,4″-dicarboxylic acid was functionalized with a –CH_2NH_2 group completely retains its CO_2 uptake capacity (12.5 wt% at 1 bar) in the

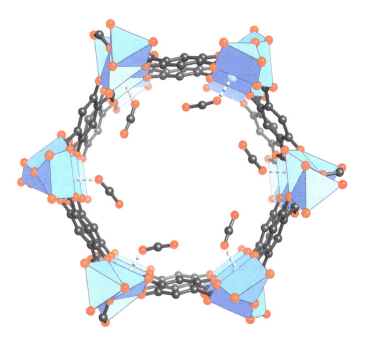

Figure 14.13: After complete activation, the hexagonal pores of M-MOF-74 are lined with open metal sites that interact strongly with CO_2 molecules; view along the crystallographic c-axis of the trigonal structure (space group $R\bar{3}$). Only one pore is shown, and all hydrogen atoms are omitted for clarity; metal-oxygen polyhedra, blue; carbon, grey; oxygen, red (reproduced from [121] with kind permission from Wiley-VCH Verlag GmbH © 2019).

presence of water although the total uptake capacity is significantly lower than that of MOF-74.

There are also some MOFs which show very high CO_2 uptake values even if they neither exhibit open metal sites nor functional groups that can bind CO_2 covalently. SIFSIX-2-Cu-i ($Cu(4,4'\text{-dpa})_2(SiF_6)$), a compound built from square $[Cu_2(H_2O)_2](dpa)_{4/2}$ (dpa = 4,4'-dipyridylacetylene) grids that are pillared by SiF_6 units to form two interpenetrated primitive cubic networks, has the highest CO_2 uptake (19.2 wt% at 298 K and 1.1 bar) for any framework material without open metal sites. The high uptake is ascribed to the strong interaction of CO_2 with the SiF_6 units.

14.2.2 Hydrogen storage

As a zero-emission fuel with a very high energy density, hydrogen is considered an ideal replacement for petrol, in particular, in automotive applications, and especially, if green hydrogen (i.e., hydrogen generated by renewable energy) is used. It should be noted that hydrogen is considered an energy carrier rather than an energy source since the production of hydrogen requires other sources of energy.

Already available physical or chemical methods for hydrogen storage have various disadvantages. The chemical storage of hydrogen using metal hydrides, boranes, or imidazolium-based ionic liquids is often associated with unfavourable kinetics, insufficient reversibility, high production costs, and a low tolerance to contaminants that may be present in hydrogen gas. The liquefaction of hydrogen and its subsequent storage at cryogenic temperatures (−252.9 °C) require high energy costs for cooling and elaborately designed and thermally insulated tank systems to minimize the leakage rate, which, however, cannot be reduced to zero. In order to achieve sufficiently high storage densities (and thus ranges), very high pressures (350–700 bar) are still required for the storage of hydrogen as a gas in the automotive sector, which in turn places high mechanical demands on the tank system.[17]

MOFs have been considered for some time as promising candidates for adsorption-based storage methods for H_2. The principle applied here is the densification of the gas by attractive interactions with the framework walls. However, it is very challenging to meet the targets, for example, that were set by the U.S. Department of Energy (DoE) for onboard hydrogen storage for light-duty fuel cell vehicles. For the year 2025 the goals are to achieve a gravimetric uptake of 0.055 kg(H_2) per kg, a volumetric capacity of 0.04 kg(H_2)/L, a refuelling time of 3–5 min, and a cycling stability of 1,500 cycles; all that at operating temperatures between −40 and +60 °C and a minimum and maximum delivery pressure of 5 and 12 bar, respectively. Note that the uptakes are values for an entire storage system (including the tank, valves, regulators, piping, insulation, etc.); hence, the storage material itself must significantly outperform these values.

Designing MOFs for hydrogen storage includes the following considerations: (i) the storage material should be as light as possible (i.e., composed of light elements), (ii) the specific surface area should be as large as possible since the amount of storage is proportional to the surface area, (iii) the pore size should be as close as possible to the kinetic diameter of the gas to be stored (for H_2: 2.89 Å), whereby either the pore size of the individual framework should be optimized or by placing several frameworks inside each other (interpenetration), (iv) and finally, the attractive interaction between the H_2 molecules and the framework atoms should be as high as possible; the ideal value of the isosteric heat of adsorption (Q_{st}) is around 20 kJ mol^{-1}.

A selection of some promising MOFs for hydrogen storage that reach at least the gravimetric target of 5.5 wt% storage capacity, albeit at 77 K and not at room temperature, are gathered in Table 14.2. As an example, the structure of UMCM-2 that reaches an uptake value of 6.9 wt% already at 46 bar is shown in Figure 14.14. UMCM-2 consists of two types of microporous cages and additionally a mesoporous cage with a size of approx. 2.6 × 3.2 Å. It is built up by two linkers, the triangular tritopic linker btb (4,4′,4″-benzene-1,3,5-triyltribenzoate) and the linear ditopic linker t^{2}dc (thieno[3,2-

---

[17] At the time of writing this book, there are just two hydrogen cars that are being produced in significant quantities: The Hyundai Nexo, and the Toyota Mirai, the latter word is Japanese and means "future".

Table 14.2: Selection of promising MOFs with regard to hydrogen storage.

Chemical formula	Common name	Surface area ($m^2 \cdot g^{-1}$)	p (bar)	T (K)	H_2 uptake (wt%)
$Be_{12}(OH)_{12}(btb)_4$	–	4,030 (BET)	20	77	6.0
$Cr_3OF(bdc)_3$	MIL-101	5,500 (Langmuir)	80	77	6.1
$Zn_4O(bbc)_2(H_2O)_3 \cdot 3\,H_2O$	MOF-200	4,530 (BET)	80	77	7.4
$Zn_4O(btb)_2$	MOF-177	4,746 (BET)	70	77	7.5
$Zn_4O(bte)_{4/3}(bpdc)$	MOF-210	6,240 (BET)	80	77	8.6
$Zn_4O(t^2dc)(btb)_{4/3}$	UMCM-2	5,200 (BET)	46	77	6.9

bbc, 4,4′,4″-(benzene-1,3,5-triyl-tris(benzene-4,1-diyl))tribenzoate; bdc, benzene-1,4-dicarboxylate; bpdc, biphenyl-1,4-dicarboxylate; btb, 4,4′,4″-benzene-1,3,5-triyltribenzoate; bte, 4,4′,4″-(benzene-1,3,5-triyl-tris(benzene-4,1-diyl))tribenzoate; t²dc, thieno[3,2-b]thiophene-2,5-dicarboxylate; MIL, Matériaux de l'Institut Lavoisier; UMCM, University of Michigan Crystalline Material.

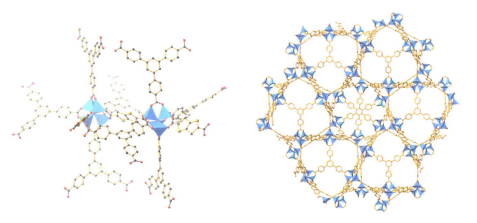

Figure 14.14: Structure of UMCM-2. Two Zn_4O clusters coordinated to three t^2dc linkers and eight btb linkers (left), structure viewed along the c-axis (right).

b]thiophene-2,5-dicarboxylate) that are linked to octahedral 6-c Zn-based SBUs identical to the one in MOF-5.

14.2.3 Methane storage

Methane is a direct fuel and occurs naturally as the main component of natural gas (>95%). It has a gravimetric heat of combustion (55.7 MJ kg^{-1}) similar to that of gasoline (46.4 MJ kg^{-1}) and is a relatively clean fuel. The two currently prevailing storage technologies are accompanied with similar problems as in the case of hydrogen storage: compressed natural gas (CNG) has a comparatively low volumetric energy density, even at the typical pressures of 200–300 bar, and liquefied natural gas (LNG) requires energy-intensive liquefaction, cooling, and storage in expensive cryogenic vessels that suffer from boil-off losses. Using adsorption-based methods, only pressures of about 65 bar are needed and the operation at ambient temperatures eliminate the need for elaborately thermally insulated tank systems.

There are similar aspects to consider when using MOFs for methane storage as in the case for hydrogen storage, namely with respect to the surface area, optimized pore diameters, and the demand of high attractive interaction of the gas with the framework. For mobile applications the so-called working capacity determines the possible driving range. The working capacity is defined as the volume of gas stored per volume of material (in v v^{-1}) in a fully loaded tank (for methane this is typically at 65 bar) minus the volume that remains in the tank when the depletion pressure (for methane typically ~5 bar) is reached.

Compared to hydrogen storage in MOFs, suffering from intrinsically low isosteric heats of adsorption, the isosteric heat of adsorption for methane is higher and often in the regime suitable for commercial applications. The initial target value of the US DoE was set to 180 (STP) cm^3 cm^{-3} at 25 °C and 35 bar (STP: standard temperature and pressure) and was updated to 263 (STP) cm^3 cm^{-3} at 65 bar in 2012. In Table 14.3 some MOFs are compiled that show storage capacities that are very close to that target value. With 259 cm^3 cm^{-3} the hitherto highest value is reached by *monolithic* HKUST-1 (Cu$_3$(btc)$_2$)[18] [122]. HKUST-1 is composed of the tritopic linker btc and a so-called dicopper paddle-wheel motif as square-planar 4-c inorganic SBU, resembling a paddle wheel with four blades. The structure of HKUST-1 is shown in Figure 14.15.

Table 14.3: Selected MOFs with a high methane storage capacity.

Chemical formula	Common name	Surface area (m$^2 \cdot$ g^{-1})	p (bar)	T (K)	CH$_4$ uptake (cm^3 cm^{-3})
Al$_8$(OH)$_8$(btb)$_4$(H$_2$btb)$_4$	MOF-519	2,400 (BET)	35	298	200
Al$_3$O(tcpt)$_{1.5}$(H$_2$O)$_3$]Cl	Al-*soc*-MOF-1	5,585 (BET)	65	298	221
Cu$_2$(adip)	PCN-14	1,753 (BET)	35	290	220
Cu$_3$(bttb)	NU-125	3,120 (BET)	58	298	228
Cu$_2$(tptc)	NOTT-101a	2,805 (BET)	35	298	237
Ni$_2$(dobdc)	Ni-MOF-74	1,593 (Langmuir)	35	298	230

adip, 5,5'-(9,10-anthracenediyl)diisophthalate; bttb, bis(μ$_{12}$-1,3,5-tris(N-(3,5-dicarboxyphenyl)-1,2,3-triazen-4-yl)benzene; btb, 4,4',4''-benzene-1,3,5-triyltribenzoate; dobdc, 2,5-dioxidobenzene-1,4-dicarboxylate; H$_4$tcpt, 3,3'',5,5''-tetrakis(4-carboxyphenyl)-*p*-terphenyl; NOTT, Nottingham; NU, Northwestern University; PCN, Porous Coordination Network.

---

[18] HKUST-1 was one of the very first MOFs and is iconic. Usually, MOFs are obtained as fine powders. The high uptake of monolithic HKUST-1 is based on a successful densification of the product.

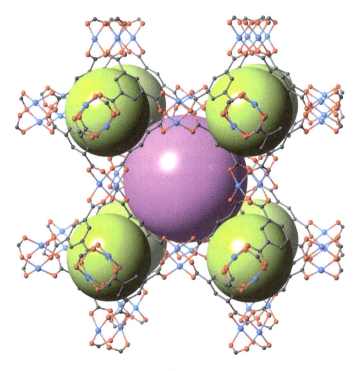

Figure 14.15: Structure of HKUST-1. The large spheres in green and pink represent the two different kinds of pores present in the framework structure; Cu, blue; O, red; C, grey; H atoms are not shown.

The chemical company BASF has developed and produced storage systems for natural gas tanks for vehicles, in which methane is stored inside an Al fumarate MOF (named Basolite A520). They had proven the technological feasibility and a number of demonstration vehicles are driving around the internal company premises; however, they can only compete economically if the oil price remains above $90 per barrel.

14.2.4 Water harvesting from air

Another currently hot topic in the area of MOFs is using them for water harvesting in arid regions in which drinking water is scarce. From laboratory testing to field trials in the driest deserts, kilogram quantities of MOFs have been tested in several generations of devices. The initial results of these experiments showed that MOFs could capture water from desert climates and deliver over 1 L/kg of MOF per day. One of the best performances is obtained by using MOF-303 – [Al(OH)(1*H*-pyrazole-3,5-dicarboxylate)] – as the active harvesting material, see Figure 14.16. A water harvester filled with MOF-303 is capable of generating 1.3 L per kg MOF per day in an indoor arid environment (32% relative humidity, 27 °C) and 0.7 L per kg MOF per day in the Mojave Desert (in conditions as extreme as 10% RH, 27 °C).

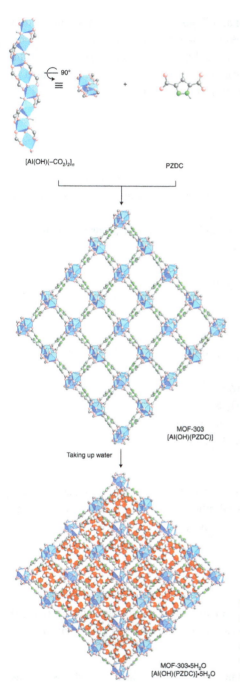

Figure 14.16: Crystal structure of MOF-303 (middle) constructed from infinite $[Al(OH)(-CO_2)_2]_n$ rod-like SBUs stitched together by V-shaped pzdc linking units (top), and the MOF-303 structure filled with water (bottom) (reproduced from [123] with kind permission by SpringerNature © 2023).

MOF-303 is constructed from one-dimensional infinite rod-like SBUs that are bridged by the V-shaped pyrazole-dicarboxylate linker. Within the SBUs, each Al(III) ion is coordinated by two bridging hydroxyl groups and four carboxyl groups to form alternating *cis–trans* corner-shared AlO_6 octahedra. The start-up company Water Harvesting Inc. – invented by Omar Yaghi – wants to bring this water harvesting technology to the market.[19]

14.2.5 Other (potential) applications

The fact that the pore size can be tailored over a wide range, the polarity of the pore walls can be varied, and that linkers with a wide variety of chemical functionalities can be incorporated, makes MOFs suitable for a whole range of other applications. In the field of sorption and separation, for example, the following applications are envisaged:
- precombustion capture of CO_2 from H_2;
- CO removal from H_2 generated by syngas;
- separation of CO_2 from CH_4 in landfill gases;
- CO purification (if CO is used as starting point for a variety of basic chemicals in C1 chemistry);
- producing pure O_2 from air, i.e., O_2/N_2 separation;
- alkane/alkene/alkyne separation, in particular, ethane/ethylene, ethylene/acetylene, and propane/propene;
- separation of noble gases, in particular, Kr/Xe;
- removal of toxic gases such as NH_3, NO_x, SO_2, and H_2S;
- removal of drugs and heavy metals in wastewater;
- indoor-quality control (moisture or odour removal); and
- adsorption-driven heat pumps.

The field of sorption and separation is the best studied and most promising field so far. But MOFs also have potential for several other fields of application, for instance, catalysis, drug delivery, and sensors, to name a few.

14.3 Networks and topology

The description and nomenclature of chemical entities in the various subfields of chemistry is developed to different degrees. In the field of molecular chemistry, the

---

[19] There are about a dozen of other MOF companies right now. How long these will last, whether more will be founded, or whether MOFs will eventually be displaced by even better materials in the various sectors on which they wanted to be applied, will only be known in later editions of this book. At least the comparatively high number of existing companies reflects the high potential of MOFs for technical-industrial applications.

description is relatively easy and the totality of all IUPAC rules for naming molecules allows drawing a unique chemical graph, i.e., a set of certain atoms that are connected in a specific way by chemical bonds, from that name.

The description of extended solids is already less trivial and several possibilities for an adequate representation arise. Beyond the purely crystallographic description (fractional coordinates of the atoms and the specification of the lattice parameters and the space group), for ionic solids it is common practice to describe the crystal structures as a packing of one type of atom where other atoms occupy the voids of the structure. A complementary approach is the description based on coordination polyhedra and their linkage that can be corner-, edge-, or face-connected. However, these descriptions are less helpful for framework-like compounds. This already becomes clear when you look at the structure of zeolites: they consist exclusively of Al/SiO_4 tetrahedra and in all of the ~250 known zeolite types all linked tetrahedra have exactly one common vertex, while shared edges or faces do not occur. Zeolites are characterized by their different *cavities*, *pores*, and *channel* systems, by their *rings* of certain sizes. Especially with the development of MOFs and related material classes, the description of framework-like substances in terms of their *underlying networks* and their *topology* (Greek τόπος meaning "place", and λόγος meaning "study") became increasingly popular. This concept allows to simplify structures by only considering the principal connections between their constituents, not their chemical nature, and how these simplified constituents are connected to each other in space. In the following, a brief introduction into networks, how chemical structures are simplified into their underlying networks and how networks in terms of topology are described should be given.

14.3.1 Graphs and nets

In mathematics, graphs are abstract objects comprising a set of vertices that are connected by a certain set of edges. In principle, there are many different kinds of graphs, however, if framework structures are considered as graphs, we are exclusively faced with a special kind of graph that are called networks (or short: nets). Nets are a subtype of graphs that have special properties:
– the graph is connected, meaning that there is a continuous path from every vertex to every other vertex;
– the graph is periodic, i.e., it has a unit cell that translates in one, two, or three directions in space;
– it has no directed edges (it is only important, *if* two vertices are connected or not);
– the graph has no multiple connections, i.e., chemically speaking, we have only single bonds between the vertices;
– there are no loops (i.e., vertices which are connected to themselves); and
– there are no "loose" ends, i.e., vertices which have only one neighbouring vertex.

In Figure 14.17 the difference between a non-simple graph and a net is visualized.

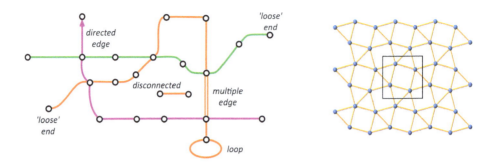

Figure 14.17: A finite, disconnected, non-simple graph with a directed edge, "loose" ends, a loop, and a multiple edge (left); a simple, infinitely extended net with translational symmetry in the plane, the black frame marks the unit cell (right) (redrawn and adapted from [124]).

14.3.2 Turning crystal structures into their underlying nets

In order to specify the topology of a crystal structure the first step is to simplify the structure and to turn it into a net. That means that the building units have to be defined that constitute the vertices and the edges of the net. Because this process is associated with disregarding the concrete chemical nature of the building units, it is also called a deconstruction process; what remains is an abstract structure of vertices and edges that only specifies which building units are connected to how many and which other building units in which way.

The deconstruction process of zeolites is rather simple and was already presented in Section 14.1.1: We just take the tetrahedral Si or Al atoms (T atoms) as vertices and connect them directly to all neighbouring T atoms that are bridged via an O atom. This means that the oxygen atoms that are only two-coordinating are omitted. That can be generalized as a rule: a two-coordinating entity is not a vertex but an edge. Furthermore, all counterions/protons are omitted as well. This leaves a network of 4-c nodes. It is already clear at this point that specifying the number of connections of the nodes is not sufficient to specify a network since all T atoms in all framework types of zeolites are four-coordinating. Additional descriptors in order to distinguish between different framework types and, hence, nets are needed, see below.

In most cases, the simplification process of MOF structures is only slightly more difficult than for zeolites. The building units that are turned into vertices are typically the SBUs and the organic linkers. For the SBUs we have to investigate the number of points of extensions as they represent the number of connections that this inorganic cluster has to other building units. For carboxylate-based MOFs these points of extension are typically placed on the carboxylate carbon atom. For the bridging organic unit we need to examine (a) how many SBUs it bridges and (b) whether it contains any branching points that represent nodes of their own, provided that the number of branches is higher than two. Applying these rules, MOF-5 can be simplified to a net that consists of only one vertex that is 6-c. To illustrate this simplification process further, let us look at another example, namely, the simplification of the MOF HKUST-1 (also compare Figure 14.15). The inorganic SBU, the "paddle-wheel" motif has four points of extensions in square-planar geometry and the linker is benzene-1,3,5-tricarboxylate and can be reduced to a (triangular) 3-c vertex (see Figure 14.18). Thus, HKUST-1 can be deconstructed to a binodal (3,4)-c net. It is a common practice to refer to a specific net with the help of a three-letter code, lower case and typeset in bold. These are net identifiers of the reticular chemistry structure resource (RCSR) database [125]. These identifiers allow for the unambiguous naming of unique nets and we discuss the nomenclature of net topologies later in more detail. The binodal (3,4)-c net of HKUST-1 is known as **tbo**.

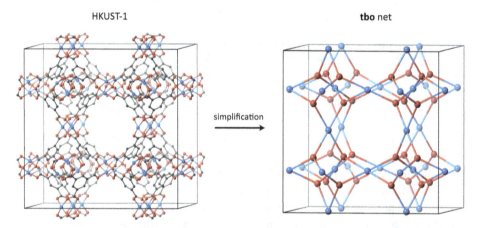

Figure 14.18: The simplification of the structure HKUST-1 to the binodal (3,4)-c net **tbo**, in which the 3-c vertices (red) represent the tritopic, triangular linker, and the 4-c vertices the square-planar inorganic SBU (blue).

14.3.3 Topology and net descriptors

In order to characterize a certain underlying net of a chemical structure that was derived by a deconstruction process, it is not sufficient to specify the coordination number of the nodes and their local coordination geometry. The important point is that topologically viewed, two nets (a set of vertices (nodes) and edges (links)) are only distinguishable or unique if they cannot be transformed into each other by bending, stretching, or tearing the vertices and edges. To express it the other way round: two nets are different if they can be transformed into each other only by breaking and (re-)connecting certain edges. For both statements an illustrative example is given in Figure 14.19.

(a)

(b)

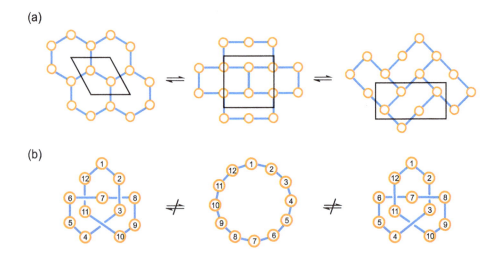

Figure 14.19: (a) Three representations of a (infinitely extended) two-periodic net having three-connecting nodes which are geometrically different but topologically identical (the unit cell is drawn in black); (b) three identical (isomorphic) graphs, in which all nodes always have an identical connection but different topological realizations (adapted and reproduced from [124] with kind permission from John Wiley & Sons © 2016).

This identity or non-identity of topological entities leads to two further questions that are closely related:
(1) How should a network be represented, if we want to, for instance, make a 2D or 3D model from which a picture can be drawn or if we want to visualize it in a software package that needs crystallographic coordinates of the nodes?
(2) On which basis can the question be answered as to whether two networks are identical?

Only when the latter question is answered, we can begin with a reasonable classification of networks and are able to identify prototypes of networks.

Embeddings

In a topological sense, vertices have no concrete location in space; they are abstract, infinitely small objects that can be placed anywhere in space, provided that their connection scheme to all other vertices remains unaltered. But if we assign the vertices coordinates in 3D space, there is a rule that we should do this in a way that the smallest number of symmetrically inequivalent vertices and also the smallest number of symmetrically inequivalent edges is used. In other words: the vertices and midpoint of the edges should be placed at coordinates with a maximum site symmetry. All vertices that are not symmetry-related to other kinds of vertices are vertices of their own, even if their coordination number and geometry is identical to those of other ones. If we have a net that consists of only 3-c nodes, but there are two kinds that are located at two different, symmetry-unrelated positions, then this is a *binodal* (3,3)-c net. The procedure to assign coordinates to the vertices (and hence edges) of an abstract graph is referred to as an *embedding* (sometimes also called realization). Then the vertices and edges (without physical properties) become nodes and links (with physical properties).

Maximizing the symmetry of a set of vertices and edges and the subsequent examination which type of net is present is typically done by specialized software packages, the most important ones are Systre [126] (a subprogram of the more comprehensive GAVROG package) and ToposPro [127].

Although this chapter deals mainly with porous crystals, the procedure to define chemical entities as vertices, to look at the neighbourhood of these vertices and to assign edges between these up to a reasonable distance, then to maximize the symmetry and finally to classify or identify this set of vertices and edges as a certain net – this is, of course, not limited to porous crystals and can be applied to any structure, even very compact ones. It is even the case that it can be very useful for deriving structural relationships. For instance, if we apply this recipe to the structure of tellurium (see Figure 14.20 and compare Section 5.1.6), we immediately recognize that it resembles the principal structure of α-Po and that it builds the same uninodal 6-c net **pcu** as MOF-5 (see Figure 14.11). This does not mean that the topological analysis should substitute the crystallographic description, but it is in many cases a useful additional information, see also Ref. [128].

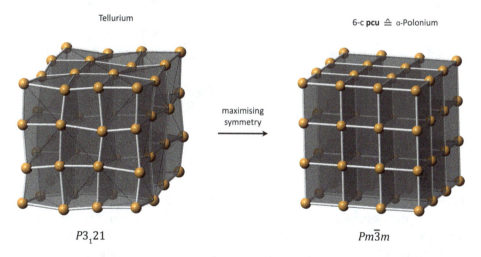

Figure 14.20: The outcome of a topological analysis of the element tellurium is the uninodal 6-c net **pcu**.

Before we will turn to the second question – what characteristics make a net unique? – we will introduce a certain technique of the so-called net decoration that makes the recognition of the "structural" features of nets much easier; it is called augmentation

Augmentation

The simplification of the chemical structure turns it into an abstract net that does not contain any chemical information anymore. Because chemists, in particular those working in the field of solid-state chemistry, are used to identify coordination polyhedra, it is helpful to turn a net into its so-called *augmented* version. Here, the vertices are replaced with their respective so-called vertex figure, i.e., "coordination polyhedra". The process of augmentation for the **tbo** net, the net representing HKUST-1, is visualized in Figure 14.21. In this way it is much easier to infer the coordination number and geometry of a node. An augmented version of a net is denoted by adding an "-a" to the identifier.

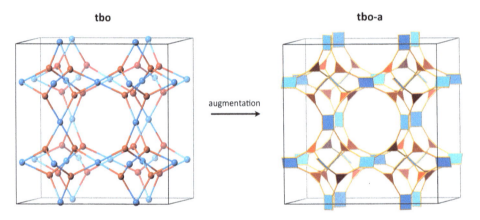

Figure 14.21: Replacing vertices with their respective vertex figure leads to the augmented version of a net. The identifier of the net gets an additional "-a". Here in the case of the **tbo** net, the 3-c vertices are replaced with triangles and the 4-c vertices with squares.

14.3.4 Characterizing and identifying nets

In Figure 14.19(b) we have already seen nets that are different, although their constituents and their connections to neighbouring vertices are identical. And we already mentioned that it is not sufficient to specify the coordination number of the vertices of the net in order to completely characterize a net. We need additional descriptors on how the vertices are spatially linked together. In this context the reader might ask, how is it possible to derive at nets that are topologically distinguishable, even if we use only nodes with identical coordination numbers and coordination geometry, because, in the first instance, the coordination geometry is not a topological distinct feature: if we consider an isolated node then we can deform a 4-c node with tetrahedral geometry into a 4-c node with square-planar geometry because bending is an allowed topological operation. However, it is indeed the local geometry of the nodes, combined with a different orientation to each other, which makes it possible to form nets from nodes of the same connectivity with globally different topologies. This is illustrated in Figure 14.22, in which the three different nets **sql**, **nbo**, and **cds** are shown that are all built from 4-c square-planar nodes.

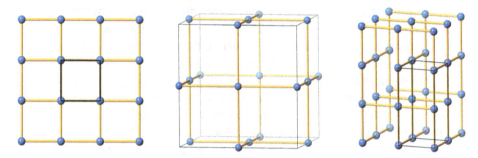

Figure 14.22: Three topologically different nets from identical square-planar 4-c nodes; the 2-periodic net **sql** (left), and the 3-periodic nets **nbo** (middle) and **cds** (right). The unit cell is shown as black frame (reprinted from [129] with kind permission from DeGryuter © 2022).

But in what way are these networks different? What properties do we need to specify in order to clearly identify them and make them distinguishable? It turns out that we have to specify exactly two characteristics to clearly distinguish one network from another: The *combination* of the coordination sequence (CS) and the vertex symbol. These are explained below.

Coordination sequence (CS) and vertex symbol

The CS specifies the number of topological neighbours of a vertex within the first, second, third ... tenth coordination sphere, this means the number of vertices which are exactly one, two, three ... ten edges away from an arbitrary chosen reference vertex (coordination sphere 0). In topological databases like the RCSR, additionally the so-called cum10 value is given, i.e., the sum over all topological neighbours from the first to the tenth coordination sphere (including the reference vertex itself). In the left part of Figure 14.23 the first three coordination shells of the uninodal 3-c honeycomb net **hcb** are illustrated, starting with a reference vertex in black (coordination shell zero) that has three neighbouring vertices (shown in blue) in the first, six vertices (shown in green) in the second, and nine vertices (shown in orange) in the third coordination shell. Note that the coordination shell of each of the different vertices has to be specified in the case that we are not dealing with uninodal nets.

The vertex symbol is the most important characteristic of a net. There are some nets which share the same vertex symbol(s) but this is extremely rare. However, as mentioned above, the combination of the CS together with the specification of the vertex symbol(s) of a net is unique, allowing an unambiguously identification of a net. Vertex symbols can be derived for nets which are periodic in two or three dimensions. The notation system is different for these two cases, however, the general recipe for deriving vertex symbols is the same: (a) you have to inspect all angles (pair of edges that meet at a vertex) and then (b) determine the ring sizes (number of vertices) of the *shortest rings* in which these angles are involved (for the difference between rings and cycles, see Ref. [130]), and (c) repeat this procedure for each different vertex type (of different symmetry or coordination number). In the two-periodic net **hcb** we have three pairs of edges that meet at each vertex (a-b, a-c, and b-c, see Figure 14.23, right). Each of those pairs is part of a six-membered ring, the vertex symbol is accordingly 6.6.6, which can be also written as 6^3. Grouping of rings in this form of a superscript is not allowed for three-periodic nets; here the number of vertices of rings is explicitly stated for every pair of edges, separated by dots. But in three-periodic nets often the pair of edges are involved in more than one shortest ring. If these shortest rings have the same size, they can be grouped with the help of the number of occurrences as subscript. To give an example, the **dia** net is a uninodal 4-c net with tetrahedral vertices; accordingly we have six pair of edges to examine, and it turns out that each pair of edges is involved in two(!) six-membered rings. The vertex symbol is, therefore, $6_2.6_2.6_2.6_2.6_2.6_2$. In computer codes the subscript numbers are given in parentheses instead: 6(2).6(2).6(2).6(2).6(2).6(2).

The simplification of MOFs and the identification of the underlying nets can be achieved with the ToposPro software, in which thousands of networks with their characteristics are stored. In the RSCR database – initiated by Michael O'Keeffe and maintained together with Olaf Delgado-Friedrichs – the most important (from their viewpoint) nets are presented, and they are given identifiers with the already mentioned three-letter codes set in bold. Currently (March 2023) the database contains 3,558 three-periodic, 203 two-periodic, 11 one-periodic, and 141 zero-periodic (polyhedra) nets. Apart from the CS and vertex symbol, numerous other specifications are given, inter alia:
- a beautiful picture of the net;
- alternative names (if they exist) for the net;
- a reference;
- the crystallographic specification of the vertices and edges;
- information about the type of embedding, the genus, and the tiling.

The author recommends studying this database over a nice glass of red wine to read more about the given specifications [124, 131] and to enjoy the beauty of chemical topology.

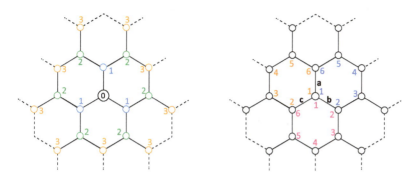

Figure 14.23: Illustration of deriving the coordination sequence of the net **hcb** that starts with 3, 6, 9, ... and the vertex symbol (6.6.6 or 6^3) (see text).

Further reading

J. Weitkamp, L. Puppe (Eds.), *Catalysis and Zeolites – Fundamentals and Applications*. Springer, Berlin, Heidelberg, **1999**.

P. A. Wright, *Microporous Framework Solids*. RSC Publishing, Cambridge, **2008**.

S. Kulprathipanja, *Zeolites in Industrial Separation and Catalysis*. Wiley-VCH Verlag, Weinheim, Germany, **2010**.

S. Kaskel (Ed.), *The Chemistry of Metal–Organic Frameworks – Synthesis, Characterization, and Applications*. Wiley-VCH Verlag, Weinheim, Germany, **2016**.

O. M. Yaghi, M. J. Kalmutzki, Ch. S. Diercks, *Introduction to Reticular Chemistry – Metal-Organic Frameworks and Covalent Organic Frameworks*. Wiley-VCH Verlag, Weinheim, Germany, **2019**.

A. Douhal, M. Anpo (Eds.), *Chemistry of Silica and Zeolite-Based Materials – Synthesis, Characterization and Applications*. Elsevier, Amsterdam, Netherlands; Cambridge, MA, United States, **2019**.

R. Freund, S. Canossa, S. M. Cohen, W. Yan, H. Deng, V. Guillerm, M. Eddaoudi, D. G. Madden, D. Fairen-Jimenez, H. Lyu, L. K. Macreadie, Z. Ji, Y. Zhang, B. Wang, F. Haase, C. Wöll, O. Zaremba, J. Andreo, S. Wuttke, C. S. Diercks, 25 Years of Reticular Chemistry, *Angew. Chem. Int. Ed. Engl.* **2021**, *60*, 23946. doi: 10.1002/anie.202101644

S. Gulati (Ed.), *Metal-Organic Frameworks (MOFs) as Catalysts*. Springer, Singapore, **2022**.

E. Pérez-Botella, S. Valencia, F. Rey, Zeolites in Adsorption Processes: State of the Art and Future Prospects, *Chem. Rev.* **2022**, *122*, 17647. doi: 10.1021/acs.chemrev.2c00140

Y. Belmabkhout, K. E. Cordova (Eds.), *Reticular Chemistry and Applications – Metal-Organic Frameworks*. DeGruyter, Boston, **2023**.

Appendix A: *Strukturbericht* designations and Pearson symbols

The *Strukturbericht* types, sometimes also called *Strukturbericht* designation is a system of classification of crystal structures. In these designations the structures are classified by *analogy* to other compounds of the same structure and each of the same type of identical compounds (i.e., with respect to the space group and sequence and occupation of Wyckoff positions) were given a symbol. While *Strukturbericht* symbols exist for many of the earliest observed and most common crystal structures, the system is not comprehensive and is no longer being updated. Modern databases such as Inorganic Crystal Structure Database (ICSD, maintained by the FIZ Karlsruhe) index thousands of structure types directly by the *prototypical* compound instead of the symbols used by the *Strukturbericht* reports. At the beginning of the year 2023 more than 272,000 inorganic compounds are gathered in the ICSD, and approx. 80% of them are assigned to about 9,300 structure types. So, for instance, what was given the symbol "B1" as a *Strukturbericht* designation is now the "NaCl structure type".

The designations were established by the journal *Zeitschrift für Kristallographie – Crystalline Materials*, which published its first round of supplemental reviews under the name *Strukturbericht* from 1913 to 1928. These reports were collected into a book published in 1931 by Paul Peter Ewald and Carl Hermann which became Volume 1 of *Strukturbericht* [132]. While the series was continued after the war under the name Structure reports, which was published through 1990,[20] the series stopped generating new symbols.

For the first volume, the designation consisted of a capital letter (A, B, C, D, ...) specifying a broad category of compounds and then a number to specify a particular crystal structure. In the second volume, subscript numbers were added, some early symbols were modified (e.g., what was initially D1 became D_{01}, D2 became D_{02}, etc.), and some more categories were added (types I, K, S). A brief historical overview of *Strukturbericht* symbols and other crystallographic classification schemes are given in the publication of Mehl [133].

The main *Strukturbericht* designation categories are as follows:
- the A compounds were reserved for structures made up of atoms of all the same chemical element. Meanwhile there are exceptions, like A_f, with γ-$HgSn_{6-10}$ being the prototypical compound.
- B designates compounds of two elements with equal numbers of atoms, i.e., those with a composition AB. Also in this category one exception is present: the structure type $B8_2$ is assigned to Ni_2In.

---

[20] International Union of Crystallography: Structure reports, https://www.iucr.org/publications/other/structure-reports, Accessed: 25 Feb 2023.

https://doi.org/10.1515/9783110657296-015

- C designates (with some exceptions) compounds of the stoichiometry AB_2.
- D designates compounds of arbitrary stoichiometry. Originally, D1–D10 were set aside for stoichiometry AB_3, D11–D20 for stoichiometry AB_n for $n > 3$, D31–D50 for $(AB_n)_2$, and D51 up for compounds $A_m B_n$ with arbitrary m and n.
- Letters between E and K designate more complex compounds.
- L designates intermetallic compounds (L from German "Legierung" = alloy).
- M designates "miscellaneous" compounds; a very vague category, that, in fact, was never used.
- O designates organic compounds.
- S designates certain silicate minerals.

Another method of describing structure types are the Pearson symbols, originated by W. B. Pearson [134]. The symbol is made up of two letters followed by a number. For example:
- diamond structure, $cF8$;
- rutile structure, $tP6$.

The two italicized letters specify the Bravais lattice. The lower-case letter specifies the crystal family/system, and the upper-case letter the type of centring. Because t is reserved for the tetragonal crystal system, the triclinic crystal system is abbreviated with a from its old name "anorthic". Furthermore, the one-side face centrings types A, B, or C are all indiscriminately marked with an S. The number at the end of the Pearson symbol gives the number of the atoms in the conventional unit cell. Because there are many possible structures that can correspond to one Pearson symbol, a prototypical compound may be useful to specify. Examples of how to write this would be $hP12$-$MgZn_2$ or $cF8$-C.

Table A.1 lists structures of the *Strukturbericht* designations. An even more comprehensive online version with the possibility to display the structure in 3D in the browser and to download the CIF file is available at the AFLOW (automatic flow for materials discovery) project website (www.aflow.org).

Table A.1: Overview of *Strukturbericht* designations/types.

Strukturbericht type	Prototype	Space group	Comment
A1	Cu	$Fm\bar{3}m$	Prototypical **ccp** (fcc) structure
A2	W	$Im\bar{3}m$	Prototypical **bcc** structure
A3	Mg	$P6_3/mmc$	Prototypical **hcp** structure
A3'	α-La	$P6_3/mmc$	ABAC... sequence of densest layers, also known as double-**hcp** packing
A4	C (diamond)	$Fd\bar{3}m$	
A5	β-Sn	$I4_1/amd$	
A6	In	$I4/mmm$	
A7	α-As	$R\bar{3}m$	
A8	Se (grey)	$P3_121$	
A9	C (graphite)	$P6_3/mmc$	
A10	α-Hg	$R\bar{3}m$	
A11	α-Ga	$Cmca$	
A12	α-Mn	$I\bar{4}3m$	
A13	β-Mn	$P4_132$	
A14	I_2	$Cmca$	
A15	β-W	$Pm\bar{3}n$	Also known as the Weaire-Phelan structure, see the box at the end of Chapter 11
A16	α-S	$Fddd$	
A17	P (black)	$Cmca$	
A20	α-U	$Cmcm$	
A_a	α-Pa	$I4/mmm$	
A_b	β-U	$P4_2/mnm$	Also known as the Frank-Kasper σ-phase, see Section 13.5.4
A_c	α-Np	$Pnma$	
A_d	β-Np	$P4/nmm$	
A_f	γ-HgSn$_{6-10}$	$P6/mmm$	Prototypical primitive hexagonal (*hP*) structure
A_g	B (tetragonal)	$P4_2/nnm$	
A_h	α-Po	$Pm\bar{3}m$	Prototypical primitive cubic structure
A_i	β-Po	$R\bar{3}m$	
A_k	γ-Se (monoclinic)	$P2_1/c$	
A_l	β-Se (monoclinic)		
B1	NaCl	$Fm\bar{3}m$	
B2	CsCl	$Pm\bar{3}m$	
B3	ZnS (zincblende)	$F\bar{4}3m$	
B4	ZnS (wurtzite)	$P6_3mc$	
B5	SiC (4*H* polytype)	$P6_3mc$	Naturally occurring as the mineral moissanite
B6	SiC (6*H* polytype)	$P6_3mc$	Naturally occurring as the mineral moissanite
B7	SiC (15*R* polytype)	$R3m$	
B8$_1$	NiAs	$P6_3/mmc$	Naturally occurring as the mineral nickeline, see also Section 5.4.1
B8$_2$	Ni$_2$In	$P6_3/mmc$	
B9	α-HgS	$P3_221$	Naturally occurring as the mineral cinnabar

Table A.1 (continued)

Strukturbericht type	Prototype	Space group	Comment
B10	PbO	$P4/nmm$	Naturally occurring as the mineral litharge
B11	γ-CuTi	$P4/nmm$	
B13	β-NiS	$R3m$	Naturally occurring as the mineral millerite
B14	FeAs	$Pnma$	Naturally occurring as the mineral westerveldite
B16	GeS	$Pnma$	Naturally occurring as the mineral herzenbergite
B17	PtS	$P4_2/mmc$	Naturally occurring as the mineral cooperite
B18	CuS	$P6_3/mmc$	Naturally occurring as the mineral covellite
B19	β'-AuCd	$Pmma$	
B20	FeSi	$P2_13$	Naturally occurring as the mineral naquite/fersilicite
B21	α-CO	$P2_13$	
B26	CuO	$C2/c$	Naturally occurring as the mineral tenorite
B27	FeB	$Pnma$	
B30	MgZn	$Imm2$	
B32	NaTl	$Fd\bar{3}m$	
B33	CrB	$Cmcm$	
B34	PdS	$P4_2/m$	
B35	CoSn	$P6/mmm$	
B37	SeTl	$I4/mcm$	
B_a	CoU	$I2_13$	
B_b	ζ-AgZn	$P\bar{3}$	
B_d	η-NiSi	$Pnma$	
B_e	CdSb	$Pbca$	
B_g	MoB	$I4_1/amd$	
B_h	WC	$P\bar{6}m2$	
B_i	AsTi	$P6_3/mmc$	
B_k	BN	$P6_3/mmc$	
B_l	AsS	$P2_1/c$	Naturally occurring as the mineral realgar
C1	CaF_2	$Fm\bar{3}m$	Naturally occurring as the mineral fluorite
$C1_b$	AgAsMg	$F\bar{4}3m$	Belongs to the class of compounds known as half Heusler phases
C2	FeS_2 (pyrite)	$Pa\bar{3}$	
C3	Cu_2O	$Pn\bar{3}m$	Naturally occurring as the mineral cuprite
C4	TiO_2 (rutile)	$P4_2/mnm$	
C5	TiO_2 (anatase)	$I4_1/amd$	
C6	CdI_2	$P\bar{3}m1$	
C7	MoS_2	$P6_3/mmc$	
C8	SiO_2 (β-quartz)	$P6_222$	
C9	SiO_2 (β-cristobalite)	$Fd\bar{3}m$	
C10	SiO_2 (β-tridymite)	$P6_3/mmc$	
$C11_a$	CaC_2	$I4/mmm$	
$C11_b$	$MoSi_2$	$I4/mmm$	

Table A.1 (continued)

Strukturbericht type	Prototype	Space group	Comment
C12	CaSi$_2$	$R\bar{3}m$	
C13	HgI$_2$	$P4_2/nmc$	Naturally occurring as the mineral coccinite
C14	MgZn$_2$	$P6_3/mmc$	One of the three Laves phases, see Section 13.5.2
C15	MgCu$_2$	$Fd\bar{3}m$	One of the three Laves phases, see Section 13.5.2
C15$_b$	AuBe$_5$	$F\bar{4}3m$	
C16	CuAl$_2$	$I4/mcm$	Naturally occurring as the mineral khatyrkite
C18	FeS$_2$ (marcasite)	$Pnnm$	
C19	α-Sm	$R\bar{3}$	ABCBCACAB... sequence of densest layers
C21	TiO$_2$ (brookite)	$Pbca$	
C23	PbCl$_2$	$Pnma$	Naturally occurring as the mineral cotunnite
C24	HgBr$_2$	$Cmc2_1$	
C29	SrH$_2$	$Pnma$	
C30	SiO$_2$ (α-cristobalite)	$P4_12_12$	
C32	AlB$_2$	$P6/mmm$	
C33	Bi$_2$Te$_3$	$R\bar{3}m$	
C34	AuTe$_2$	$C2/m$	Naturally occurring as the mineral cavalerite
C35	CaCl$_2$	$Pnnm$	Naturally occurring as the mineral hydrophilite
C36	MgNi$_2$	$P6_3/mmc$	One of the three Laves phases, see Section 13.5.2
C37	Co$_2$Si	$Pnma$	
C38	Cu$_2$Sb	$P4/nmm$	
C40	CrSi$_2$	$P6_222$	
C42	SiS$_2$	$Ibam$	
C43	ZrO$_2$	$P2_1/c$	Naturally occurring as the mineral baddeleyite
C44	GeS$_2$	$Fdd2$	
C45	CuCl$_2 \cdot 2$ H$_2$O	$Pmna$	Naturally occurring as the mineral eriochalcite
C46	AuTe$_2$	$Pma2$	Naturally occurring as the mineral krennerite
C47	SeO$_2$	$P4_2/mbc$	Naturally occurring as the mineral downeyite
C49	ZrSi$_2$	$Cmcm$	
C50	α-PdCl$_2$	$Pnnm$	
C52	β-TeO$_2$ (tellurite)	$Pbca$	
C53	SrBr$_2$	$P4/n$	
C54	TiSi$_2$	$Fddd$	
C$_a$	Mg$_2$Ni	$P6_222$	
C$_b$	Mg$_2$Cu	$Fddd$	
C$_c$	α-ThSi$_2$	$I4_1/amd$	
C$_e$	PdSn$_2$	$Aba2$	
C$_g$	ThC$_2$	$C2/c$	
C$_h$	Cu$_2$Te	$P6/mmm$	
C$_k$	LiZn$_2$	$P6_3/mmc$	

Table A.1 (continued)

Strukturbericht type	Prototype	Space group	Comment
$D0_1$	NH_3	$P2_13$	
$D0_2$	$CoAs_3$	$Im\bar{3}$	Naturally occurring as the mineral skutterurite
$D0_3$	BiF_3	$Fm\bar{3}m$	
$D0_4$	$CrCl_3$	$P3_112$	
$D0_5$	BiI_3	$R\bar{3}$	
$D0_8$	MoO_3	$Pnma$	Naturally occurring as the mineral molybdite
$D0_9$	$\alpha\text{-}ReO_3$	$Pm\bar{3}m$	
$D0_{11}$	Fe_3C	$Pnma$	Naturally occurring as the mineral cementite
$D0_{12}$	FeF_3	$R\bar{3}c$	
$D0_{14}$	AlF_3	$R32$	
$D0_{16}$	NH_4I_3	$Pnma$	
$D0_{17}$	BaS_3	$P\bar{4}2_1m$	
$D0_{18}$	Na_3As	$P6_3/mmc$	
$D0_{19}$	Ni_3Sn	$P6_3/mmc$	
$D0_{20}$	$\varepsilon\text{-}NiAl_3$	$Pnma$	
$D0_{21}$	Cu_3P	$P\bar{3}c1$	
$D0_{22}$	Al_3Ti	$I4/mmm$	
$D0_{23}$	Al_3Zr	$I4/mmm$	
$D0_{24}$	Ni_3Ti	$P6_3/mmc$	
$D0_a$	$\beta\text{-}TiCu_3$	$Pmmn$	
$D0_c$	SiU_3	$I4/mcm$	
$D0_d$	Mn_3As	$Cmcm$	
$D0_e$	Ni_3P	$I\bar{4}$	
$D1_1$	SnI_4	$Pa\bar{3}$	
$D1_2$	SiF_4	$I\bar{4}3m$	
$D1_3$	Al_4Ba	$I4/mmm$	
$D1_a$	Ni_4Mo	$I4/m$	
$D1_b$	Al_4U	$Imma$	
$D1_c$	$PtSn_4$	$Aba2$	
$D1_d$	$PtPb_4$	$P4/nbm$	
$D1_e$	ThB_4	$P4/mbm$	
$D1_f$	Mn_2B	$Fddd$	
$D1_g$	$B_{13}C_2$	$R\bar{3}m$	
$D2_1$	CaB_6	$Pm\bar{3}m$	
$D2_3$	$NaZn_{13}$	$Fm\bar{3}c$	
$D2_a$	$TiBe_{12}$	$P6/mmm$	
$D2_b$	$Mn_{12}Th$	$I4/mmm$	
$D2_c$	U_6Mn	$I4/mcm$	
$D2_d$	$CaCu_5$	$P6/mmm$	
$D2_e$	$BaHg_{11}$	$Pm\bar{3}m$	
$D2_f$	UB_{12}	$Fm\bar{3}m$	
$D2_g$	Fe_8N	$I4/mmm$	
$D2_h$	$MnAl_6$	$Cmcm$	
$D3_1$	Hg_2Cl_2	$I4/mmm$	Naturally occurring as the mineral calomel

Table A.1 (continued)

Strukturbericht type	Prototype	Space group	Comment
$D5_1$	Al_2O_3 (corundum)	$R\bar{3}c$	
$D5_2$	La_2O_3	$P\bar{3}m1$	
$D5_3$	Mn_2O_3	$Ia\bar{3}$	Naturally occurring as the mineral bixbyite
$D5_5$	Mg_3P_2	$Pn\bar{3}m$	
$D5_6$	Al_2O_3 (β-alumina)	$P6_3/mmc$	
$D5_7$	γ-Fe_2O_3	$P4_332$	Naturally occurring as the mineral maghemite
$D5_8$	Sb_2S_3	$Pnma$	Naturally occurring as the mineral stibnite
$D5_9$	Zn_3P_2	$P4_2/nmc$	
$D5_{10}$	Cr_3C_2	$Pnma$	Naturally occurring as the mineral tongbaite
$D5_{11}$	Sb_2O_3	$Pccn$	Naturally occurring as the mineral valentinite
$D5_{12}$	β-Bi_2O_3	$P\bar{4}b2$	
$D5_{13}$	Al_3Ni_2	$P\bar{3}m1$	
$D5_a$	Si_2U_3	$P4/mbm$	
$D5_b$	Pt_2Sn_3	$P6_3/mmc$	
$D5_c$	Pu_2C_3	$I\bar{4}3d$	
$D5_e$	Ni_3S_2	$R32$	Naturally occurring as the mineral hazelwoodite
$D5_f$	As_2S_3	$P2_1/c$	Naturally occurring as the mineral orpiment
$D6_1$	Sb_2O_3	$Fd\bar{3}m$	Naturally occurring as the mineral senarmontite
$D7_1$	Al_4C_3	$R\bar{3}m$	
$D7_2$	Co_3O_4	$Fd\bar{3}m$	Co_3O_4 has a spinel structure
$D7_3$	Th_3P_4	$I\bar{4}3d$	
$D7_a$	δ-Ni_3Sn_4	$C2/m$	
$D7_b$	Ta_3B_4	$Immm$	
$D8_1$	Fe_3Zn_{10}	$Im\bar{3}m$	
$D8_2$	Cu_5Zn_8	$I\bar{4}3m$	The γ-brass structure
$D8_3$	Cu_9Al_4	$P\bar{4}3m$	
$D8_4$	$Cr_{23}C_6$	$Fm\bar{3}m$	
$D8_5$	Fe_7W_6	$R\bar{3}m$	
$D8_6$	$Cu_{15}Si_4$	$I\bar{4}3d$	
$D8_8$	Mn_5Si_3	$P6_3/mcm$	Naturally occurring as the mineral mavlyanovite
$D8_9$	Co_9S_8	$Fm\bar{3}m$	
$D8_{10}$	Cr_5Al_8	$R3m$	
$D8_{11}$	Co_2Al_5	$P6_3/mmc$	
$D8_a$	Th_6Mn_{23}	$Fm\bar{3}m$	
$D8_b$	σ-CrFe	$P4_2/mnm$	
$D8_c$	Mn_2Zn_{11}	$Pm\bar{3}$	
$D8_d$	Co_2Al_9	$P2_1/c$	
$D8_e$	$Mg_{32}(Al,Zn)_{49}$	$Im\bar{3}$	
$D8_f$	Ir_3Ge_7	$Im\bar{3}m$	
$D8_g$	Ga_2Mg_5	$Ibam$	
$D8_h$	W_2B_5	$P6_3/mmc$	
$D8_i$	Mo_2B_5	$R\bar{3}m$	
$D8_k$	Th_7S_{12}	$P6_3/m$	

Table A.1 (continued)

Strukturbericht type	Prototype	Space group	Comment
$D8_l$	Cr_5B_3	$I4/mcm$	
$D8_m$	W_5Si_3	$I4/mcm$	
$D10_1$	C_3Cr_7	$Pnma$	
$D10_2$	Fe_3Th_7	$P6_3mc$	
$E0_1$	PbFCl	$P4/nmm$	Naturally occurring as the mineral matlockite
$E0_2$	AlOOH	$Pnma$	Naturally occurring as the mineral diaspore
$E0_3$	Cd(OH)Cl	$P6_3mc$	
$E0_4$	γ-FeO(OH)	$Cmcm$	Naturally occurring as the mineral lepidocrocite
$E0_5$	FeOCl	$Pmmn$	
$E0_6$	γ-MnO(OH)	$P2_1/c$	Naturally occurring as the mineral manganite
$E0_7$	FeAsS	$P2_1/c$	Naturally occurring as the mineral arsenopyrite
$E1_1$	$CuFeS_2$	$I\bar{4}2d$	Naturally occurring as the mineral chalcopyrite
$E1_2$	$Zn(NH_3)_2Cl_2$	$Imma$	
$E1_3$	$Mg(NH_3)_2Cl_2$	$Cmmm$	
$E1_4$	$PNCl_2$	$P4_2/n$	
$E1_a$	$MgCuAl_2$	$Cmcm$	
$E1_b$	$AgAuTe_4$	$P2/c$	Naturally occurring as the mineral sylvanite
$E2_1$	$CaTiO_3$	$Pm\bar{3}m$	The structure type is referred to the cubic phase of $CaTiO_3$
$E2_2$	$FeTiO_3$	$R\bar{3}$	Naturally occurring as the mineral ilmenite
$E2_4$	NH_4CdCl_3	$Pnma$	
$E2_5$	NH_4HgCl_3	$P4/mmm$	
$E2_6$	$KMg(H_2O)_6(Cl,Br)_3$	$P4/n$	Naturally occurring as the mineral bromocarnallite
$E3$	$CdAl_2S_4$	$I\bar{4}$	
$E3_2$	B_2CaO_4	$Pbcn$	
$E3_3$	$FeSb_2S_4$	$Pnma$	Naturally occurring as the mineral berthierite
$E3_4$	$K_2HgCl_4 \cdot H_2O$	$Pbam$	
$E3_5$	$K_2SnCl_4 \cdot H_2O$	$Pnma$	
$E4_1$	Fe_2TiO_5	$Cmcm$	Naturally occurring as the mineral pseudobrookite
$E5_1$	$FeNb_2O_6$	$Pbcn$	Naturally occurring as the mineral columbite
$E8_1$	$Eu_2Ir_2O_7$	$Fd\bar{3}m$	The structure of cubic pyrochlore
$E9_1$	$Ca_3Al_2O_6$	$Pm\bar{3}m$	
$E9_2$	$NaBe_4SbO_7$	$P6_3mc$	Naturally occurring as the mineral swedenborgite
$E9_3$	Fe_3W_3C	$Fd\bar{3}m$	
$E9_4$	Al_5C_3N	$P6_3mc$	
$E9_a$	$FeCu_2Al_7$	$P4/mnc$	
$E9_b$	π-$FeMg_3Al_8Si_6$	$P\bar{6}2m$	
$E9_c$	Al_9Mn_3Si	$P6_3/mmc$	
$E9_d$	$AlLi_3N_2$	$Ia\bar{3}$	
$E9_e$	$CuFe_2S_3$	$Pnma$	Naturally occurring as the mineral cubanite

Table A.1 (continued)

Strukturbericht type	Prototype	Space group	Comment
$F0_1$	NiSSb	$P2_13$	Naturally occurring as the mineral ullmannite
$F0_2$	COS	$R3m$	
$F1_1$	$Hg(CN)_2$	$I\bar{4}2d$	
$F2_1$	$K_4[Mo(CN)_8] \cdot 2H_2O$	$Pnma$	
$F4_1$	$Fe_2(CO)_9$	$P6_3/m$	
$F5_1$	$CrNaS_2$	$R\bar{3}m$	Naturally occurring as the mineral caswellsilverite
$F5_2$	KHF_2	$I4/mcm$	
$F5_5$	$NaNO_2$	$Imm2$	Low-temperature phase
$F5_6$	$CuSbS_2$	$Pnma$	Naturally occurring as the mineral chalcostibite
$F5_7$	$NH_4H_2PO_2$	$Cmma$	
$F5_8$	NH_4HF_2	$Pmna$	
$F5_9$	$KCNS$	$Pbcm$	
$F5_{10}$	$KAg(CN)_2$	$P\bar{3}1c$	
$F5_{12}$	$AgNO_2$	$Imm2$	
$F5_{13}$	KBO_2	$R\bar{3}c$	
$F5_{14}$	NH_4ClBrI	$Pnma$	
$F5_a$	$KFeS_2$	$C2/c$	
$G0_1$	$CaCO_3$ (calcite)	$R\bar{3}c$	
$G0_2$	$CaCO_3$ (aragonite)	$Pnma$	
$G0_3$	$NaClO_3$	$P2_13$	
$G0_6$	$KClO_3$	$P2_1/m$	
$G0_7$	$KBrO_3$	$R3m$	
$G0_8$	NH_4NO_3	$Pm\bar{3}m$	Polymorph I
$G0_9$	NH_4NO_3	$P4bm$	Polymorph II
$G0_{10}$	NH_4NO_3	$Pnma$	Polymorph III
$G0_{11}$	NH_4NO_3	$Pmmn$	Polymorph IV
$G0_{12}$	$NaHCO_3$	$P2_1/c$	
$G1_1$	$MgCa(CO_3)_2$	$R\bar{3}$	Naturally occurring as the mineral dolomite
$G2_1$	$Pb(NO_3)_2$	$Pa\bar{3}$	
$G2_2$	$Nd(BrO_3)_3 \cdot 9 H_2O$	$P6_3mc$	
$G3_2$	Na_2SO_3	$P\bar{3}$	
$G5_1$	H_3BO_3	$P\bar{1}$	
$G5_2$	$Al(PO_3)_3$	$I\bar{4}3d$	
$G7_1$	$CeF(CO_3)$	$P\bar{6}2c$	Naturally occurring as the mineral bastnäsite
$G7_2$	$Be_2BO_3(OH)$	$Pbca$	Naturally occurring as the mineral hambergite
$G7_4$	$Cu_3(CO_3)_2(OH)_2$	$P2_1/c$	Naturally occurring as the mineral azurite
$H0_1$	$CaSO_4$	$Cmcm$	Naturally occurring as the mineral anhydrite
$H0_2$	$BaSO_4$	$Pnma$	Naturally occurring as the mineral barite
$H0_4$	$CaWO_4$	$I4_1/a$	Naturally occurring as the mineral scheelite
$H0_5$	$KClO_4$	$F\bar{4}3m$	High-temperature modification; turns into the $H0_2$ structure below 299.5 °C
$H0_6$	$MgWO_4$	$P2/c$	Naturally occurring as the mineral huanzalaite

Table A.1 (continued)

Strukturbericht type	Prototype	Space group	Comment
$H1_1$	Al_2MgO_4	$Fd\bar{3}m$	Naturally occurring as the mineral spinel
$H1_5$	K_2PtCl_4	$P4/mmm$	
$H1_6$	K_2SO_4	$Pnma$	Naturally occurring as the mineral arcanite
$H1_7$	Na_2SO_4	$Fddd$	Naturally occurring as the mineral thenardite
$H2_1$	Ag_3PO_4	$P\bar{4}3n$	
$H2_4$	Cu_3S_4V	$P\bar{4}3m$	Naturally occurring as the mineral sulvanite
$H2_5$	$AsCu_3S_4$	$Pmn2_1$	Naturally occurring as the mineral enargite
$H2_6$	Cu_2FeS_4Sn	$I\bar{4}2m$	Naturally occurring as the mineral stannite
$H2_7$	$Zn_2(AsO_4)(OH)$	$Pnnm$	Naturally occurring as the mineral adamite
$H3_2$	$KAl(SO_4)_2$	$P321$	naturally occurring as the mineral steklite
$H4_5$	α-$NiSO_4 \cdot 6\,H_2O$	$P4_12_12$	naturally occurring as the mineral retgersite
$H4_6$	$CaSO_4 \cdot 2\,H_2O$	$C2/c$	Naturally occurring as the mineral gypsum
$H4_7$	$CaSO_4(H_2O)_{0.5}$	$C2$	Naturally occurring as the mineral bassanite
$H4_{10}$	$CuSO_4 \cdot 5\,H_2O$	$P\bar{1}$	Naturally occurring as the mineral chalcanthite
$H4_{12}$	$NiSO_4 \cdot 7\,H_2O$	$P2_12_12_1$	Naturally occurring as the mineral morenosite
$H5_7$	$Ca_5F(PO_4)_3$	$P6_3/m$	Naturally occurring as the mineral fluorapatite
$H5_8$	$Na_6ClF(SO_4)_2$	$Fm\bar{3}m$	Naturally occurring as the mineral sulphohalite
$J1_1$	K_2PtCl_6	$Fm\bar{3}m$	
$J1_7$	$MgCl_2 \cdot 6\,H_2O$	$C2/m$	Naturally occurring as the mineral bischofite
$J2_6$	Na_3AlF_6	$P2_1/c$	Naturally occurring as the mineral cryolite
$K0_1$	$K_2S_2O_5$	$P2_1/m$	
$K0_2$	KSO_3	$P321$	
$K1_2$	$CsSO_3$	$P\bar{6}2c$	
$K3_5$	$KB_5O_8 \cdot 4\,H_2O$	$Aba2$	Naturally occurring as the mineral santite
$K7_5$	$Na_5Al_3F_{14}$	$P4/mnc$	Naturally occurring as the mineral chiolite
$K7_6$	$AuCsCl_3$	$I4/mmm$	
$L1_0$	$CuAu$	$P4/mmm$	
$L1_1$	$CuPt$	$R\bar{3}m$	Rhombohedral polymorph
$L1_2$	Cu_3Au	$Pm\bar{3}m$	Naturally occurring as the mineral bogdanovite
$L1_3$	$CuPt$	$Fd\bar{3}m$	Cubic polymorph
$L2_1$	Cu_2MnAl	$Fm\bar{3}m$	Heusler phase
$L2_2$	Sb_2Tl_7	$Im\bar{3}m$	
$L'3_2$	β-V_2N	$P\bar{3}1m$	
$S0_1$	Al_2SiO_5 (kyanite)	$P\bar{1}$	
$S0_2$	Al_2SiO_5 (andalusite)	$Pnnm$	
$S0_3$	Al_2SiO_5 (sillimanite)	$Pnma$	
$S0_5$	$Al_2SiO_4(F,OH)_2$	$Pnma$	Naturally occurring as the mineral topaz
$S0_6$	$CaTiSiO_5$	$C2/c$	Naturally occurring as the mineral titanite
$S0_7$	$Mg(F,OH)_2 \cdot Mg_2SiO_4$	$Pnma$	Naturally occurring as the mineral norbergite

Table A.1 (continued)

Strukturbericht type	Prototype	Space group	Comment
$S1_1$	$ZrSiO_4$	$I4_1/amd$	Naturally occurring as the mineral zircon
$S1_2$	$Mg2SiO4$	$Pnma$	Naturally occurring as the mineral forsterite
$S1_3$	$Be2SiO4$	$R\bar{3}$	Naturally occurring as the mineral phenakite
$S1_4$	$Ca_3Al_2(SiO_4)_3$	$Ia\bar{3}d$	Naturally occurring as the mineral garnet
$S3_1$	$Be_3Al_2Si_6O_{18}$	$P6/mcc$	Naturally occurring as the mineral beryl
$S4_1$	$CaMg(SiO_3)_2$	$C2/c$	Naturally occurring as the mineral diopside
$S4_3$	$MgSiO_3$	$Pbca$	Naturally occurring as the mineral enstatite
$S6_1$	$NaAlSi_2O_6 \cdot H_2O$	$I4_1/acd$	Naturally occurring as the mineral analcime
$S6_2$	$Na_4(AlSiO_4)_3Cl$	$P\bar{4}3n$	Naturally occurring as the mineral sodalite
$S6_3$	$CaB_2Si_2O_8$	$Pnma$	Naturally occurring as the mineral danburite
$S6_5$	$NaAlSiO_4$	$P2_13$	Naturally occurring as the mineral α-carnegieite
$S6_6$	Na_2CaSiO_4	$P2_13$	
$S6_7$	$KAlSi_3O_8$	$C2/m$	Naturally occurring as the mineral sanidine
$S6_8$	$NaAlSi_3O_8$	$P\bar{1}$	Naturally occurring as the mineral albite
$S6_{10}$	$Na_2Al_2Si_3O_{10} \cdot 2 H_2O$	$Fdd2$	Naturally occurring as the mineral sodalite

References

[1] E. Colacino, G. Ennas, I. Halasz, A. Porcheddu, A. Scano (Eds.), *Mechanochemistry*. De Gruyter, Berlin/Boston, **2021**.
[2] P. A. Heiney, J. E. Fischer, A. R. McGhie, W. J. Romanow, A. M. Denenstein, J. P. McCauley, A. B. Smith, D. E. Cox, *Phys. Rev. Lett.* **1991**, *66*, 2911. doi: 10.1103/PhysRevLett.66.2911
[3] M. O'Keeffe, S. Andersson, *Acta Crystallogr. A* **1977**, *33*, 914. doi: 10.1107/S0567739477002228
[4] U. Müller, *Inorganic Structural Chemistry*, 2nd ed. Wiley, Chichester, **2006**.
[5] L. Pauling, *J. Am. Chem. Soc.* **1929**, *51*, 1010. doi: 10.1021/ja01379a006
[6] H. Bärnighausen, *MATCH Commun. Math. Comput. Chem.* **1980**, *9*, 139. doi: https://match.pmf.kg.ac.rs/electronic_versions/Match09/match9_139-175.pdf
[7] J. George, D. Waroquiers, D. Di Stefano, G. Petretto, G.-M. Rignanese, G. Hautier, *Angew. Chem. Int. Ed. Engl.* **2020**, *59*, 7569. doi: 10.1002/anie.202000829
[8] G. O. Brunner, D. Schwarzenbach, *Zeitschrift Für Kristallographie – Cryst. Mater.* **1971**, *133*, 127. doi: 10.1524/zkri.1971.133.16.127
[9] J. Lima-de-faria, E. Hellner, F. Liebau, E. Makovicky, E. Parthé, *Acta Crystallogr. A* **1990**, *46*, 1. doi: 10.1107/S0108767389008834
[10] U. Müller, *Symmetry Relationships Between Crystal Structures: Applications of Crystallographic Group Theory in Crystal Chemistry*. Oxford University Press, Oxford, **2013**.
[11] R. H. Mitchell, M. D. Welch, A. R. Chakhmouradian, *Mineral. Mag.* **2017**, *81*, 411. doi: 10.1180/minmag.2016.080.156
[12] H. D. Megaw, *Crystal Structures – A Working Approach*. W.B. Saunders, London, **1973**.
[13] M. J. Buerger, *J. Chem. Phys.* **1947**, *15*, 1. doi: 10.1063/1.1746278
[14] A. F. Wells, *Three-Dimensional Nets and Polyhedra*. John Wiley & Sons, Inc., New York, **1977**.
[15] A. F. Wells, *Structural Inorganic Chemistry*, 5th ed. Oxford University Press, Washington, **1984**.
[16] https://dictionary.iucr.org/Crystal_system
[17] A. Bravais, *J. École Polytech.* **1850**, *19*, 1.
[18] S. F. A. Kettle, L. J. Norrby, *J. Chem. Educ.*, **1993**, *70*, 959.
[19] A. Schoenflies, *Krystallsysteme Und Krystallstructur*. B.G. Teubner, Leizpig, **1891**.
[20] M. O'Keeffe, B. G. Hyde, *Crystal Structures: Patterns and Symmetry*. Dover Publ. Inc., New York, p 11, **2020**.
[21] W. T. Carnall, S. Siegel, J. R. Ferraro, B. Tani, E. Gebert, *Inorg. Chem.* **1973**, *12*, 560.
[22] E. S. Fedorov, *Symmetry of Crystals*. (1891) Translated from Russian by David und Katherine Harker, New York, American Crystallographic Association Monograph No. 7. American Crystallographic Association, New York, **1971**.
[23] I. M. Aroyo (Ed.), *International Tables for Crystallography, Volume A: Space-Group Symmetry*, 6th ed. John Wiley & Sons, Hoboken, **2016**.
[24] R. W. G. Wyckoff, *The Analytical Expression of the Results of the Theory of Space Groups*. Publication no. 318. Carnegie Institution of Washington, Washington, **1922**.
[25] T. Hales, M. Adams, G. Bauer, T. D. Dang, J. Harrison, L. T. Hoang, C. Kaliszyk, V. Magron, S. Mclaughlin, T. T. Nguyen, Q. T. Nguyen, T. Nipkow, S. Obua, J. Pleso, J. Rute, A. Solovyev, T. H. A. Ta, N. T. Tran, T. D. Trieu, J. Urban, K. Vu, R. Zumkeller, *Forum Math., Pi* **2017**, *5*, e2. doi: 10.1017/fmp.2017.1
[26] T. C. Hales, *Notices AMS*, **2000**, *47*, 440.
[27] T. C. Hales, *Ann. Math.* **2005**, *162*, 1065. doi: 10.4007/annals.2005.162.1065
[28] A. R. West, *Solid State Chemistry and Its Applications*, 2nd ed., student ed. John Wiley & Sons, Hoboken, p 23, **2014**.
[29] W. Barlow, *Nature* **1883**, *29*, 186. doi: 10.1038/029186a0
[30] H. Jagodzinski, *Acta Crystallogr.* **1949**, *2*, 201. doi: 10.1107/S0365110X49000552

[31] P. T. B. Shaffer, *Acta Crystallogr. B* **1969**, *25*, 477. doi: 10.1107/S0567740869002457
[32] U. Häussermann, S. I. Simak, *Phys. Rev. B: Condens. Matter.* **2001**, *64*, 245114.
[33] F. Hoffmann, *Introduction to Crystallography*. Springer, Cham, **2020**.
[34] A. Walsh, G. W. Watson, *J. Solid State Chem.* **2005**, *178*, 1422. doi: 10.1016/j.jssc.2005.01.030
[35] D. J. Payne, R. G. Egdell, A. Walsh, G. W. Watson, J. Guo, P.-A. Glans, T. Learmonth, K. E. Smith, *Phys. Rev. Lett.* **2006**, *96*, 157403. doi: 10.1103/PhysRevLett.96.157403
[36] J. Breternitz, S. Schorr, *Adv. Energy Mater.* **2018**, *8*, 1802366. doi: 10.1002/aenm.201802366
[37] N. Mercier, *Angew. Chem. Int. Ed. Engl.* **2019**, *131*, 18078. doi: 10.1002/anie.201909601
[38] V. M. Goldschmidt, *Naturwissenschaften* **1926**, *14*, 477. doi: 10.1007/BF01507527
[39] A. M. Glazer, *Acta Crystallogr. B* **1972**, *28*, 3384. doi: 10.1107/S0567740872007976
[40] C. J. Howard, H. T. Stokes, *Acta Crystallogr. B* **1998**, *54*, 782. doi: 10.1107/S0108768198004200
[41] C. J. Howard, H. T. Stokes, *Acta Crystallogr. B* **2002**, *58*, 565. doi: 10.1107/S010876810200890X
[42] L. Wei, A. Stroppa, Z.-M. Wang, S. Gao, *Hybrid Organic-Inorganic Perovskites*. Wiley-VCH, Weinheim, Germany, **2020**.
[43] N. R. Wolf, B. A. Connor, A. H. Slavney, H. I. Karunadasa, *Angew. Chem. Int. Ed. Engl.* **2021**, *60*, 16264. doi: 10.1002/anie.202016185
[44] S. N. Ruddlesden, P. Popper, *Acta Crystallogr.* **1957**, *10*, 538. doi: 10.1107/S0365110X57001929
[45] S. N. Ruddlesden, P. Popper, *Acta Crystallogr.* **1958**, *11*, 54. doi: 10.1107/S0365110X58000128
[46] Y. Shiomi, *Anomalous and Topological Hall Effects in Itinerant Magnets*. Springer, Tokyo, pp 72 ff, **2013**.
[47] W. Fischer, *Zeitschrift Für Kristallographie* **1991**, *194*, 67. doi: 10.1524/zkri.1991.194.1-2.67
[48] W. Fischer, *Zeitschrift Für Kristallographie – Cryst. Mater.* **1991**, *194*, 87. doi: 10.1524/zkri.1991.194.14.87
[49] W. H. Baur, *Mater. Res. Bull.* **1981**, *16*, 339. doi: 10.1016/0025-5408(81)90051-9
[50] M. O'Keeffe, *Mater. Res. Bull.* **1984**, *19*, 1433. doi: 10.1016/0025-5408(84)90255-1
[51] W. H. Baur, *Crystallogr. Rev.* **2007**, *13*, 65. doi: 10.1080/08893110701433435
[52] H. P. Beck, *Zeitschrift Für Kristallographie – Cryst. Mater.* **2012**, *227*, 843. doi: 10.1524/zkri.2012.1550
[53] F. A. Kröger, H. J. Vink. In: F. Seitz, D. Turnbull (Eds.) *Solid State Physics*. Academic Press, pp 307–435, **1956**. doi: 10.1016/S0081-1947(08)60135-6
[54] R. W. Pohl, *Kolloid-Zeitschrift* **1935**, *71*, 257. doi: 10.1007/BF01423782
[55] M. Nespolo, *J. Appl. Crystallogr.* **2019**, *52*, 451. doi: 10.1107/S1600576719000463
[56] W. D. Callister Jr, *Materials Science and Engineering: An Introduction*, 7th ed. John Wiley & Sons, Inc, New York, USA, **2007**.
[57] P. E. Tomaszewski, R. W. Cahn, *MRS Bull.* **2004**, *29*, 348. doi: 10.1557/mrs2004.105
[58] S. Takeuchi, T. Hashimoto, *J. Mater. Sci.* **1990**, *25*, 417. doi: 10.1007/BF00714049
[59] N. Daneu, A. Rečnik, W. Mader, *Am. Mineral.* **2014**, *99*, 612. doi: 10.2138/am.2014.4672
[60] W. D. Callister Jr, *Materials Science and Engineering: An Introduction*. John Wiley & Sons, Inc., New York, **2007**.
[61] R. Hoffmann, *Solids and Surfaces: A Chemist's View of Bonding in Extended Structures*. Wiley-VCH, Weinheim, Germany, **1989**.
[62] R. Dronskowski, *Computational Chemistry of Solid State Materials: A Guide for Materials Scientists, Chemists, Physicists and Others*. Wiley-VCH, Weinheim, Germany, **2005**.
[63] F. Bloch, *Zeitschrift Für Physik* **1929**, *52*, 555. doi: 10.1007/BF01339455
[64] C. Kittel, *Introduction to Solid State Physics*, 8th ed. John Wiley & Sons, Inc, Hoboken, USA, **2005**.
[65] D. Treves, *J. Appl. Phys.* **1965**, *36*, 1033. doi: 10.1063/1.1714088
[66] P. M. Price, W. E. Mahmoud, A. A. Al-Ghamdi, L. M. Bronstein, *Front. Chem.* **2018**, *6*, 619. doi: 10.3389/fchem.2018.00619
[67] D. Chang, M. Lim, J. A. C. M. Goos, R. Qiao, Y. Y. Ng, F. M. Mansfeld, M. Jackson, T. P. Davis, M. Kavallaris, *Front. Pharmacol.* **2018**, *9*, 831. doi: 10.3389/fphar.2018.00831
[68] T. H. Maiman, *Nature* **1960**, *187*, 493. doi: 10.1038/187493a0
[69] D. Van Delft, P. Kes, *Phys. Today* **2010**, *63*, 38. doi: 10.1063/1.3490499

[70] W. Meissner, R. Ochsenfeld, *Naturwissenschaften* **1933**, *21*, 787. doi: 10.1007/BF01504252
[71] J. Bardeen, L. N. Cooper, J. R. Schrieffer, *Phys. Rev.* **1957**, *106*, 162. doi: 10.1103/PhysRev.106.162
[72] C. Buzea, K. Robbie, *Supercond. Sci. Technol.* **2004**, *18*, R1. doi: 10.1088/0953-2048/18/1/R01
[73] A. W. Sleight, J. L. Gillson, P. E. Bierstedt, *Solid State Commun.* **1975**, *17*, 27. doi: 10.1016/0038-1098(75)90327-0
[74] C. Michel, L. Er-Rakho, B. Raveau, *Mater. Res. Bull.* **1985**, *20*, 667. doi: 10.1016/0025-5408(85)90144-8
[75] J. G. Bednorz, K. A. Müller, *Z. Phys. B: Condens. Matter.* **1986**, *64*, 189. doi: 10.1007/BF01303701
[76] K. A. Müller, J. G. Bednorz, *Science* **1987**, *237*, 1133. doi: 10.1126/science.237.4819.1133
[77] R. Chevrel, M. Sergent, J. Prigent, *J. Solid State Chem.* **1971**, *3*, 515. doi: 10.1016/0022-4596(71)90095-8
[78] H. W. Kroto, J. R. Heath, S. C. O'Brien, R. F. Curl, R. E. Smalley, *Nature* **1985**, *318*, 162. doi: 10.1038/318162a0
[79] Y. Takabayashi, K. Prassides, *Philos. Trans. A Math. Phys. Eng. Sci.* **2016**, *374*, 20150320. doi: 10.1098/rsta.2015.0320
[80] J. Nagamatsu, N. Nakagawa, T. Muranaka, Y. Zenitani, J. Akimitsu, *Nature* **2001**, *410*, 63. doi: 10.1038/35065039
[81] V. Moshchalkov, M. Menghini, T. Nishio, Q. H. Chen, A. V. Silhanek, V. H. Dao, L. F. Chibotaru, N. D. Zhigadlo, J. Karpinski, *Phys. Rev. Lett.* **2009**, *102*, 117001. doi: 10.1103/PhysRevLett.102.117[001]
[82] A. P. Drozdov, P. P. Kong, V. S. Minkov, S. P. Besedin, M. A. Kuzovnikov, S. Mozaffari, L. Balicas, F. F. Balakirev, D. E. Graf, V. B. Prakapenka, E. Greenberg, D. A. Knyazev, M. Tkacz, M. I. Eremets, *Nature* **2019**, *569*, 528. doi: 10.1038/s41586-019-1201-8
[83] C. Mielke 3rd, D. Das, J.-X. Yin, H. Liu, R. Gupta, Y.-X. Jiang, M. Medarde, X. Wu, H. C. Lei, J. Chang, P. Dai, Q. Si, H. Miao, R. Thomale, T. Neupert, Y. Shi, R. Khasanov, M. Z. Hasan, H. Luetkens, Z. Guguchia, *Nature* **2022**, *602*, 245. doi: 10.1038/s41586-021-04327-z
[84] V. A. Drits, B. B. Zviagina, D. K. McCarty, A. L. Salyn, *Am. Mineral.* **2010**, *95*, 348. doi: 10.2138/am.2010.3300
[85] B. Butz, R. Schneider, D. Gerthsen, M. Schowalter, A. Rosenauer, *Acta Mater.* **2009**, *57*, 5480. doi: 10.1016/j.actamat.2009.07.045
[86] S. T. Norberg, N. Ishizawa, S. Hoffmann, M. Yoshimura, *Acta Crystallogr. E* **2005**, *61*, i160. doi: 10.1107/S1600536805021331
[87] M. J. Li, L. P. Xu, K. Shi, J. Z. Zhang, X. F. Chen, Z. G. Hu, X. L. Dong, J. H. Chu, *J. Phys. D Appl. Phys.* **2016**, *49*, 275305. doi: 10.1088/0022-3727/49/27/275305
[88] R.-J. Xie, N. Hirosaki, *Sci. Technol. Adv. Mater.* **2007**, *8*, 588. doi: 10.1016/j.stam.2007.08.005
[89] G. E. R. Schulze, *Metallphysik*. Akademie-Verlag, Berlin, Germany, **1967**.
[90] G. Sauthoff, *Intermetallics*. VCH, Weinheim, Germany, **1995**.
[91] W. Steurer, J. Dshemuchadse, *Intermetallics – Structures, Properties, and Statistics*. Oxford University Press, Oxford, United Kingdom, **2016**.
[92] R. Pöttgen, D. Johrendt, *Intermetallics – Synthesis, Structure, Function*, 2nd ed., De Gruyter, Berlin, Germany, **2018**.
[93] J. Mohd Jani, M. Leary, A. Subic, M. A. Gibson, *Mater. Des.* **2014**, *56*, 1078. doi: 10.1016/j.matdes.2013.11.084
[94] H. G. Grimm, A. Sommerfeld, *Zeitschrift Für Physik* **1926**, *36*, 36. doi: 10.1007/BF01383924
[95] I. Žutić, J. Fabian, S. Das Sarma, *Rev. Mod. Phys.* **2004**, *76*, 323. doi: 10.1103/RevModPhys.76.323
[96] M. Z. Hasan, C. L. Kane, *Rev. Mod. Phys.* **2010**, *82*, 3045. doi: 10.1103/RevModPhys.82.3045
[97] M. Mekata, *Phys. Today* **2003**, *56*, 12. doi: 10.1063/1.1564329
[98] N. F. Mott, H. Jones, *The Theory of the Properties of Metals and Alloys*. Clarendon Press, Oxford, UK, **1936**.
[99] U. Mizutani, *Introduction to the Electron Theory of Metals*. Cambridge University Press, Cambridge, UK, **2001**. doi: 10.1017/CBO9780511612626
[100] R. Nesper, *Z. Anorg. Allg. Chem.* **2014**, *640*, 2639. doi: 10.1002/zaac.201400403

[101] F. C. Frank, J. S. Kasper, *Acta Crystallogr.* **1958**, *11*, 184. doi: 10.1107/S0365110X58000487
[102] F. C. Frank, J. S. Kasper, *Acta Crystallogr.* **1959**, *12*, 483. doi: 10.1107/S0365110X59001499
[103] M. J. Murray, J. V. Sanders, *Philos. Mag. A* **1980**, *42*, 721. doi: 10.1080/01418618008239380
[104] H. Hartmann, F. Ebert, O. Bretschneider, *Z. Anorg. Allg. Chem.* **1931**, *198*, 116. doi: 10.1002/zaac.19311980111
[105] M. Costa, A. T. Costa, J. Hu, R. Q. Wu, R. B. Muniz, *J. Phys.: Condens. Matter.* **2018**, *30*, 305802. doi: 10.1088/1361-648X/aacc08
[106] T. C. Hales, *Discrete Comput. Geom.* **2001**, *25*, 1. doi: 10.1007/s004540010071
[107] W. Thomson, *The London, Edinburgh, and Dublin Philosophical Magazine and Journal of Science* **1887**, *24*, 503. doi: 10.1080/14786448708628135
[108] D. Weaire, R. Phelan, *Philos. Mag. Lett.* **1994**, *69*, 107. doi: 10.1080/09500839408241577
[109] M. V. Stackelberg, H. R. Müller, *Zeitschrift für Elektrochemie, Berichte der Bunsengesellschaft für physikalische Chemie*, **1954**, *58*, 25. doi: 10.1002/bbpc.19540580105
[110] A. F. Masters, T. Maschmeyer, *Microporous Mesoporous Mater.* **2011**, *142*, 423. doi: 10.1016/j.micromeso.2010.12.026
[111] D. S. Coombs, A. Alberti, T. Armbruster, G. Artioli, C. Colella, E. Galli, J. D. Grice, F. Liebau, J. A. Mandarino, H. Minato, E. H. Nickel, E. Passaglia, D. R. Peacor, S. Quartieri, R. Rinaldi, M. Ross, R. A. Sheppard, E. Tillmanns, G. Vezzalini, *Mineral. Mag.* **1998**, *62*, 533. doi: 10.1180/002646198547800
[112] S. Kulprathipanja (Ed.), *Zeolites in Industrial Separation and Catalysis*. Wiley-VCH, Weinheim, **2010**. doi: 10.1002/9783527629565
[113] N. A. Anurova, V. A. Blatov, G. D. Ilyushin, D. M. Proserpio, *J. Phys. Chem. C* **2010**, *114*, 10160. doi: 10.1021/jp1030027
[114] G. T. Kokotailo, S. L. Lawton, D. H. Olson, W. M. Meier, *Nature* **1978**, *272*, 437. doi: 10.1038/272437a0
[115] E. M. Flanigen, J. M. Bennett, R. W. Grose, J. P. Cohen, R. L. Patton, R. M. Kirchner, J. V. Smith, *Nature* **1978**, *271*, 512. doi: 10.1038/271512a0
[116] D. Dubbeldam, S. Calero, T. J. H. Vlugt, *Mol. Simul.* **2018**, *44*, 653. doi: 10.1080/08927022.2018.1426855
[117] H. Li, M. Eddaoudi, M. O'Keeffe, O. M. Yaghi, *Nature* **1999**, *402*, 276. doi: 10.1038/46248
[118] H. Jiang, D. Alezi, M. Eddaoudi, *Nat. Rev. Mater.* **2021**, *6*, 466. doi: 10.1038/s41578-021-00287-y
[119] W. Xuan, C. Zhu, Y. Liu, Y. Cui, *Chem. Soc. Rev.* **2012**, *41*, 1677. doi: 10.1039/c1cs15196g
[120] I. Hönicke, I. Senkovska, V. Bon, I. Baburin, S. Raschke, J. D. Evans, S. Kaskel, *Angew. Chem. Int. Ed.* **2018**, *57*, 13780. doi: 10.1002/anie.201808240
[121] O. M. Yaghi, C. S. Diercks, M. J. Kalmutzki, *Introduction to Reticular Chemistry: Metal-Organic Frameworks and Covalent Organic Frameworks*. Wiley-VCH Verlag GmbH, Weinheim, Germany, **2019**.
[122] T. Tian, Z. Zeng, D. Vulpe, M. E. Casco, G. Divitini, P. A. Midgley, J. Silvestre-Albero, J.-C. Tan, P. Z. Moghadam, D. Fairen-Jimenez, *Nat. Mater.* **2018**, *17*, 174. doi: 10.1038/nmat5050
[123] Z. Zheng, H. L. Nguyen, N. Hanikel, K. K.-Y. Li, Z. Zhou, T. Ma, O. M. Yaghi, *Nat. Protoc.* **2023**, *18*, 136. doi: 10.1038/s41596-022-00756-w
[124] F. Hoffmann, M. Fröba, *Network topology*. In: S. Kaskel (Ed.) *The Chemistry of Metal-Organic Frameworks*. Wiley-VCH, Weinheim, Germany, pp 5–40, **2016**. doi: 10.1002/9783527693078.ch2
[125] M. O'Keeffe, M. A. Peskov, S. J. Ramsden, O. M. Yaghi, *Acc. Chem. Res.* **2008**, *41*, 1782. doi: 10.1021/ar800124u
[126] (a) O. Delgado-Friedrichs, M. O'Keeffe, *Acta Crystallogr. A* **2003**, *59*, 351. doi: 10.1107/S0108767303012 [017]. (b) http://gavrog.org/ (accessed: 03/21/2023)
[127] (a) V. A. Blatov, A. P. Shevchenko, D. M. Proserpio, *Cryst. Growth Des.* **2014**, *14*, 3576. doi: 10.1021/cg500498k. (b) https://topospro.com/ (accessed: 03/21/2023)
[128] A. P. Shevchenko, V. A. Blatov, *Struct. Chem.* **2021**, *32*, 507. doi: 10.1007/s11224-020-01724-4
[129] F. Hoffmann, *Phys. Sci. Rev.* **2022**. doi: 10.1515/psr-2019-0073

[130] O. Delgado-Friedrichs, M. O'Keeffe, *J. Solid State Chem.* **2005**, *178*, 2480. doi: 10.1016/j.jssc.2005.06.011
[131] O. M. Yaghi, M. J. Kalmutzki, C. S. Diercks, *Topology*. In: *Introduction to Reticular Chemistry – Metal-Organic Frameworks and Covalent Organic Frameworks*. Wiley-VCH Verlag, Weinheim, Germany, pp 431–452, **2019**.
[132] P. P. Ewald, C. C. Hermann (Eds.), *Strukturbericht 1913–1928*. Akademische Verlagsgesellschaft, Leipzig, **1931**.
[133] M. J. Mehl, *J. Phys. Conf. Ser.* **2019**, *1290*, 012016. doi: 10.1088/1742-6596/1290/1/012016
[134] W. B. Pearson, *A Handbook of Lattice Spacings and Structures of Metals and Alloys*, Vol. 2. Pergamon Press, Oxford, **1967**.

Subject index

μ-phase 290
σ-phase 289
123 oxide 234
(8−N) rule 277

A15 phase 290
A15 phases 232
ABC-6 family 307
abherent 258
abrasives 247
Abrikosov, Alexei 229
Abrikosov lattice 230
Abrikosov vortices 229
Acheson, Edward Goodrich 255
Acheson furnace 255
Acheson process 255
actinoids 63
activation
– of MOFs 313
activators 214
adamite 342
AFT framework 307
ageing 152
γ-AlON 262
aircraft construction 265
albite 30, 343
$AlCl_3$ structure type 78, 112
alkali metal chalcogenides 80
alkali metal fullerides 237
alkali metal halides 7
alloys 128, 134
– magnetic properties 197
– ordered 265
all-silica zeolites 297
α-Al_2O_3 structure type 78, 112
alpha cage 301
AlPO-52 307
AlPOs 298
aluminium nitride 259
aluminium oxide 121
aluminium oxynitride 262
aluminium titanate 249
aluminophosphates 298
aluminosilicates 297
amine scrubbing 315
analcime 343
anatase 9, 34

anglesite 32
anhydrite 341
annealing 123
anti-CaF_2 structure type 80
anti-$CdCl_2$ type 77
anti-CdI_2 structure type 107
anticuboctahedron 51, 62
antiferromagnetic 104
anti-PbO structure type 81
anti-perovskite structure type 96
anti-ReO_3 structure type 99
apatite 35
aperture 302
aragonite 32
arcanite 342
Aristotle 279
aristotype 13, 91, 111
arsenopyrite 340
ATI 249
atomic force microscopy (AFM) 252
augmentation 328
axis of rotation 27
azeotrope 157
azurite 341

baddeleyite 247, 337
BAM 216
band dispersion 172
band gap
– tunability 270
band structure 163, 168, 197
band theory 163
bandwidth 172
barcode scanners 221
Bardeen, John 230
barioferrite 209
barite 341
barium titanate 96, 250
Barlow packings 58
Barlow, William 58
Bärnighausen, Hartmut 10
Bärnighausen tree 111
Bärnighausen trees 13
Barrer 296
baryte 32
base 3
base-centring 21

BASF 320
basic structures 13
Basolite A520 320
bassanite 342
bastnäsite 341
Baur, Werner H. 111
BCS theory 230
Bednorz, Johannes Georg 234
benitoite 36
berthierite 340
beryl 343
beryll 36
beryllium 62, 215
Bethe-Slater curve 196
bew net 272
BiI_3 structure type 112
binders 242
bischofite 342
bixbyite 339
blacklight lamps 217
Bloch, Felix 166
Bloch functions 166
Bloch phase factors 169
Bloch walls 199
Bloch waves 167
Bloch's theorem 166
blocking temperature 203
blood-brain barrier 203
Blu-ray discs 221
body-centred cubic 4
body-centring 21
bogdanovite 342
Bohr magneton 189
bond valence strength 8
BORAZON 259
boron carbide 254
boron nitride 258
borosilicates 299
branching points 325
brass 273
β-brass 128
brass instruments 128
brass-like phases 273
β-brass phase 275
β'-brass phase 269
γ brass 275
γ-brass 339
Bravais, Auguste 22
Bravais lattice 15

Bravais lattices 22
bricks 245
Brillouin, Léon Nicolas 168
Brillouin zone 168
bromocarnallite 340
Bronze Age 208
brookite 9
brucite 106
Brunner, Georg O. 11
Buckminsterfullerene 3
Buerger, Martin J. 13
bunsenite 134
Burgers circuit 140
Burgers, Martinus 139
Burgers vector 139

$CaCl_2$ structure type 108
cadmium 62
cadmium iodide 106
CaF_2 structure type 79
β-cage 293
cahnite 33
calcite 35, 76
calomel 338
CAN framework 307
cancrinite 307
carbon 13
carbon black 147
carborundum 256
carbothermic reduction 255
carnegieite 343
cast iron 137
caswellsilverite 341
cathode-ray tube (CRT) 212
cation-anion radius ratio 6
cavalerite 337
cavities 296
$CdCl_2$ structure type 77
CdI_2 structure type 105
CDs 221
cds net 329
cementite 338
centre of inversion 25
centring 15, 21
ceramic capacitors 252
ceramics 242
– boride 253
– carbide 253
– household 245

– piezoelectric 252
– SiALON 261
Ceran 263
cha cavity 307
chabazite 296, 302
chalcanthite 30, 342
chalcopyrite 34, 84, 340
chalcostibite 341
channel system
– dimensionality 301
chemical topology 330
chemiluminescence 212
Chevrel phases 236
China clay 247
chiolite 342
chrome green 134
chrome magnesia bricks 249
chrome yellow 134
chromium 121
chromium oxide green 134
cinnabar 335
clay minerals 242
clinker 245
clinohedrite 31
clinoptilolite 296, 302
CMAT 216
coarse ceramics 245
coccinite 337
coercivity field strength 201
coherence length 218
coloradoite 37
colour rendering index (CRI) 215
Colour TVs 217
columbite 340
colusite 37
compass needle 185
composite building units 299
compressed natural gas (CNG) 318
condensed phase rule 155
conduction band 183
cone cells 215
congruent melting 159
Cooper, Leon Neil 230
Cooper pairs 230, 237
cooperite 336
coordination number 6
– effective 11
coordination polyhedra 3, 5, 11
coordination sequence 329

coordinatively unsaturated metal sites (CUS) 315
copper 3, 60
corundum 35, 114, 134, 220, 249
cotunnite 337
coumarins 222
covellite 336
Cr_3Si structure type 232, 290
$CrCl_3$ structure type 78
cristobalite 136
critical electric current flux density 229
critical magnetic field strength 229
critical radius ratio 6
critical temperature 229
Cronstedt 295
cryolite 342
crystal classes 24
crystal system 15, 19
crystallographic point group 47
crystallographic restriction theorem 27
crystallographic shear planes 144
CS planes 144
CsCl structure type 119, 288
cubanite 340
cubic boron nitride (CBN) 259
cubic closest packing 4, 50
cubic spinel ferrites 208
cuboctahedra 68
cuboctahedron 51, 61
$CuFeS_2$ structure type 84
cuprite 336
Curie paramagnetism 185
Curie temperature 193
Curie-Weiss law 188
Curie's law 187
Czochralski, Jan 142
Czochralski method 142

DABCO 97
danburite 343
De Broglie 183
deconstruction process 324
deflocculants 242
Delgado-Friedrichs, Olaf 330
density of states (DOS) 173, 197
dentistry 208
derivative structures 13
dermatology 208
dewaxing 309
diabolite 34

diagram of symmetry elements 47
diamagnetism 185
diamond 13, 83, 147
diamond net 89
diaspore 340
dielectric ceramics 96
dihexagonal-dipyramidal crystal class 36
dihexagonal-pyramidal crystal class 35
diopside 343
diploidal crystal class 37
direct band gap 183
discharge lamp 215
dislocations 138
dispersion forces 3
disphenoidal crystal class 32
ditetragonal-dipyramidal crystal class 34
ditetragonal-pyramidal crystal class 34
ditrigonal-dipyramidal crystal class 36
ditrigonal-pyramid crystal class 35
ditrigonal-scalenohedral crystal class 35
dolomite 341
dopant atoms 123
double exchange
– magnetic 208
double helices 45
double perovskites 97
double-eight ring 300
double-four ring 299
double-six ring 299
downeyite 337
drug delivery 203
DUT-60 313
DVDs 221
dye lasers 222

earthenware 245
Earth's lower mantle 130
eddy current 209
edge dislocation 138, 243
Einstein coefficients 218
electric resistance 227
electrical conductivity 167
electroluminescence 212
electron gas 191
electron microscopes 253
electroneutrality 7
electron-phonon collisions 227
electron-phonon coupling 184

electrons
– quasi-free 167
electrostatic bond strength 8
electrostatic interactions 3
electrostatic repulsion 9
electrostatic valence rule 8
embedding 327
emission bands 214
enantiomorphy 37
enargite 117, 342
energy band 165, 214
energy gap 176, 180
enstatite 343
epsomite 32
eriochalcite 337
eucryptite 136
europium 64
eutactic sphere packings 70
eutectic line 158
eutectic mixture 157
eutectic point 157
exchange energy 196
exchange integral 196
excimer 222
excimer lasers 222
exhaust gas purification 136
extrinsic defects 121

face-centring 21
face-linked octahedra 8
face-linked tetrahedra 8
faujasite 304
fayalite 117, 155
F-centre 127
Fe_2C structure type 109
Fedorov, Evgraf Stepanovich 46
feldspar 30, 247
fergusonite 33
Fermi level 173, 191
Fermi temperature 192
ferrimagnetic 104
ferrimagnetism 207
ferroelectrics 250
ferromagnetism 185, 199
ferropericlase 130
fersilicite 336
$FeTiO_3$ structure type 115
fine ceramics 245

Fischer-Tropsch synthesis 305
flue gas desulfurization (FGD) 297
fluid catalytic cracking 296
fluorapatite 215
fluorescence 212
fluorescence spectroscopy 212
fluorescent lamp 215
fluorite 37, 79, 212, 336
fluorite structure type 80
forsterite 117, 155, 343
FRAM modules 252
framework types 296
framework-type code (FTC) 299
Frank-Kasper phases 279
Frank-Kasper polyhedron 68
Frank-Kasper σ phase 335
Frenkel defects 124
Frenkel, Yakov Il'ich 123
fresnoite 34
Friauf polyhedron 280
frontier orbital 173
full Heusler compounds 269
fullerene 237
furnace linings 247

g factor 189
gadolinium 199
gain medium 218–219
galenite 37
gallophosphates 299
GaPOs 299
garnet 343
garnets 208
GAVROG 327
$GdFeO_3$ structure type 93
gemstones
– synthetic 208
general position diagram 48
generators 47
Ginzburg, Vitaly 230
Ginzburg-Landau theory 232
glass-ceramics 263
Glazer notation 93
glide plane
– diagonal 41
– diamond-like 41
– double 41
glide planes 40
"glow-in-the-dark" toys 212

Goldschmidt tolerance factor 92
Gor'kov, Lev Petrovich 232
Graham, Thomas 137
grain boundaries 149
graphene sheet 3
graphite 3, 13, 36
graphs 323
green body 243
greenockite 35
gypsum 31, 342
gyroidal crystal class 37

Haber-Bosch process 130
Hales, Thomas C. 50
half Heusler phases 336
half-Heusler compounds 269
halide double perovskites 98
halite 37
halophosphate 215
hambergite 341
hard disk drives 203
hard magnetic materials 286
hazelwoodite 339
hcb net 329
H-centre 128
heats of adsorption 319
helimagnetism 105, 193
hemimorphite 32
He-Ne laser 220
Hermann & Mauguin notation 30
Hermann, Carl 25
herzenbergite 336
heterovalent 135
hettotype 13, 93
Heusler alloys 269
Heusler, Friedrich 269
Heusler phase 342
hexagonal closest packing 4, 50
hexagonal tungsten bronzes 100
hexagonal-dipyramidal crystal class 35
hexagonal-pyramidal class 35
hexagonal-trapezohedral crystal class 35
hexakisoctahedral crystal class 36
hexoctahedral crystal class 37
hextetrahedral crystal class 37
HgI_2 structure type 86
high-quartz 263
HKUST-1 319, 325
holohedry 34, 36

homovalent 135
HPSiC 255
huanzalaite 341
Hückel approximation 165
Hückel theory 165
human skin 215
Hume-Rothery phases 273
Hume-Rothery rules 279
Hume-Rothery solubility rules 134
Hume-Rothery, William 134, 273
Hund's first rule 196
Hund's rule 204, 207
hybrid organic-inorganic materials 310
hybrid organic-inorganic perovskites 96
hydrocracking 304
hydrogen storage 137, 317
hydrophilite 337
hypermorphy 24, 37
hypomorphy 24
hysteresis 201

ice 147
icosahedron 68
identity 25
illite 246
ilmenite 115, 340
ilmenite structure type 92
incandescent lamp 215
incongruent melting 160
indirect band gap 183
induced defects 123
Inorganic Crystal Structure Database (ICSD) 333
intercalation 147
intermetallic compounds 128, 265
intermetallic phases
– classification 266
– definition 265
internal conversion 213
International Mineralogical Association (IMA) 296
International Tables for Crystallography 47
International Union of Crystallography 15
International Zeolite Association (IZA) 296
interstitial 122
interstitial site 50, 69, 122
interstratification 147
intersystem crossing (ISC) 214
intrinsic defects 121
ion radius 70
ionic radius 7

IRASER 218
IRMOF-74 315
iron 137
iron(II,III) hexacyanoferrate(II,III) 310
isomorphic graphs 326
isotypic 13
isovalent 135

Jablonski, Alexander 212
Jablonski diagram 212
Jagodzinski, Heinz Ernst 58
Jahn-Teller distortion 81, 94, 106, 234
Johnson solid 222

k space 167
K_2NiF_4 structure type 100
kagome net 271, 282, 285
Kagome superconductors 240
Kamerlingh Onnes, Heike 227
kaolin 247
kaolinite 247
keatite 263
Kelvin conjecture 292
Kelvin foam 293
Kepler conjecture 50
Kepler, Johannes 50
Kerameikos 242
kgm net 271
khatyrkite 337
kitchenware 247
Koch-Cohen cluster 131
krennerite 337
krh net 273
Kröger-Vink notation 125, 135
kyanite 30

labrophones 129
labrosones 129
lambda sensor 136
Landé g factor 190
langistite 104
lanthanoids 63
– magnetic properties 198
lanthanum 58, 64
LAP 216
LAS glass-ceramics 263
laser interferometers 221
laser medium 218
laser scalpel 221

laser transitions 220
lasers 97, 208, 218
lattice 17
lattice defects 227
lattice parameters 3, 16
lattice points 18
lattice vectors 18
lattice vibrations 184
Laves phases 281
lead chalcogenide 13
lead zirconate titanate 96, 250
Leggett, Anthony James 230
lepidocrocite 340
Li$_2$O structure type 80
ligand field 190
light bulb 215
light-emitting diode (LED) 97, 184, 224
limit of stability 161
Linde Y 304
line defects 138
linker 310
links 327
liquefied natural gas (LNG) 318
liquidus line 156
litharge 81, 336
lithium-ion batteries 91
lodestone 206
London penetration depth 229
lone pair
– stereochemically active 81
lonsdaleite 284
Lord Kelvin 292
Lorentz force 230
Löwenstein's rule 298
LPSSiC 255
LS coupling 189
lta cavity 304
LTA framework 299
lubricant 120, 258
luminescence 212

M phase 290
Madelung constant 89
Madelung energy 63
maghemite 339
Magnéli, Arne 144
Magnéli phases 144
magnesia 249
magnesiowüstite 130

magnesium 36, 62
magnesium tungstate 215
magnet 185
magnetic double exchange 205
magnetic flux 186
magnetic hyperthermia therapy 203
magnetic materials
– hard 201
– soft 201
magnetic resonance imaging 232
magnetic superexchange 203
magnetic susceptibility 185
magnetism 185
magnetite 37, 130, 206
magnetisation axes 199
magnetoplumbites 209
manganese 67
manganite 340
marcasite 109
MASERs 218
matlockite 340
Matthiessen, Augustus 227
Mauguin, Charles-Victor 25
mavlyanovite 339
M-centre 128
MeAPOs 299
mechanochemistry 2
Meissner, Walther 228
Meissner-Ochsenfeld effect 228
melting points 63
mercury 227
mesoporous carbons 295
metal aluminophosphates 299
metal trihalides 113
metallic bonds 3
metal-metal bonds 9
metal-metal interactions 103
metal-organic chemical vapour deposition (MOCVD) 259
metal-organic frameworks 14, 310
metal-to-insulator transition 176
methane hydrate 294
methane storage 319
methanol to gasoline (MTG) 297
methanol to olefin 297
metric 16
MFI framework 308
mica 247
microporous materials 313, 296

microstructure 201
millerite 336
mirror plane 26
mirror symmetry 26
miscibility 133
miscibility gap 133, 158
MOF-303 320
MOF-5 312, 325, 327
moissanite 335
MO-LCAO approach 164
molecular orbital (MO) theory 163
molecular sieves 295
molybdenite 120
molybdenum disilicide 256
molybdenum oxide 144
molybdite 338
monoclinic-domatic crystal class 31
monoclinic-prismatic crystal class 31
monoclinic-sphenoidal crystal class 31
montmorillonite 246
mordenite 296, 305
morenosite 342
morphology 15, 24
morphotropic phase boundary (MPB) 252
MoS structure type 119
mosaic blocks 148
mosaic spread 148
mosaicity 148
motif 3, 18
– chiral 48
Müller, Karl Alexander 234
multiplicity 3, 49
mutinaite 310

NaCl structure type 72
naquite 336
native defects 123
natural tilings 301
nbo net 329
Néel relaxation 202
Néel temperature 192–193
Néel wall 200
nesosilicate 208, 215
nesosilicates 9, 246
net decoration 328
nets 323
network 3, 14, 323
network types 312
neutron diffraction 204

neutron moderator 249
NiAs structure type 102
niccolite 102
nickeline 102, 335
Niggli formula 12
Niggli, Paul 12
N-iodosuccinimide 33
nitride ceramics 257
noble gases 3
nodes 327
non-stoichiometric compounds 130
non-stoichiometric defects 121
norbergite 342
NTSC 217
n-type conduction 130

Ochsenfeld, Robert 228
octahedral voids 55
olivine 9, 33, 118
open metal sites 315
ophthalmology 208
optical pumping 220
orpiment 339
orthoclase 31
orthorhombic-dipyramidal crystal class 32
orthorhombic-pyramidal crystal class 32
overlap integral 165, 172
oxalic acid dihydrate 31
oxide ceramics 247
oxocuprates 235
O'Keeffe, Michael 70, 111, 330

P phase 290
packing density 4
packing direction 57
packing-dominated phases 280
paddle-wheel motif 325
PAL 217
paraelectric phase 251
parallelepiped 16
paramagnetism 185
partial density of states (PDOS) 173
Pauli paramagnetism 185, 192
Pauli principle 231
Pauling, Linus Carl 5
Pauling rules 5
Pauli's principle 191, 204
PbFCl structure type 82
α-PbO structure type 81

pcu net 327
Pearson symbols 22
pedial crystal class 30
Peierls distortion 174
Peierls, Rudolf 174
pentagonal dodecahedron 37
pentasils 308
periclase 130, 134
peritectic point 160
peritectic reaction 161
permanent magnet 185, 201
permittivity 250
perovskite 91
– stacking variants 94
perovskite structure type 91
perovskite supergroup 97
perovskites 13, 234
– oxygen-deficient 234
phase diagrams 154
phase rule 154
phenakite 343
phillipsite 296, 302
phonons 184
phosphorescence 212
phosphors 212
– inorganic 214
photodetectors 97
photoluminescence 212
photoluminescence spectrum 212
phylloalumosilicates 246
phyllosilicates 147
physical vapour deposition (PVD) 259
piezoelectricity 252
pinacoidal crystal class 30
pinning centres 230
plane groups 3
plane waves 167
plastic deformation 141
plasticizers 242
Plateau's laws 292
p-n junction 225
Pohl, R.W. 126
point defect 121, 122, 243
point group 15
point reflection 26
points of extension 313, 324
polar axis 31
polonium 66
polymorphs 154

polytypes 59
polytypism 147
Popper, P. 101
population inversion 219
porcelain 245
potassium tetrathionate 31
precipitates 149, 151, 243
precipitation hardening 151
primitive unit cell 20
principle of symmetry 10
product selectivity 309
prototypes 12
prototypical compounds 3
Prussian blue 134, 310
pseudo symmetry 24
pseudobrookite 340
PtS structure type 85
p-type conduction 130
pumping mechanism 218
Purcell, Edward Mill 166
pyrite 37
pyritohedra 293
pyrochlore 340
pyrrhotite 104
PZT ceramics 250

quartz 35, 136, 247
– high-temperature modification 35
β-quartz 263

R phase 290
RAM modules 252
rare earth oxyfluorides 80
RCSR database 325
realgar 336
realization 327
remanence 201
ReO_3 structure type 99
retgersite 33, 342
reticular chemistry 312
RGB-(red-green-blue)-LEDs 225
$RhBr_3$ structure type 78
rhenium(VI)trioxide 180
rhodamines 222
rhombohedral centring 21
rhombohedral crystal class 35
rhombohedral lattice 20
rhombohedral unit cell 20
rhombohedron 20

right-hand rule 139
rings 301
rock salt 3, 12
rod packings 4
rod-like SBUs 322
rotary-reflection axis 29
rotoinversion axis 28
ruby 121, 134
ruby laser 220
Ruddlesden, S.N. 101
Ruddlesden-Popper phases 101
Russel-Saunders coupling 189
rutile 9, 109, 144, 150, 249
Ryabinin, Y.N. 232

sanidine 343
santite 342
SAPO-34 307
SAPOs 299
SAT framework 307
saturation magnetisation 197, 201
scapolite group 33
scheelite 33, 341
Schoenflies, Arthur Moritz 25, 46
Schoenflies notation 29
Schottky defect 124
Schottky, Walter Hans 123
Schrieffer, John Robert 230
Schrödinger equation 164
Schwarzenbach, Dieter 11
screw axes 40
screw axis 67
screw dislocation 140, 243
screw rotations 42
SECAM 217
secondary building unit (SBU) 299, 310
second-harmonic generation (SHG) 222
semiconductor 121
semi-Meissner state 239
SF convention 139
shape-selective catalyst 309
Shubnikov phase 229
SiALON 261
silicalite 309
silicoaluminophosphates 299
silicogermanates 298
silicon carbide 59, 254
silicon nitride 260
similarity principle 12

simplification process 324
single crystal 121
sintering 142
SiS_2 structure type 87
SiSiC 255
site symmetry 3
skutterurite 338
snub cube 288
snub cuboctahedron 288
snub disphenoid 222
soap bubble foam 292
sod cages 304
sodalite 343
sodalite cage 293, 301
sodium chlorate 37
sodium tungsten bronzes 130
solar cells 97
solid electrolyte 136
solid oxide fuel cells 136, 248
solid solutions 132, 268
– phase diagram 156
solid-solid reaction 242
solidus line 156
solvothermal synthesis 313
sorption dimensionality 301
space group 3, 15, 46
specific surface area 313
sphalerite 37, 82
sphere packings 4, 50
– densest 50
spin quantum number 189
spinel 37, 208, 262, 342
spinel structure type 88
spinel structure
– inverse 90
spin-only formula 191
spintronics 269, 291
spontaneous magnetisation 199, 201
sql net 271, 329
SSiC 255
SSZ-13 307
stacking disorder 147
stacking faults 146
stacking variants 58, 63, 120
stannite 34, 342
steel 137
steklite 342
stellerite 295
stibnite 339

stilbenes 222
stimulated emission 218
stishovite 111
stoichiometric defects 121
stoichiometry 130
Stokes 212
Stokes shift 212
Stoner-Wohlfarth particles 202
stoneware 245
structure types 3, 13
struvite 32
stuffed silica 136
stumpflite 104
subgrains 148
subgroup 129
sublattice 129
substitutional solid solutions 134
sucrose 31
sulphohalite 342
sulvanite 342
supercage 304
supercavity 304
supercell 129
superconductivity 227
superconductor
– paramagnetic 237
– re-entrant 237
superhydride 240
superstructure 129, 268
superstructure reflections 129
swedenborgite 340
sylvanite 340
symmetry elements 3
symmetry rules 32
Systre 327

tanning lamps 217
tartaric acid 31
TBCCO 235
tbo net 325, 328
tellurium 67, 327
tenorite 336
tephroite 117
terephthalic acid 312
tetartoidal crystal class 37
tetragonal sphere packing 111
tetragonal tungsten bronze 100
tetragonal-dipyramidal crystal class 33
tetragonal-disphenoidal crystal class 33

tetragonal-pyramidal crystal class 33
tetragonal-scalenohedral crystal class 34
tetragonal-trapezohedral crystal class 33
tetrahedral voids 55
tetrahedrally close packings 279
tetrahedrite 37
thenardite 342
tialite 249
β-$TiCl_3$ structure type 113
tilt boundary 148
tin 66
TiO_2 structure type 109
titanite 342
titanium carbide 253
titanium nitride 257
tongbaite 339
topaz 33, 342
topicity 312
topological dimensionality 301
topological insulators 269
topology 14, 323
ToposPro 327, 330
tourmaline 35
transformer cores 209
transistors 97
transition metals
– magnetic properties 198
translation operations 16
translation principle 18
tri-band spectrum 216
trichromatic theory 215
tridymite 136
trigonal dodecahedron 222
trigonal-dipyramidal crystal class 35
trigonal-pyramidal crystal class 35
triphosphor lamp 216
triple points 155
trirutile compounds 111
troilite 104
TS-1 309
tungsten 11, 65
β-tungsten 293
tungsten bronzes 100
tungsten carbide 253
tungsten oxide 144
tungsten trioxide 100
turbostratic-like disorder 147
twin boundaries 150
twist boundary 148

type I superconductors 229
type II superconductors 229
type-1.5 superconductors 239

ullmannite 37, 341
ultra-stable zeolite Y 305
UMCM-2 317
unit cell 3, 16
– magnetic 204
UTD-1 302

V_2N structure type 107
vacancies 243
vacancy 122
valence bond (VB) theory 163
valence electron concentration (VEC) 273
valentinite 339
Van Vleck paramagnetism 185
V-centre 128
Vegard's rule 134
vertex symbol 330
viewing directions 24
voids 4, 55
volatile organic compounds (VOCs) 306
vortices 229

Wadsley defects 144
water harvesting 320
wave function
– periodic 167
wave vector 167
Weaire-Phelan structure 291, 293, 335
Weiss constant 188
Weiss domains 199
Wells, Alexander F. 14
westerveldite 336
white graphite 258
wüstite 130
Wyckoff letter 3, 49
Wyckoff positions 13, 49

xylene 309

YAG 208, 222
Yaghi, Omar 313, 322
YBCO 234
YIG 208
Young-Helmholtz theory 215
YOX 216
YSZ 248
ytterbium 64
yttrium aluminium garnet 208
yttrium iron garnet 208
yttrium oxysulfide 217
yttrium-stabilized zirconia 135

zeolite A 303
zeolite Alpha 304
zeolite topologies 296
zeolite Y 304
zeolites 13, 295
– natural 296
zeolitic tuff 306
zinc 36
zinc orthosilicate 215
zincblende structure type 82
zincite 35
Zintl, Eduard 277
Zintl phases 277
Zintl substructure 105
Zintl-Klemm concept 277
Zintl-Klemm-Busmann concept 277
zircon 34, 343
zirconia 135, 208, 247
zirconium nitride 257
α-$ZnCl_2$ structure type 87
γ-$ZnCl_2$ structure type 86
ZnS (wurtzite) structure type 116
ZSM-5 308

Formula index

Ag 4, 37, 107, 275
Ag_2F 107
Ag_2O 107
Ag_3Al 275
$AgBiS_2$ 75
$AgBiSe_2$ 75
AgBr 73, 126
AgCl 73
AgF 73
AgI 73
β-AgI 117
$Ag(I)Ag(III)O_2$ 86
$AgInS_2$ 117
AgMgAs 269
$AgMgF_3$ 96
AgO 86
$Al_{23}O_{27}N_5$ 262
$Al_2(F,OH)_2SiO_4$ 33
Al_2O_3 121, 134, 220, 261
$Al_2Si_2O_5(OH)_4$ 247
Al_2SiO_5 30
Al_2TiO_5 249
$AlCl_3$ 78, 112
AlF_3 99
AlN 117, 249, 259
AlP 83
AlSb 83
$AlSbO_4$ 111
$AlTaO_4$ 111
$Al_xNi_yCo_z$ 202
α-Al_2O_3 114, 247
γ-Al_2S_3 114
AmO 73, 80
AmO_2 80
AsI_3 112
AsLiMg 270
Au 4, 37, 270, 275
Au_3Al 275
Au_3Cu 270
$AuOCs_3$ 96

B_4C 254
$B_{13}C_2$ 254
Ba_2PbO_4 101
Ba_2SnO_4 101
$Ba_2TiSi_2O_8$ 34
BaF_2 80

$BaFe_{12}O_{19}$ 209
$BaMgAl_{10}O_{17}$ 216
$BaNiO_3$ 95
$BaSb_2O_6$ 108
$BaSi_2$ 277
$BaSi_2O_5$ 217
$BaSnO_3$ 95
$BaSO_4$ 32
$BaTiO_3$ 95–96, 126, 250
$BaTiS_3$ 96
$BaTiSi_3O_9$ 36
$BaZrS_3$ 96
Be_2C 81
$Be_3Al_2Si_6O_{18}$ 36
BeO 117, 249
BeS 83
$BiBr_3$ 78
α-BiF_3 269
$BiFeO_3$ 95
BiI_3 112
$BiInO_3$ 95
BiOCl 82
BiScPt 270
BiTeBr 106
BiTeI 106
$BiUO_4$ 80
α-BN 258
β-BN 83, 258
γ-BN 117, 258

$C_2O_4 \cdot 2\,H_2O$ 31
$Ca_2B(OH)_4AsO_4$ 33
$Ca_5(PO_4)_3(Cl,F)$ 215
$Ca_5(PO_4)_3(Cl,F,OH)$ 35
CaAs 108, 278
$CaAs_2O_6$ 108
$CaBr_2$ 109, 111
CaC_2 256
$CaCl_2$ 108, 111
$CaCO_3$ 32, 35, 76
$CaCu_5$ 285
CaF_2 37, 79, 185
CaI_2 106
$CaIn_2$ 278
$CaIrO_3$ 95
$CaMg_2$ 156
CaO 134–135, 248, 263

Ca(OH)$_2$ 106
CaSb$_2$O$_6$ 108
CaSi 278
CaSi$_2$ 278
CaSO$_4$ · 2 H$_2$O 31
CaTiO$_3$ 91, 93
CaTiS$_3$ 96
CaUO$_4$ 80
CaWO$_4$ 33
CaZnSiO$_4$ · H$_2$O 31
Cd 4
CdAs$_2$S$_6$ 108
CdBr$_2$ 77
CdCl$_2$ 8, 77, 106
CdCO$_3$ 76
CdCr$_2$Se$_4$ 91
CdF$_2$ 80
CdI$_2$ 8, 106
CdO 73
Cd(OH)$_2$ 106
CdS 35, 73
CdSb$_2$O$_6$ 108
CdSe 73
CdSe (red) 117
CdSnO$_3$ 95
CdTe 73, 83
Ce$_{0.67}$Tb$_{0.33}$MgAl$_{11}$O$_{19}$ 216
CeNi$_3$ 286
CeO$_2$ 80, 124
CfBr$_3$ 78
CfOF 80
CH$_3$NH$_3$PbCl$_3$ 98
CmH$_2$ 80
CmO$_2$ 80
Co 4, 91, 109, 114, 315
CO$_2$ 315
Co$_2$As$_3$ 114
Co$_2$C 109
Co$_2$N 109
Co$_3$O$_4$ 91
Co$_3$S$_4$ 91
CoAs 104, 108, 278
CoAs$_2$O$_6$ 108
CoAs$_3$ 278
CoBr$_2$ 106
CoCl$_2$ 77, 195
CoCO$_3$ 76
CoF$_2$ 111
CoFe$_2$O$_4$ 91

CoI$_2$ 106
CoIn$_2$S$_4$ 91
CoO 192, 195, 203
Co(OH)$_2$ 106
CoS 104
CoSb 104
CoSe 104
CoTe 104
CoTe$_2$ 104
CoZn$_3$ 275
Cr 4, 107, 114, 134, 195, 232, 291, 293
Cr$_2$N 107
Cr$_2$O$_3$ 114, 134, 195
Cr$_3$Si 232, 291, 293
CrAl$_2$S$_4$ 91
CrBr$_2$ 106
CrCl$_3$ 78, 112
CrI$_2$ 106
CrI$_3$ 78
CrN 73
CrNbO$_4$ 111
CrO$_2$ 111, 195
CrS 104
CrSb 104
CrSbO$_4$ 111
CrSe 104
CrTaO$_4$ 111
CrTe 104
Cs$_2$O 77
Cs$_3$C$_{60}$ 237
CsAuCl$_3$ 96
CsBr 73, 119
CsCdBr$_3$ 96
CsCdCl$_3$ 96
CsCl 73, 118, 288
CsCN 119
CsF 7
CsGaCl$_3$ 96
CsH 73
CsHgBr$_3$ 96
CsHgCl$_3$ 96
CsI 73, 119
CsPbI$_3$ 96
CsSH 119
CsSnI$_3$ 94, 96
Cu 4, 37, 84, 99, 269–270, 274–275
Cu$_2$FeSnS$_4$ 34
Cu$_{10}$(Fe,Zn)$_2$Sb$_4$S$_{13}$ 37
Cu$_{13}$VAs$_3$S$_{16}$ 37

Cu_2MnAl 269
Cu_2MX_3 84
Cu_3Au 270
Cu_3N 99
Cu_5Si 275
Cu_5Zn_8 274
CuAu 270
CuBr 37, 81, 83, 106
$CuBr_2$ 106
CuCl 37, 83
$CuCr_2Te_4$ 91
$CuFe_2O_4$ 91
$CuFeS_2$ 34, 84
$CuGe_2P_3$ 84
CuI 37, 73, 83, 117
$CuInS_2$ 117
$CuNMn_3$ 96
CuO 86
$CuSi_2P_3$ 84
$CuSO_4 \cdot 5\,H_2O$ 30
CuZn 128

DyH_2 80
DyOF 80

$ErCoO_3$ 95
ErOF 80
EuO 73
EuOF 80

α-Fe 4, 114, 129, 200
γ-Fe 137
(Fe,Mg)O 130
$(Fe,Mg)SiO_3$ 95
$Fe_{0.48}Pt_{0.52}$ 202
$Fe_{10}S_{11}$ 104
$Fe_{11}S_{12}$ 104
$Fe_{1-x}O$ 130, 144
Fe_2C 109
Fe_2N 107
Fe_2O_3 195
α-Fe_2O_3 114
Fe_2SiO_4 117
Fe_3Al 269
Fe_3O_4 37, 91, 130, 195, 205
Fe_7S_8 104
Fe_9S_{10} 104
$FeBr_2$ 106

$FeBr_3$ 112
$FeCl_2$ 77
$FeCl_3$ 112
$FeCO_3$ 76
FeF_2 111
FeI_2 106, 195
$FeNi_4$ 202
FeO 73, 130, 192, 195, 203
FeOF 111
$Fe(OH)_2$ 106
FeS 37, 104
FeS_2 37
FeSb 104
$FeSbO_4$ 111
FeSe 104
$FeTaO_4$ 111
FeTe 104
$FeTiO_3$ 115

Ga_2Au 81
GaAs 37, 83
GaN 117
α-Ga_2O_3 114
GaP 83
GaSb 83
$GaSbO_4$ 111
$GdCl_3$ 78
$GdFeO_3$ 93, 95
GdOF 80
GeO_2 111

H_2S 322
$HgAs_2O_6$ 108
$HgCr_2S_4$ 91
HgF_2 80
HgI_2 86
HgTe 37, 269
$HoCrO_3$ 95
HoHO 80
$HoMn_5$ 286
$HoMo_6S_8$ 237
HoOF 80

In_2Au 81
In_2Pt 81
InAs 37, 73
InBi 104
$InBO_3$ 76

$InBr_3$ 78
$In(OH)_3$ 99
$InSb$ 37, 83
$InTe$ 73
Ir 4, 81
Ir_2P 81
$IrBr_3$ 78
$IrCl_3$ 78
IrO_2 111
$IrSb$ 104
$ISAg_3$ 96

$K_{0.26}WO_3$ 101
K_2CuF_4 101
K_2NiF_4 100, 234
K_2O 80, 263
K_2S 31, 80
$K_2S_4O_6$ 31
K_2Se 80
K_2Te 80
K_3C_{60} 237
$K_3Pr_2(NO_3)_9$ 37
$KAlSi_3O_6$ 31
$KaTaO_3$ 95
$KCeF_4$ 80
KCl 7, 127
$KCoF_3$ 96
$KCrF_3$ 96
$KCuF_3$ 96
KF 7
$KFeF_3$ 96
KGe 278
KH 73
$KLaF_4$ 80
$KMgF_3$ 96
$KMnCl_3$ 96
$KMnF_3$ 96
$KNbO_3$ 95
$KNiF_3$ 96
KV_3Sb_5 240
$KZnF_3$ 96

$La_{1.85}Ba_{0.15}CuO_4$ 102
$LaAlO_3$ 95
$LaAs$ 73
LaB_6 253
$LaBi$ 73
$LaCoO_3$ 95
$LaFeO_3$ 195

LaH_{10} 240
LaH_2 80
$LaMnO_3$ 95
$LaMo_6S_8$ 236
LaN 73
$LaNiO_3$ 234
LaP 73
$LaPO_4$ 216–217
LaS 73
$LaSb$ 73
$LaSe$ 73
$LaTe$ 73
$LaTiO_3$ 95
$LaTiSbO_6$ 108
Li_2HfF_6 108
Li_2NbF_6 108
Li_2NbOF_5 108
Li_2O 80
Li_2PbF_6 108
Li_2S 80
Li_2Se 80
Li_2Te 80
Li_2ZrF_6 108
$(LiAl_3)Al_2O_8$ 91
$LiAlSiO_4$ 136
$LiAs$ 278
$LiBiS_2$ 74
$LiBr$ 7
$LiCoO_2$ 91
α-$LiFeO_2$ 74
γ-$LiFeO_2$ 75
LiH 73
LiI 7
$LiMgN$ 81, 270
$LiMn_2O_4$ 91
$LiMo_2F_6$ 111
$LiNiO_2$ 75
$LiNO_3$ 76
$LiOH$ 81
γ-$LiTlO_2$ 74
$LiYS_2$ 74

$MAPbCl_3$ 97
Mg 4, 63, 81, 117, 160
$(Mg,Fe)_2SiO_4$ 9
$(Mg,Fe)SiO_4$ 33
Mg_2Ca 160
Mg_2Ge 81
Mg_2Pb 81

Mg_2Si 81
Mg_2SiO_4 117
Mg_2Sn 81
$MgAl_2O_4$ 37, 88
MgB_2 239
$MgBr_2$ 106
$MgCl_2$ 77
$MgCO_3$ 76
$MgCu_2$ 282
MgF_2 109, 111
$MgFe_2O_4$ 91, 208
MgH_2 111
MgI_2 106
$MgIn_2O_4$ 91
$MgNi_2$ 285
MgO 126, 130, 134–135, 248–249
$Mg(OH)_2$ 106
$MgSO_4 \cdot 7 H_2O$ 32
$MgSrAl_{10}O_{17}$ 217
$MgTe$ 73
$MgWO_4$ 215
$MgZn_2$ 284
α-Mn 67
Mn_2SiO_4 117
Mn_3O_4 91
$MnAs$ 104, 108
$MnAs_2O_6$ 108
$MnBi$ 104
$MnBr_2$ 106
$MnCl_2$ 77
$MnCO_3$ 76
MnF_2 111
$MnFe_2O_4$ 208
MnI_2 106
MnO 73, 111, 192, 195, 203, 205
MnO_2 111
$Mn(OH)_2$ 106
MnP 104
MnS 195
$MnSb$ 104
$MnSe$ 104
$MnTe$ 104
Mo 4, 290
$Mo_{21}Cr_9Ni_{20}$ 290
$Mo_3Cr_2Co_5$ 290
$MoBr_3$ 114
$MoCl_3$ 78
MoF_3 99
$MoNa_2O_4$ 91

$MoNi_4$ 273
MoO_2 111
MoO_3 144
MoS_2 119, 181
$MoSi_2$ 256

$Na_{0.48}WO_3$ 101
$Na_2Al_2Si_3O_{10} \cdot 2 H_2O$ 298
Na_2O 80, 263
Na_2S 80
Na_2Se 80
Na_2Te 80
Na_3As 278
Na_3C_{60} 238
Na_3N 99
$NaAl_{11}O_{17}$ 209
$NaAlSi_3O_8$ 30
$NaCl$ 12, 37, 71–72, 127, 185
$NaClO_3$ 37
$NaErF_4$ 80
$NaFeF_3$ 96
α-$NaFeO_2$ 75
NaH 73
$NaHoF_4$ 80
$NaLaS_2$ 74
$NaMgF_3$ 96
$NaNO_3$ 76
$NaTl$ 277
α-$NaTlO_2$ 74
Na_xWO_3 130
$NaYbF_4$ 80
$NaYF_4$ 80
$NaZn_{13}$ 288
Nb 4, 145–146, 232, 290, 292
$Nb_{10}Ni_9Al_3$ 290
$Nb_{14}W_3O_{44}$ 146
Nb_2O_5 145
$Nb_3(Al,Ge)$ 292
Nb_3Ge 232, 292
Nb_3Sn 292
NbF_3 99
NbO_2 111
$NbOCl_3$ 12
NbS_2 181
$NbVH_4$ 80
$Nd_2Fe_{14}B$ 195, 202, 253
$NdAlO_3$ 95
$NdOF$ 80
NH_3 322

NH$_4$Br 73, 119
NH$_4$Cl 73, 119
NH$_4$I 73, 119
NH$_4$MgPO$_4$ · 6 H$_2$O 32
Ni 4, 202
Ni$_{0.75}$Fe$_{0.2}$Mo$_{0.05}$ 202
NiAs 102, 104, 108
NiAs$_2$O$_6$ 108
NiBi 104
NiBr$_2$ 77, 106
NiCl$_2$ 77, 195
NiF$_2$ 109, 111
NiFe$_2$O$_4$ 91
NiI$_2$ 77
NiLi$_2$F$_4$ 91
NiMn$_2$O$_4$ 90
NiO 73, 134, 192, 195, 203
Ni(OH)$_2$ 106
NiS 104
NiSb 104
NiSbS 37
NiSe 104
NiTe 104
NO$_x$ 322
NpO$_2$ 80

Os 4
OsO$_2$ 111

Pb$_2$CuCl$_2$(OH)$_4$ 34
PbCrO$_4$ 134
β-PbF$_2$ 80
PbFCl 82
PbFe$_{12}$O$_{19}$ 209
PbI$_2$ 106
PbMo$_6$S$_8$ 236
PbNCa$_3$ 96
PbO 13, 81, 111
PbO$_2$ 111
PbS 13, 37
PbSb$_2$O$_6$ 108
PbSe 13
PbSO$_4$ 32
PbTe 13
PbTiO$_3$ 95
PbZrO$_3$ 95, 252
Pb(Zr$_x$Ti$_{1-x}$)O$_3$ 96, 250
Pd 4, 109

Pd$_2$B 109
PdAs$_2$O$_6$ 108
PdF$_2$ 111
PdH$_x$ 137
PdO 86
PdS 86
PdSb 104
PdTe 104
α-Po 4, 67
PrH$_2$ 80
PrOF 80
Pt 4, 37
PtBi 104
PtO 86, 109
PtO$_2$ 109
PtS 86, 106
PtS$_2$ 106
PtSb 104
PtSe$_2$ 106
PtTe$_2$ 106
PuH$_2$ 80
PuNi$_3$ 286
PuO 73, 80
PuO$_2$ 80
PuOF 80

Rb$_2$O 80
Rb$_2$S 80
Rb$_3$C$_{60}$ 237
RbCs$_2$C$_{60}$ 237
RbF 7
RbGdF$_4$ 80
RbH 73
Re 4
ReO$_3$ 99, 180
ReSi$_2$ 256
Rh 4, 81, 109, 114
Rh$_2$C 109
Rh$_2$O$_3$ 114
Rh$_2$P 81
RhBi 104
RhBr$_3$ 78
RhCl$_3$ 78
RhNbO$_4$ 111
RhSbO$_4$ 111
RhTaO$_4$ 111
RhVO$_4$ 111
Ru 4

RuBr$_3$ 114
RuCl$_3$ 78
RuO$_2$ 109, 111

SbI$_3$ 112
SbScPt 270
Sc$_2$O$_3$ 8
ScBO$_3$ 76
ScCl$_3$ 112
ScF$_3$ 99
Sc(OH)$_3$ 99
Se 66
Si$_3$N$_4$ 260
α-Si$_3$N$_4$ 260
β-Si$_3$N$_4$ 260
SiC 59, 254
α-SiC 117
β-SiC 83
SiO$_2$ 35, 111
β-SiO$_2$ 35
SiS$_2$ 87
SmCo$_5$ 202, 286
SmH$_2$ 80
SmHO 80
α-Sn 66
β-Sn 66
SnCo$_2$O$_4$ 91
SnMg$_2$O$_4$ 91
SnMn$_2$O$_4$ 91
SnMo$_6$S$_8$ 236
SnO 81, 111
SnO$_2$ 111
SnS$_2$ 106
SnZn$_2$O$_4$ 91
SO$_2$ 322
Sr$_2$FeMoO$_6$ 98
Sr$_3$Ti$_2$O$_7$ 102
Sr$_4$Ti$_3$O$_{10}$ 102
SrAl$_{11}$O$_{18}$ 217
SrAs$_2$O$_6$ 108
SrB$_4$O$_7$ 217
SrCoO$_3$ 95
SrF$_2$ 80
SrGa$_2$ 277
SrMoO$_3$ 95
SrRu$_2$O$_6$ 108
SrRuO$_3$ 95
SrSb$_2$O$_6$ 108
SrSi$_2$ 277

SrTiO$_3$ 8, 91, 99
SrTiS$_3$ 96
SrZrS$_3$ 96

Ta 4
TaC 253
TaF$_3$ 99
TaO$_2$F 99
TbOF 80
β-TcCl$_3$ 78
Te 66–67
TeO$_2$ 111
Th$_2$Ni$_{17}$ 286
Th$_2$Zn$_{17}$ 286
ThO$_2$ 80
Ti 4, 114
Ti$_2$O$_3$ 114
TiAl 265, 272
TiAl$_3$ 272
TiAs 104
TiB$_2$ 253
TiBr$_2$ 106
TiBr$_3$ 112
TiC 71, 73, 253
α-TiCl$_3$ 112
β-TiCl$_3$ 113
TiCo$_2$O$_4$ 91
TiFe$_2$O$_4$ 91
TiI$_2$ 106
TiMg$_2$O$_4$ 91
TiMn$_2$O$_4$ 91
TiN 73, 257
TiNbO$_4$ 111
TiO 9, 34, 71, 73, 109, 111, 126, 144, 150, 192
TiO$_2$ 9, 34, 109, 111, 126, 144, 150
TiOF$_2$ 99
TiP 104
TiS 104
TiS$_2$ 106
TiSb 104
TiSe 104, 106
TiSe$_2$ 106
TiTaO$_4$ 111
TiTe 104, 106
TiTe$_2$ 106
TiVH$_4$ 80
TiVO$_4$ 111
TiZn$_2$O$_4$ 91
TlBiF$_4$ 80

TlBiTe$_2$ 75
TlBr 119
TlCl 119
TlCN 119
TlSbTe$_2$ 75

UCr$_2$O$_6$ 108
UO 73, 80, 99
UO$_2$ 80
UO$_3$ 99
UV$_2$O$_6$ 108

V 4, 107, 114, 144, 292
V$_2$N 107
V$_2$O$_3$ 114
V$_2$O$_5$ 144
V$_3$Si 292
VBr$_2$ 106
VC 73
VCo$_2$O$_4$ 91
VF$_2$ 111
VH$_2$ 80
VI$_2$ 106
VI$_3$ 78, 112
VMg$_2$O$_4$ 91
VO 73, 111, 192
VO$_2$ 111
VP 104
VS 104
VSb 104
VSbO$_4$ 111
VSe 104
VTaO$_4$ 111
VTe 104
VZn$_2$O$_4$ 91

W 4, 290
W$_7$Fe$_6$ 290
WC 253
WNa$_2$O$_4$ 91
WO$_2$ 111
WO$_3$ 100, 144
WSi$_2$ 256

Y$_2$O$_2$S 217
Y$_2$O$_3$ 135, 217, 248
Y$_3$Al$_2$(AlO$_4$)$_3$ 208

Y$_3$Al$_5$O$_{12}$ 222
Y$_3$Fe$_2$(FeO$_4$)$_3$ 208
Y$_3$Fe$_5$O$_{12}$ 195
YBa$_2$Cu$_3$O$_7$ 234
YbF$_2$ 80
YbMnO$_3$ 95
YBO$_3$ 76
YBr$_2$ 109
YCl$_3$ 78
YH$_2$ 80
YI$_3$ 112
YMn$_5$ 286
YNbO$_4$ 33
YOF 80
YPO$_4$ 217
YVO$_4$ 217

Zn$_2$SiO$_4$ 215
Zn$_3$(PO$_4$)$_2$ 217
Zn$_4$Si$_2$O$_7$(OH)$_2$ · H$_2$O 32
ZnBr$_2$ 77
α-ZnCl$_2$ 87
γ-ZnCl$_2$ 86
ZnCO$_3$ 76
ZnCr$_2$Se$_4$ 91
ZnF$_2$ 111
ZnFe$_2$O$_4$ 208
ZnK$_2$(CN)$_4$ 91
ZnO 35, 73, 117
ZnS 35, 37, 82, 217
α-ZnS 116
β-ZnS 116
ZnS (wurtzite) 116
ZnSe 83
ZnSiO$_4$ 217
ZnTe 83
Zr 4
ZrAl$_3$ 272
ZrAu$_4$ 273
ZrI$_3$ 9, 113
ZrN 257
ZrO$_2$ 135, 247
ZrS$_2$ 106, 181
ZrSe$_2$ 106
ZrSiO$_4$ 34
ZrTe$_2$ 106

Printed in the USA
CPSIA information can be obtained
at www.ICGtesting.com
JSHW061945140324
59250JS00008B/97